# The CRF Signal

# The CRF Signal
## Uncovering an Information Molecule

Jay Schulkin

OXFORD
UNIVERSITY PRESS

OXFORD
UNIVERSITY PRESS

Great Clarendon Street, Oxford, OX2 6DP,
United Kingdom

Oxford University Press is a department of the University of Oxford. It furthers the University's objective of excellence in research, scholarship, and education by publishing worldwide. Oxford is a registered trade mark of Oxford University Press in the UK and in certain other countries



First Edition published in 2017



Published in the United States of America by Oxford University Press
198 Madison Avenue, New York, NY 10016, United States of America

British Library Cataloguing in Publication Data

Data available

Library of Congress Control Number: 2016958446

ISBN 978–0–19–879369–4

Joe Herbert
Lou Muglia

# Foreword

I was there!

The thesis of this book is that the neuropeptide corticotrophin releasing factor (CRF) serves as an information signal for coordinating challenges to the body most often described in the form of the general term *stress. Stress* here can be defined from Selye, as "the nonspecific (common) result of any demand upon the body" (Selye's Guide to Stress Research, Van Nostrand, New York, 1990). CRF was long hypothesized by endocrinologists to exist, and it finally was discovered in 1981 by Wylie Vale, Catherine Rivier, Joachim Speiss, and their teams at The Salk Institute. In a series of elegant classical physiological studies, they quickly confirmed that this 41 amino acid peptide controlled the hypothalamic-pituitary-adrenal response to stress. Indeed, I was there. I, as well as my colleagues in the Arthur Vining Davis Center for Behavioral Neurobiology, was the first to inject CRF into the brain of rats and systematically observe behavioral effects. CRF at low doses produced arousal, but at high doses produced anxiety-like responses that led to our general hypothesis for its action: "CRF has a direct neurotropic role in the central nervous system to mediate behavioral responses to adaptive challenge particularly those involving activation and stress." For about 10 years, this hypothesis remained unproven, with much supporting evidence from neuroanatomical studies and from direct neuropharmacological studies where CRF itself exaggerated behavioral responses to stress. However, with the discovery of the effectiveness of CRF peptide antagonists and then small molecule antagonists ultimately one could accept the general hypothesis that CRF has a neurotropic role in the brain that parallels its peripheral actions to drive physiological responses to environmental challenge. Modern molecular genetic techniques are now targeting specific CRF pathways to confirm its neurotropic actions in specific neurocircuits that mediate behavioral responses to environmental challenge and the fascinating bidirectional interaction between hormonal consequences of CRF activation (glucocorticoid physiology) and brain function.

This book by Jay Schulkin is a comprehensive, scholarly tour de force of the physiology (using the broad sense of the word) of CRF, giving insights into its role in maintaining homeostasis and driving allostasis and pathology. Well written and well researched, it flows logically and pulls together diverse literature into common themes. Chapters explore the discovery of CRF, its evolution, its role throughout the body, how it is conserved evolutionarily, the role of CRF in the hypothalamic-pituitary-adrenal axis and the critical classical negative feedback mechanism, how CRF plays a key role in fetal maturation and parturition, and its role in the central nervous system to mediate fear and anxiety, but also social contact, learning, and memory. An overarching theme is how CRF functions in response to challenge but also can drive pathophysiology such as anxiety and depression, addiction and addiction-like disorders, and irritable bowel syndrome. He ends the book with a compelling treatise on its role as an information molecule that promotes change in almost all of the body's organ systems, not just in behavior (see above, the original hypothesis regarding the neurotropic actions of CRF). Perhaps more importantly, he argues how such a fundamental peptide system may contribute to individual differences in how the body develops resilience and self-regulation, and—dare I be even more provocative than Dr. Schulkin—happiness.

George F. Koob, Ph. D.

# Preface

My career in science began with the study of one molecule: aldosterone, a steroid secreted by the adrenal gland that regulates fluid balance and sodium in the body, but that also acts in the brain to help insure salt ingestion. It was a simple model, given to me by my science teacher George Wolf, in which to study whole-body regulation, and furthered by many collaborations with a very rich set of colleagues, one of whom, Alan Epstein, had the idea that adrenal steroids facilitate angiotensin in the brain to help generate water and salt ingestion. This idea, held by others (James Fitzsimmons) and furthered by many people (e.g. Steven J. Fluharty), would set a context for an early program project grant from National Institute of Mental Health.

I left New York to follow my wife to Washington, DC: best decision in my life—to marry April Oliver. I joined Philip Gold at the National Institute of Mental Health. I took with me this basic idea of a steroid facilitating a neuropeptide, a focus on a neural circuit that included the amygdala, and a whole-body regulation orientation to mammalian physiology, but now also mental health.

Philip Gold introduced me to the study of CRF after he hired me in 1992 as the Associate Director of the Clinical Neuroendocrinology Branch at the National Institute of Mental Health. Until that time, I had worked on basic motivational systems in which steroids facilitate neuropeptides in the regulation of behavior and physiological viability. It occurred to me that the focus of our branch, which was anxiety, depression, and trauma, might similarly fall under that purview.

What dominated the branch and the field was the negative restraint of CRF by glucocorticoids, as well as its devolution of the normal hypothalamic function under conditions of sustained and relenting duress. An alternative view was that glucocorticoids might facilitate CRF in regions that are now being tied to fear-related events. With a postdoctoral fellow in our branch, Shinya Makino, we went about looking to see if it were possible. Three colleagues were very supportive: Mitch Kling in the branch at NIMH, Jeffrey Rosen in the Biological Psychiatry Branch, and Bruce McEwen at Rockefeller University.

We found that it was, and at least one other scientist had found something similar, but the findings were buried in a paper (Swanson and Simmons 1989) devoted to the hypothalamus; since then, many others have found variants on the theme. This book reflects a piece of the literature.

After all, a steroid facilitating a neuropeptide was not uncommon. What was common was only considering CRF for its role in dysregulation of normal hypothalamic function. And indeed, along with post-traumatic distress disorder and a variety of conditions of adversity, the steroid cortisol, all purpose and all pervasive in coping with events along with the more narrowly focused neuropeptide CRF, emphasized just one form of regulation: we know now that the universe of CRF and glucocorticoid interaction is very rich.

What the literature says is that CRF is first diversely distributed across species (invertebrates and vertebrates), and most of it exists outside the hypothalamus. Indeed, CRF is so richly outside that it interacts with multiple forms of regulation by adrenal steroids in different regions and across different end organ systems (e.g., placenta).

CRF, as you will see, is involved in the great array of biological phenomenon. And I have approached the subject in this brief book under that searchlight of discovery.

I have approached CRF in a very similar manner to my book on sodium hunger published in 1991. I approached the hunger for sodium in as many ways as I could. It was both a narrow subject, a specific hunger, and a broad one, sodium and fluid homeostasis.

CRF is a narrow subject, one information molecule among many others; narrow, therefore, because it is one object, but CRF is also broadly involved in many biological functions. So the aim of this book is to show both sides and from a particular point of view. CRF and its associated intellectual partner glucocorticoids are part of a larger family of chemical signaling systems in the body.

The book covers a small part of the CRF story; the literature is quite extensive. I have selected some of the more dominant, obscure, and unusual features of CRF. I apologize in advance for the many that are left out. This is but a small window into the diverse roles and expressions of CRF and those of us who have studied it.

I thank the many people who have allowed me to collaborate with them. My science teacher George Wolf long ago told me to find a place to be part of contributing to the science, enjoy being part of the world of scientific inquiry, and collaborate with good people. I hope I have done just that.

The oxygen of intellectual exchange is the collaborations and friendships that emerge in the context of inquiry. Many, many thanks for the many people who made this book possible.

# Table of Contents

Chapter 1

# Introduction

Life began on Earth approximately 3.8 billion years ago in the form of single-celled prokaryotes. Initial prokaryotic life required carbon dioxide to survive; however, at approximately 3.0 billion years ago, photosynthesizing cyanobacteria evolved on Earth, thus requiring oxygen, water, and sunlight to survive. Regardless of the introduction of photosynthesizing bacteria, life continued to persist in variable forms with variable requirements. Following the initial emergence and proliferation of life 3.8 billion years ago, the rate of change appears to have been slow for the first several billion years of life's history. However, within the half a billion years comprising the Cenozoic, Mesozoic, and Paleozoic periods, life began to change with a voracity that produced a great variety of living things, from new species of protozoa and bacteria to multicellular organisms like fungi, plants, and invertebrates. Despite such a variety in life forms, one common feature of all of these organisms was the expression of peptides and steroid hormones, and most, for instance, had endorphins, insulin, or adrenocorticotropic hormone (ACTH), to name a few.

Our evolutionary history is very much tied to the expression of—as I will call them throughout this book—information molecules. These information molecules are chemical signaling agents that are important for the functioning not only of human beings but of all living things. Information molecules convey signals between bodily systems and provide a rich history of the ever-developing role of these signals in diverse species throughout time.

Information molecules are structures of beauty. This book is about living tissue and the information molecules that participate in the origins of life and regulate, sustain, and build biological tissue. In particular, this book focuses on a molecule called corticotrophin releasing factor (CRF). An information molecule, by metaphorical expression, anchored in biology. I describe in this book the diverse function of CRF: its many roles in signaling, facilitating change, and responding to danger throughout its history.

## CRF

CRF is a peptide hormone produced in diverse tissues in the body. CRF, which is also often referred to as *corticotrophin releasing hormone* or *CRH*, is most commonly linked to the hypothalamic pituitary adrenal (HPA) axis. It was named CRF for its role as the releasing factor of ACTH, which in turn triggers the release of cortisol (Hauger et al. 2003).

A core idea was posited with some evidence in the late 1930s (Harris 1937) and then more specifically in the 1940s and 1950s of diverse releasing factors from the hypothalamus (e.g., thyroid releasing factor) to the pituitary gland (e.g., thyroid stimulating factor), (e.g., Green and Harris 1946; Guillemin and Rosenberg 1955; Saffran and Schally 1955; Saffran et al. 1955; Sawyer et al. 1949a, b). Neural control of pituitary function became, over a 30-year period, a well-established piece of our understanding of neuroendocrine function (see Fink 2015; Watts 2015).

CRF is well known and studied as a hypothalamic releasing factor (Schally et al. 1958; Saffran et al. 1958; Vale et al. 1972; Guillemin and Schally 1963). It, along with ACTH and cortisol, are mobilized under diverse conditions, including adversity, and are often thought of as part of the stress axis. CRF, however, is much more than this. One aim of this book is pushing the conception of CRF beyond the HPA axis and what most people know about CRF. Since its original discovery, research has unearthed that this molecule is much broader than a hypothalamic releasing factor. And it took a while to discern CRF and its properties outside of its role as a releasing factor of ACTH. The scientific community now knows that CRF is a dynamic and diversely widespread peptide hormone that plays many roles and has many functions, in addition to its role as a releasing factor in the brain, as is detailed throughout this book.

Within the scientific community, there is some variation in nomenclature. Some scientists refer to CRF as CRH; therefore, throughout this book, references to figures from outside sources may refer to CRF as CRH. While the object is the same, and the term *CRF* is used in this text, many references to CRH have been maintained out of consideration for other authors' perspectives and interpretation of the object. This author is of the mindset that *CRH* is a fitting term for the molecule, and all of its functions, as it exists today. But that said, I understand and respect the historical and prominent point of view (i.e., referring to the molecule as CRF). It is up to each reader and scientist to determine which term is the best fit given their understanding and perspective of its many roles.

Wylie Vale, a student of Roger Guillemin, was one of the most important, if not the most important researchers in the study of CRF: Guillemin was a student of Hans Selye, and, notably, Guillemin, Selye, and Andrew Schally won a Noble Prize for their research on hypothalamic releasing factors (see Guillemin 1977; Schally 1977). Initially, Vale arrived at Guillemin's laboratory at the Baylor School of Medicine to work on TSH-releasing factor, but eventually he turned to the study of CRF at the Salk Institute, where he had a great many students. Vale and many colleagues (e.g., Rivier et al. 1983) and students did perhaps more in the initial focus on and study of CRF than any other group, and CRF was in part further described and then investigated in great detail by his lab. His group will be well represented across many pages in this book, but, still, these studies are only a small part of the investigations into this important information molecule (see Bale and Chen 2012).

CRF is a peptide and thus is one example of an information molecule. Peptides are strings of amino acids that can be short (as in oxytocin and amino acids) or long (as in CRF and forty-one amino acids) (Table 1.1).

For instance, everyone seems to have heard of endorphins associated with the regulation of pleasure and pain, leptin associated with fat and food regulation, or, perhaps, oxytocin associated with breastfeeding or social contact. CRF is well known as a neuropeptide, and is expressed across diverse regions of the brain. Perhaps less well known, however, is CRF's tie to diverse forms of behavior, including appetitive behaviors and avoidance behaviors (Koob and Bloom 1985). As shown in Table 1.2, in various behavioral paradigms (e.g., elevated plus-maze, cued electric shock and taste conditioning), centrally administered CRF peptides induce behavioral effects that can be reversed by CRF blockers.

Indeed, the idea of neuroendocrine systems as information molecules that capture classes of related molecules in form and function are common currency in neuroscience (e.g. Starling 1923; Pavlov 1928). Diverse peptides and steroids (e.g., cortisol, estrogen, thyroxin, and vitamin D [a steroid misnamed as a vitamin]) are part of a super family of steroid and steroid receptors. The conceptual construction of related peptides and steroids (or transmitters) is a common approach to understanding the diverse forms of cross talk or activation of receptor sites.

As we will see throughout the text, the effect of CRF is much broader than may be generally thought, and it is tied to many forms of adaptation. What is an information molecule in this age of information? CRF is a signal, among other things, that alerts one to danger. A key aim of this book is clarify under what conditions and by what mechanism CRF is triggered as a signal; how it is transmitted; and how it is delivered across receptor sites in different parts of the body.

**Table 1.1** Alignment of amino acid sequences of members of the CRF peptide family.

| **Corticotrophin Releasing Factor Sequences** | | |
|---|---|---|
| Perciformes | Tilapia | SEDPPISLDLTFHLLREMMEMSRAEQLAQQAQNNRRMMELF |
| Salmoniformes | Rainbow trout | SDDPPISLDLTFHMLRQMMEMSRAEQLQQQAHSNRKMMEIF |
| Siluriformes | Bullhead | SEDPPISLDLTFHLLREMMEMSRAEQLAQQAQNNRRMMELF |
| Cypriniformes | Sucker | SEEPPISLDLTFHLLREVLEMARAEQLVQQAHSNRKMMEIF |
| | Carp | SEEAPISLDLTFHLLREVLEMARAEQMAQQAHSNRKMMEIF |
| | Goldfish | SEEPPISLDLTFHLLREVLEMARAEQMAQQAHSNRKMMEIF |
| Amphibia | Xenopus | AEEPPISLDLTFHLLREVLEMARAEQIAQQAHSNRKLMDII |
| Mammalia | Sheep | SQEPPISLDLTFHLLREVLEMTKADQLAQQAHSNRKLLDIA |
| | Cow | SQEPPISLDLTFHLLREVLEMTKADQLAQQAHNNRKLLDIA |
| | Human | SEEPPISLDLTFHLLREVLEMARAEQLAQQAHSNRKLMEII |
| | Dog | SEEPPISLDLTFHLLREVLEMPGAEQLAQQAHSNRKLMEII |

**Table 1.2** Behavioral effects of centrally administered CRF peptides.

| **CRF/Behavioral** |
|---|
| Suppresses exploration of unfamiliar environments |
| Facilitates startle |
| Induces conditioned fear |
| Enhances stress-induced freezing |
| Decreases food intake |
| Produces aversion |
| Enhances sensitization |
| Enhances defensive burying |

*Source*: Adapted from George F. Koob, Stephen C. Heinrichs, Emilio Merlo Pich, Frédérique Menzaghi, Helen Baldwin, Klaus Miczek, and Karen T. Britton, *The Role of Corticotropin-Releasing Factor in Behavioural Responses to Stress*, *Corticotropin-Releasing Factor*, 172, pp. 277–95, doi: 10.1002/9780470514368.ch14, Copyright © Ciba Foundation 1993.

In this book, I will also discuss just how diverse a peptide CRF is, as demonstrated by its presence in various tissues in the body that range from the skin to the placenta, and to various regions of the brain. As Dobzhansky (1962) noted, in light of Darwin (1874), and then looking beyond Darwin, CRF must be placed in the larger world of regulatory biology. Evolutionary trends do not proceed in a continuous one-dimensional direction; there are starts, turns, and abrupt ends. The study of CRF or other information molecules is mostly about diverse functions in physiological and behavioral regulation of the internal milieu and adapting to an ecological and or social context.

The CRF hormone family was found to be expressed in peripheral tissues (e.g., CRF in skin, colon, kidney, liver, and adipose tissue; urocortins in the heart and stomach) in rodents, as well as primates, implying that the CRF hormone-family expression in peripheral tissues is probably widespread among mammalian species and thus is likely an ancient adaptation. One example is the expression of CRF in the colon (Yuan et al. 2010). Double labeling of rat proximal colon tissue shows that CRF is present in all ganglia in myenteric plexuses and most of the enterochromaffin cells in mucosal epithelia. What is notable is first that it is detected and second that it is increased dramatically under conditions of adversity, in this by the insertion of endotoxins.

## CRF and Fearful Danger

One example of CRF's function is its role in the brain when responding to fearful stimuli. As we will see, CRF expression underlies fear not only in humans but also across many species. CRF is elevated under unfamiliar or unpredictable contexts of potential danger. In this way, the molecule acts as a signal about potential danger, which organisms perceive as fear (Davis et al. 1997, 2010).

Fear is costly to health, and this is partly because the excessive expression of CRF is costly. One result of prolonged overexposure to CRF is the deterioration of neural structures. In other words, the elevation of CRF or receptor activation in specific regions of the brain is essential for responding to potential danger that evokes fear, but a continuous state of fear is damaging to bodily capability. Understanding the link between adaptive functions of CRF, as well as pathology, is a continuous theme of this book.

Some important issues that emerge when considering information molecules such as CRF are production, receptor activation, and maintenance, all within the context of biological viability. Maintenance and regulation of peptides like CRF are essential to sustaining internal viability. Regulation is important in the diverse tissue in the body including, in the case of detection of danger, neural tissue. Again, however, the same signaling system that can indicate danger can itself cause danger when overexpressed for an inappropriate time and can begin to induce neural deterioration. In this way, CRF can move from protective signaling to compromising tissue. Diverse forms of bodily vulnerability and viability result in increased expression; for instance, hypothalamic CRF is augmented with multiple sclerosis (Erkut et al. 1995).

This book will explore the expression and putative role of CRF in diverse end organ systems tied to pathology. In these cases, the pathologies resulting from CRF function may present an easier mechanism by which we can understand the role of CRF in the body. CRF is tied to diverse forms of visceral distress, including, for instance, irritable bowel syndrome. CRF along with immunological proteins (e.g., cytokines) are embedded in inflammatory tissue.

CRF is almost always tied to inflammatory signals, such as those seen in bacterial vaginosis, colonic mucosa with colitis, liver fibrosis, melanoma, and chronic inflammatory disorders (Sato et al. 2002; Chatzaki et al. 2013; Yuan et al. 2012; Meng et al. 2015; Wang et al. 2016; Liu et al. 2016; Gay et al. 2008); and psychosocial adversity, which enhances CRF expression, increases colitis, and reduces recovery (Reber et al. 2006). In fixed melanoma tissues, CRF antibodies were expressed in a highly similar pattern to alpha melanocyte-stimulating hormone (a-MSH) antibodies when detected by the avidin-biotin-peroxidase system. Areas of positive melanoma cells expressed both a-MSH and CRF antibodies (Sato et al. 2002).

These CRF-related inflammatory signals result from the body being under siege in one way or another. Bacterial vaginosis, for instance, is a signal of danger and during pregnancy is tied to the vulnerability of giving birth to low birth weight babies (see Chapter 4).

Elevated levels of CRF in mammary glands can accelerate the growth of breast cancer, or in the ovaries and ovarian cancer. Inflammation in the tissue is one clear expression of CRF or immunological responses to tissue danger. In general, this information molecule goes from a signal of danger to being part of the process that results in tissue deterioration.

Physiological concepts like inflammation or metabolism are not as simple as they seem at first glance, nor are other behavioral concepts such as fear and depression. Signals of danger, for instance, are diverse. They may include the expression of fear at an unexpected sight, or the signal from bleeding after a predator attack, or alertness for danger when approaching a water hole to drink. CRF elevation is tied to signals of danger in all of these cases and more (Anisman et al. 2007; Merali et al. 2001). Figure 1.1 demonstrates the instances when CRF is inferred to be elevated in nature, based on analogous laboratory demonstrations (Figure 1.1).

For instance, in laboratory-bred rodents for which water or food is provided, in several different contexts, CRF expression is elevated in several regions of the brain, including the hypothalamus, amygdala, and frontal cortex. In nature, a trip to the water hole is not a free ride for most animals as the event is typically rich in predatory possibilities and conflict; approach and avoidance behaviors permeate the event. What is at stake is large: survival or death.

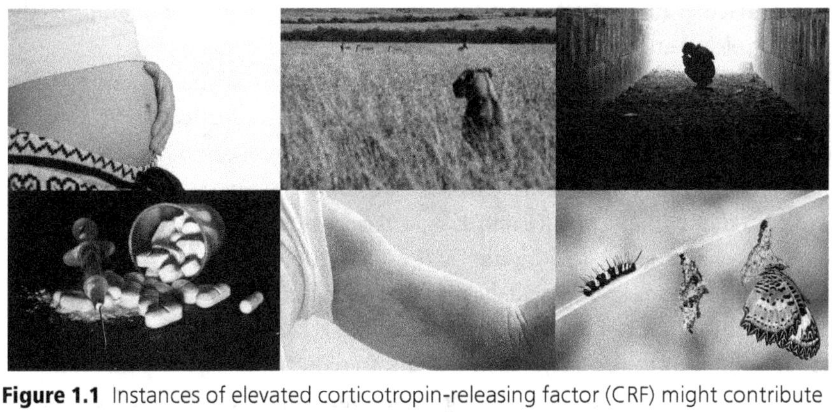

**Figure 1.1** Instances of elevated corticotropin-releasing factor (CRF) might contribute to: pregnancy and parturition, presence of predators, fear and anxiety, addiction, inflammation and infection, and metamorphosis.

Indeed, under conditions of predatory threat (e.g., hawks) a number of non-predatory bird species produce fewer hatchlings. In laboratory models, such conditions of predator threat may also result in elevated levels of CRF in the brain (Clinchy et al. 2011).

CRF evolution is knotted to other sets of information molecules such as cytokines, and is tied to inflammation and the immune signaling system. CRF is tied to inflammation such that it is expressed in tissue (e.g., in infections of the brain, placenta, skin, liver, or kidney), and it signals danger or bodily harm. As in other expressions of CRF, it is initially adaptive upon activation; however, it becomes maladaptive if it persists beyond biological capability. CRF is important for growth and bodily composition and capability (Murâni et al. 2006), but also pathology. In other words, ongoing infections, injuries, or psychosocial stress can lead to the induction of inflammation and an influx of the information molecules associated with CRF. Over time, inflammation can pose a danger to bodily tissue.

## Diversity of Expression of CRF

One apparent fact about CRF is that it can be produced in various tissues and has receptor sites across many regions of the body. These regions include the skin, gastrointestinal tract, brain, heart, kidney, and placenta to name a few. Vitamin D, which is produced in the skin by ultraviolet light, can influence CRF gene expression in diverse tissue (e.g., skin). Ultraviolet radiation initiates within the skin a response system, which can activate the HPA and CRF production, by producing humoral signals that travel to the brain, directly to the paraventricular nucleus (PVN) or pituitary, or to the adrenal gland; ultimately, cortisol release and steroidogenesis induction is stimulated.

Indeed the skin, the largest organ in the body, is a massive endocrine gland; many steroids are produced in the skin, including glucocorticoids, Vitamin D, androgens, and estrogens (Slominski et al. 2013; Viau et al. 2001); it is a working factory of endocrine events. Ultraviolet light, which induces vitamin D, also results in the synthesis of CRF and a wide variety of hypothalamic pituitary and extrahypothalamic-related events (Zbytek et al. 2006).

CRF is a peptide that is widely distributed in the body and the brain, and much of this book will be about understanding relationships between location and function. At the outset, it is important to reflect that variation in an information molecule is a little like the variation found in speciation. In these cases, the expression of CRF is diverse in not only its site but also its purpose.

Two recurrent features linked to CRF are movement and change. CRF expression in regions of the brain is tied to paying attention to novel events and invoking movement in response to those events. Indeed, CRF provokes simple organized rhythmic behavior (e.g., Rosen et al. 1994). Depending upon the time and level, its provoked behaviors can range from startled responses to seizure. CRF-induced seizures are tightly linked to a G-protein coupled type CRF 1 receptor (Narla et al. 2016). CRF is also tied to movement under normal and pathological conditions, largely based on a dose response. As a demonstration of this function, movement in diverse species has been noted when CRF is directly administered to the brain.

CRF is also tied to the physiology of change, such as in metamorphosis or birth events. In several species tested in the laboratory (Figure 1.2), CRF is associated with metamorphosis, and infusions of CRF facilitate the induction of metamorphosis. Elevated levels of CRF can be beneficial, for instance, by promoting metamorphosis to adulthood in newts or toads, and in its production in the placenta to facilitate primate parturition and birth. CRF, however, is not

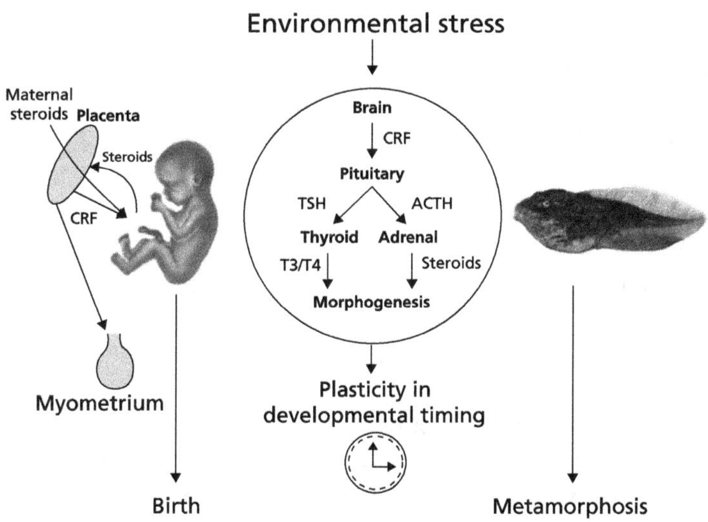


**Figure 1.2** Environmental conditions affect the timing of life history transitions through the evolutionarily conserved actions of stress hormones in humans and amphibians. Unlike the tadpole, which is a free-swimming animal, the human fetus is influenced by hormones of maternal and placental origin. Conversely, fetal hormones can alter physiology of the mother (and the placenta); for example, maternal cortisol is transferred to the fetal compartment and can alter fetal development. Adrenal steroids of fetal origin exert a positive feedback on placental corticotrophin-releasing factor (CRF) production and promote uterine contraction.

Reproduced from Erica J. Crespi and Robert J. Denver, Ancient origins of human developmental plasticity, *American Journal of Human Biology*, 17 (1), pp. 44–54, DOI: 10.1002/ajhb.20098, Copyright © 2004 Wiley-Liss, Inc.

the sole player in these cases as thyrotropin-releasing hormone, thyroxine, and adrenal steroids all play important roles (Denver 1999, 2009).

Environmental conditions and adversity can influence timing of metamorphosis and development across species (e.g., humans and amphibians.) The human fetus is influenced by both maternal and placental hormones, and fetal hormones can have an effect on the mother and placenta. Metamorphosis is clearly not a stagnant, standard process; it is influenced by many factors. CRF is thus tied to promoting change: internal end organ change in these instances.

Cortisol facilitates CRF gene expression in the placenta and several extrahypothalamic regions (e.g., the amygdala, see Chapters 3 and 5). There is normal parturition in which cortisol promotes fetal development (e.g., lung formation) and promotes peptides such as CRF to facilitate parturition. Excessive production of CRF or its analogs under conditions of adversity can—as in other areas—have negative consequences on growth and change. For instance, under conditions of duress, prolonged CRF exposure in utero can promote development more quickly, resulting in preterm low-birth weight babies and compromised reproductive fitness (Majzoub et al. 1999; Smith 2007; Robinson et al. 1989; Torricelli et al. 2011a).

Importantly, CRF is influenced by diverse steroids including vitamin D, progesterone, estrogen, and testosterone, as well as cortisol. Estrogen, for example, can increase differential CRF gene expression in the brain. CRF in the amygdala can be increased without affecting CRF in hypothalamic regions by estrogen. Testosterone can increase CRF gene expression in the brain and facilitate aggressive and fear-related behaviors (Wood and Stanton 2012).

Elevated estrogen also increases CRF gene expression in the placenta, and thereby helps facilitate parturition. Progesterone, which is sometimes referred to as having a quiescent effect on tissue (e.g., cervix), has the opposite effect of cortisol. Progesterone decreases CRF gene expression and therefore participates in delaying parturition; thus, in other words, in the placenta, progesterone decreases CRF (Challis 2001; Norwitz and Caughey 2011).

Hypothalamic sources of CRF are the primary source of CRF in the regulation of the HPA axis. Differential regulation of CRF by steroids in the brain and other bodily tissues is a common theme (Frim et al. 1990; Swanson and Simmons 1989). This reflects, in part, an adaptive design integral to our evolutionary capabilities, and a more flexible regulation of both the internal milieu and the social/ecological context. Greater regulatory flexibility is one critical feature in our evolutionary ascent. While CRF is a major focus, the overall inquiry of this book will draw on the regulatory and evolutionary links of diverse information molecules (neurotransmitters, steroids, other peptides, and neuropeptides, in addition to CRF) in understanding the regulation of the internal and social/ecological milieu, as well as the organization of action.

CRF will be used as a window into an evolutionary perspective (Lovejoy et al. 2009). This book will trace CRF through its evolutionary past, hundreds of millions of years. It begins with the origins of information molecules and can be traced back nearly to the origins of life on the planet through both vertebrates and invertebrates. The evolutionary history of CRF is perhaps tied to circadian rhythmicity and the light–dark cycle through the origins of mammals, the placenta, and lactation.

Figure 1.3 shows a phylogenetic tree of the number of genes for CRF, which helps visualize the evolutionary history of CRF (Alsop and Vijayan 2009). Species like the carp, goldfish, white sucker, and sockeye salmon have two genes for CRF, and they all descend from the shared Ostariophysi. However, the green

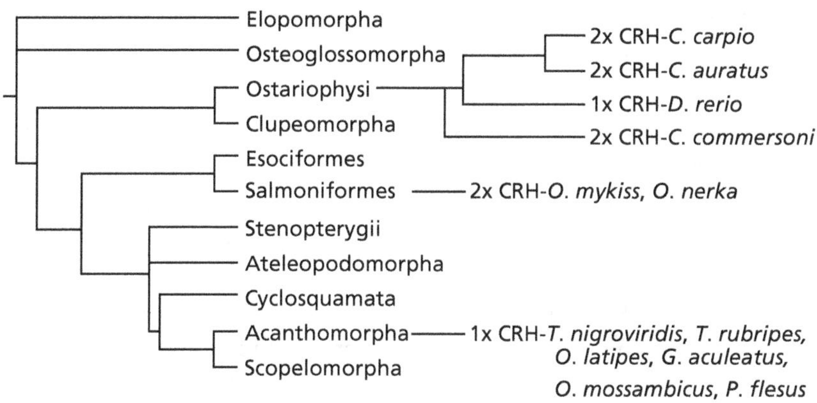


**Figure 1.3** The phylogeny of the number of genes for corticotrophin-releasing factor (CRF). Please note that although zebrafish have two POMC genes, ACTH is derived from only one of the duplicate copies. Trees were redrawn from The Tree of Life Web Project. The additional cyprinid (Ostariophysi) phylogeny was based on trees constructed from sequences of GR and the mitochondrial cytochrome b. C. carpio (carp), C. auratus (goldfish), D. rerio (zebrafish), C. commersoni (white sucker), O. mykiss (rainbow trout), O. nerka (sockeye salmon), T. nigroviridis (green spotted puffer), T. rubripes (fugu), O. latipes (medaka), G. aculeatus (stickleback), O. mossambicus (tilapia), P. flesus (flounder), O. keta (chum salmon), D. labrax (European sea bass), Pimephales promelas (fathead minnow), and H. burtoni (Burton's mouth brooder).

spotted puffer, fugu, medaka, stickleback, tilapia, and flounder only have one gene, and they all descend from the shared ancestral clade, Acanthomorpha.

In this inquiry, one will see just how far beyond the HPA axis CRF goes. Modern genetic tools such as optogenetics (Deisseroth 2015; Asok et al. 2016a; Sanford et al. 2016) allow the neuroscience researcher to investigate regional specificity of CRF neurons; those CRF neurons in regions of the brain tied to memory, or diverse forms of distress (gastrointestinal), fear, or anxiety (Chen et al. 1992). Changes in CRF can result in differential survival responses, to avoid events or approach events, or have a mixture of the two responses.

This book also emphasizes the evidence collected over the last few decades suggesting that one important function of the glucocorticoids is to facilitate CRF expression and not necessarily to restrain CRF under diverse regulatory conditions that are broadly tied to forms of behavioral and physiological adaptation (e.g., adaptation to uncertain and fear-provoking events, to parturition, and to birth). The other theme considered herein is the diverse genetic and epigenetic forms of regulation that underlie adaptation.

There are 7 chapters in the book in addition to the introduction and conclusion:

*Chapter 2.* Evolution and Diversification of Function of an Information Molecule: Corticotrophin-Releasing Factor

Chapter 2 begins with a depiction of the evolutionary origins of CRF in living things. CRF appears to date back hundreds of millions of years. It is found in diverse invertebrates, including flies and bees. Invertebrates' brains look nothing like those of vertebrates, except for the diverse information molecules that underlie both brain systems. There is no clear anatomical organ like the HPA axis in invertebrates, yet information molecules, including CRF, are just as important to invertebrate functioning as they are to vertebrates. CRF in invertebrates is linked to basic regulatory functions such as osmotic regulation, food intake, learning, and circadian rhythmicity.

*Chapter 3.* Conservation of CRF in Brains and Its Regulation by Adrenal Steroids

The regulation of the HPA axis has been categorized for over 40 years as the classical mechanism of slow-acting genomic regulation of gene products, which has given way to both slow and fast regulation of the HPA axis. Of course we still do not know how cortisol restrains the production of CRF in the paraventricular nucleus and thereby directly decreases ACTH and subsequently decreases cortisol; the classical negative-feedback regulatory system that we know exists with plenty of examples provides a mechanism, but how it works well, that is another thing.

As we will see, the HPA axis is restrained by glucocorticoids; but other regions of the brain are not, such as the central nucleus of the amygdala and bed nucleus of the amygdala. But we now know that both chemically and electrically, these regions are not the same or are not "created equal" (Dabrowska et al. 2011, 2013). As shown in Table 1.3, PVN CRF neurons may be more heavily anchored to glutamate, whereas neurons of the bed nucleus of the stria terminalis (BNST) are to GABAergic. This differential regulation of CRF neurons may be a core feature in design.

CRF neurons have distinct membrane properties when in the PVN and the BNST (Figure 1.4). Neurons of the PVN (neurosecretory neurons: magnocellular and parvocellular) exhibit significantly lower resting membrane potentials than BNST neurons do, as well as a higher threshold for spike firing, higher input resistances, and larger fAHP than CRF neurons in the BNST. These differential neural relationships of CRF populating cells are further differentiated by distinct electrophysiological responses to neural stimulation (Dabrowska et al. 2015).

And one mechanism for which CRF is regulated is via cannabinoid-related mechanisms. An example of this cannabinoid-related mechanism is rapid glucorticoid signaling in PVN parvocellular neuroendocrine cells. When glucocorticoids bind to membrane-associated receptors that activate G-protein-dependent intracellular signaling pathways and produce cAMP, this results in endocannabinoid (CB) synthesis and retrograde release. Endocannabinoids then bind to presynaptic CB1 receptors and suppress glutamate release onto the PVN nucleus (Hill and McEwen 2009; Hill et al. 2013).

*Chapter 4*. Radical Change: Metamorphosis and Parturition

CRF is involved in two quite striking processes: the first is metamorphosis and the second is parturition. Both of these processes revolve around the wondrously vital feature of radical change. In both instances, we are discussing the changing of form and the maturation of an organism. While one change is ancient and one more modern, both continue to be important features of life on this planet and both represent the importance of development and its responsiveness to changing environments.

The events take place in environments that affect the events and alter the internal milieu. Challenging environments (temperature and food resource availability) can accelerate this change prematurely. Hostile or less friendly environments (e.g. nutrition, drug use, metabolism, and fear) can accelerate or delay fetal maturation and parturition. Under these conditions, CRF is one signal in the body for normal development or the induction of prematurity under conditions of adversity.

**Table 1.3** Distinct membrane properties of CRF neurons in the PVN and BNST.

| CRF Neurons | Spike | | | | | |
|---|---|---|---|---|---|---|
| | **Amplitude (mV)** | **Half Width (ms)** | **Threshold (mV)** | **Rise Time (ms)** | **Decay Time (ms)** | **fAHP (mV)** |
| PVNpc (12) | 70.1 ± 2.4 | 1.003 ± 0.072 | −33.1 ± 0.89 | 0.36 ± 0.022 | 0.98 ± 0.094 | −14.5 ± 0.9 |
| PVNmc (4) | 73.3 ± 4.5 | 1.01 ± 0.042 | −33.2 ± 1.12 | 0.37 ± 0.024 | 0.89 ± 0.030 | −19.2 ± 2.6* |
| Type III BNST (19) | 76.9 ± 1.4 | 1.04 ± 0.024 | −39.0 ± 0.75*,# | 0.44 ± 0.016 | 1.12 ± 0.045 | −4.9 ± 0.5**,## |

*,**, $p < 0.05$, and 0.01 respectively, vs. PVNpc; #,##, $p < .05$ and .01 respectively, vs. PVNmc. PVNpc, parvocellular neurons of the PVN; PVNmc, magnocellular neurons of the PVN.

Adapted from J. Dabrowska, R. Hazra, J-D. Guo, S. DeWitt, and DG Rainnie, Central CRF neurons are not created equal: phenotypic differences in CRF-containing neurons of the rat paraventricular hypothalamus and the bed nucleus of the stria terminalis, *Frontiers in Neuroscience*, **7**:156, Table 2, doi: 10.3389/fnins.2013.00156, 

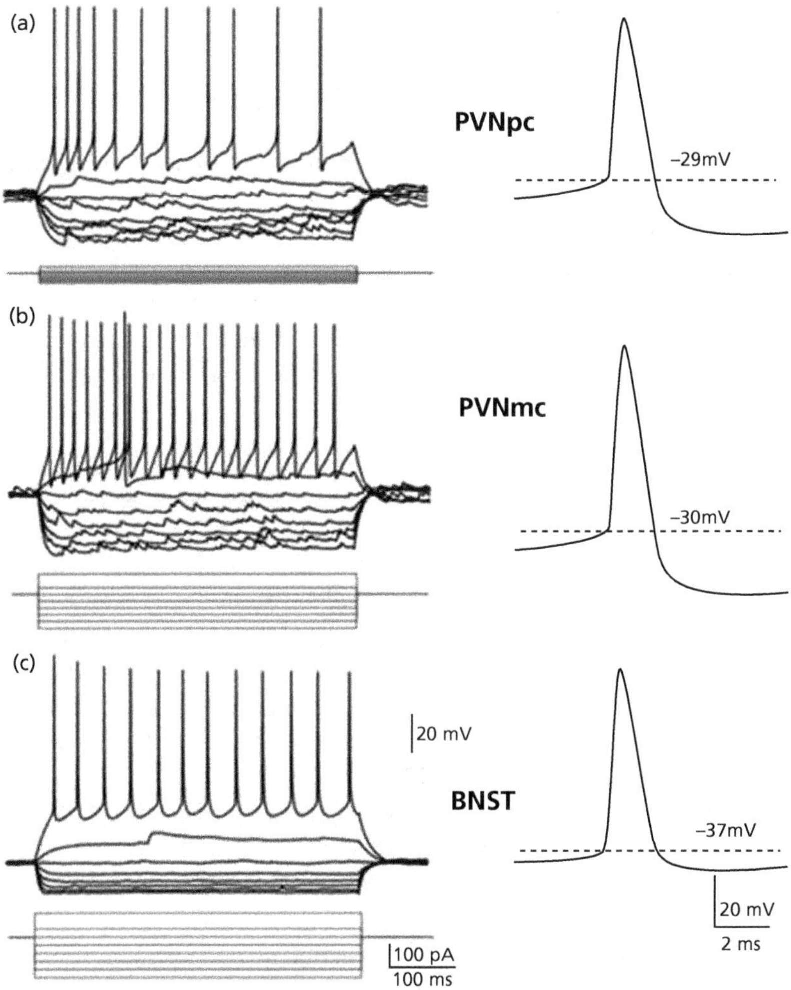


**Figure 1.4** Distinct membrane properties of putative corticotrophin-releasing factor (CRF) neurons in the PVN and type III CRF neurons in the BNSTALG. Representative traces demonstrate the membrane responses of these neurons to membrane current injections and typical spikes. (a) PVNpc neurons have extremely high input resistance and high threshold for spike firing. When hyperpolarized, these neurons exhibited an inward rectification current. No outward rectification was observed during depolarization current injections. The right trace shows a typical spike that has a high firing threshold and a prominent fAHP. (b) PVNmc neurons have smaller input resistance than that of PNVpc neurons and have high threshold for spike firing. Single spike trace also shows this neuron has a high spike threshold and prominent fAHP. (c) In comparison to the PVN CRF neurons, CRF neurons in the bed nucleus of the stria terminalis (BNST) have lower input resistances, lower spike threshold, and smaller fAHP. PVNmc and PVNpc represent magnocellular and parvocellular neurons of PVN, respectively.

Adapted from J. Dabrowska, R. Hazra, J-D. Guo, S. DeWitt, and D.G. Rainnie, Central CRF neurons are not created equal: phenotypic differences in CRF-containing neurons of the rat paraventricular hypothalamus and the bed nucleus of the stria terminalis, *Frontiers in Neuroscience*, **7**:156, Figure 4, doi: 10.3389/fnins.2013.00156, © 2013 Dabrowska, Hazra, Guo, DeWitt and Rainnie.

In mammals, glucocorticoids both stimulate and inhibit CRF promoter sites in placental tissue (Robinson et al. 1988, 1989). The induction of CRF in the placenta by glucocorticoids has been linked to cAMP mediated mechanisms; cAMP activates CRF promoter sites (Nicholson et al. 2004). The regulation of the CRF gene by glucocorticoids that has been described in placental syncytiotrophoblast cells by rising glucocorticoids underlies parturition and is mediated, in part, by a CRF type 1 receptor (Smith 2007; Majzoub 2006; Challis 2001).

Thus this chapter will focus on these two phenomena both under normal conditions and under adverse circumstances in which the role of CRF may be more pronounced.

*Chapter 5*. CRF: Contributor to Fear and Anxiety

Fear is an adaptive signal about danger. For instance, fox-odor exposure to potential prey is tied to CRF gene expression in several regions of the brain of the prey, including the amygdala (Rosen et al. 1996; Takahashi et al. 2005). These effects have been shown in other models within the amygdala and the bed nucleus of the stria terminalis as well (Rajbhandari et al. 2015; Butler et al. 2016); the olfactory mechanism through which this occurs is still being studied (Kondoh et al. 2016). Moreover, the odors of stressed conspecifics result in withdrawal and wariness, responses that are attenuated by CRF antagonists (Takahashi et al. 1990; Roltsch et al. 2014).

Excessive fear is tied to anxiety disorders and vulnerability to the breakdown of mental and physical health. CRF in the brain is tied to these events. CRF for instance may be constrained by the neurotransmitter gamma-Aminobutyric acid (GABA) in key regions of the forebrain and is mobilized by brainstem catecholaminergic neurons that are critical in coping with and adapting to everyday life; and of course, one is less able to do so when these information molecules are compromised by genetic predispositions and social duress. One hypothesis about CRF and the brain is that at least two forebrain sites are differentially involved in regulating both adaptive fear and deleterious chronic anxiety (Kolber et al. 2010; Davis 1998; Davis and Shi 1999). A great variety of events can cause fear in individuals: anything from downsizing at work to acts of terrorism and crime.

The breakdown is the unrelenting sense of adversity and fear of loss. CRF is involved in both the adaptive side and the devolution of function side. For instance, CRF is disregulated in post-traumatic stress disorder and psychotic depression (Nemeroff et al. 1984; Gold et al. 1988; Binneman et al. 2008; Holsboer and Ising 2008; Waters et al. 2015).

This chapter will focus first on the diverse forms of fear (e.g., fear of the unfamiliar, fear about a specific object, and conditioned fear) and then on the mobilization of motivated behavior to combat and cope with fear. Biological responses to adversity of fear-related events, including the biology of temperament in humans, reflect the genetics and epigenetic regulation of information molecules. Resilience is one key factor in our everyday and evolutionary success. CRF is part of that story.

*Chapter 6.* Social Contact, Gonadal Steroids, and CRF

CRF is vital for rapid-fire responses to a vast array of events. These events are often social in nature; and they are much greater than that having to do with fear. Indeed one result of the rapid expression of CRF in different regions of the brain is perhaps to serve the large social/ecological demands in adapting to changing contexts. Parental behavior is perhaps one good example.

This chapter begins with a brief discussion of CRF in approach/avoidance behaviors across pre- and postnatal events. What will follow is the description of diverse steroids, in particular gonadal steroids (e.g., testosterone and estrogen) and their impact on CRF and other peptides expression, and, finally, sex differences in the expression of CRF in the brain.

As shown in all chapters of this book, CRF is richly expressed across evolution and across the brain within many different species. As this chapter shows, CRF also plays diverse roles in behavioral and physiological adaptation, depending upon context, capability, and end organ systems. For instance, CRF is differentially expressed in the brains of different strains of the same species, different species, and sexual differences in the same species, which affects behavior and social contact, and underlies social approach and avoidance (Young 2002). Sex differences are apparent in CRF receptors, as well as in heightened attention and responsiveness (Bangasser et al. 2013, 2015; Van Kempen et al. 2015; Lukkes et al. 2016, Beery et al. 2016). It has been shown that activation of the type 1 receptor within the BNST in some contexts can lead to maternal neglect in lactating mothers (Klampfl et al. 2016).

*Chapter 7.* Surviving, Remembering, Adversity

While CRF is intimately involved in organ development, it is also linked to devolution of function and conditions of danger.

CRF expression itself reveals developmental changes, particularly, in the brain. CRF is linked to diverse forms of learning and timing of events. But CRF may either enhance or degrade learning and memory. CRF tends to enhance salience and visibility, and, therefore, learning and memory consolidation may be enhanced. However, excessive CRF expression begins to compromise these essential capabilities, and it promotes neural atrophy deterioration (Goebel et al. 2010).

Moreover, there are sexual dimorphic differences in CRF neuronal regions. These brain regions include the amygdala, the bed nucleus of the stria terminalis in the forebrain, and regions in the brainstems such as the locus coeruleus (Valentino et al. 1993). These regions of the brain are tied to the regulation of the HPA axis.

There are sexual differences in the responses to inflammatory expression, for which CRF appears to be playing a role, in addition to metabolic function. One further issue is the possible role of CRF in the vulnerability of affective disorders in women, for which the reported incidence when compared can be twice as high as in men (Bale 2011; Valentino et al. 2013).

*Chapter 8*. Cravings and Addictions: The Contributions of CRF to Approach, Ingestion, and Withdrawal

CRF is tied both to the ingestion of diverse drugs and to food withdrawal (Koob 2009). This process, however, is little understood. Indeed, one of the most important discoveries in the addiction research field was that for all addictive drugs that have been tested, this dual phenomenon of ingestion and withdrawal has been expressed; this includes cocaine, heroin, alcohol, and cannabis. The brain is active in all stages of addiction (preoccupation/anticipation, binge/intoxication, withdrawal/negative affect, and psychic pain), and differentially regulated (Koob 2010).

This phenomenon occurs in at least three prominent regions of the brain: the PVN, amygdala, and bed nucleus of the stria terminalis. Two important suggestions about the bed nucleus is first that it is tied to nonspecific anxiety and second that it is tied to the vulnerability to relapse and the reinstatement of drug use and drug abuse. CRF infusions elicit both forms of behavior.

In the bed nucleus of the stria terminalis, there is a network of afferent and efferent connections that interact with brain response effector systems, and through these systems, the extended amygdala and its connections are modulated via both the brain arousal or buffer systems. These neural systems either are local circuits or derived from cell bodies and terminals existing in the hypothalamus or brainstem. Brain arousal system components, for instance, include norepinephrine, vasopressin, and hypocretin. Brain buffer components include nociception cell bodies and terminals, NPY terminals, and endocannabinoids.

CRF and its regulation have also been linked to stress-induced obesity (Dallman et al. 2003), a vulnerability, and a result of our affluent society infused with less exercise and big proportions of junk food. CRF expression in different regions of the brain is tied to the endless barrage of incentives or junk foods or sweet rewards that have little to no nutritional value (Power and Schulkin 2013).

One region of the brain tied to incentive salience is the nucleus accumbens. CRF infusions in this region of the brain can amplify the CRF-related vulnerability to obesity and attention to junk food rewards (Berridge et al. 2004). The interaction of corticosteroids and CRF function together to enhance these events.

*Chapter 9.* Conclusion: The Study of CRF—an Epistemic and Medical Endeavor

This chapter describes key conceptual and historical events in thinking about information molecules that have shaped discovery and understanding in the biological sciences, in general, and the behavioral and neural sciences, in particular. CRF is an historical object; a part of our conceptual, scholarly, experimental, and social landscape.

The conclusion again addresses the question of what constitutes an information molecule, in general, and CRF, in particular. Our age is the age of information; CRF is now set in the larger context of the information age that is the last hundred years. Thus, this conclusion will summarize the preceding chapters, and look at the large social context and the great range of scientific fields to understand more fully this peptide, its structure, and its history. Our understanding of this one information molecule is expanding, as we learn more about its expression or overexpression beyond what is adaptive, as well as its moderation by other information molecules (e.g., GABA and oxytocin) and various contexts (e.g., species, age, gender, setting, or incentive). Information molecules cannot be looked at as independent entities, isolated from their context and other molecules, factors that contribute to diverse outcomes.

Again, the focus is narrow on the one hand and broad on the other. The role of CRF is highlighted throughout the book against the backdrop of the idea of information molecules in the body and their diverse roles in biological viability and evolution; CRF is just one among many others.

Chapter 2

# Evolution and Diversification of Function of an Information Molecule: Corticotrophin-Releasing Factor

## Introduction

Diverse information molecules have a long evolutionary history. Peptides, steroids, and neurotransmitters can be found in plants and insects (Strand 1999). For example, at night, plants secrete melatonin, an information molecule that carries diverse forms of instructions that are linked to circadian rhythms (e.g., Wehr et al. 1993). Thus in plants, invertebrates, and vertebrates, melatonin acts as part of the transduction mechanism in adjusting to daylight and nighttime (Strand 1999). In a diverse array of animals, melatonin is linked to a number of physiological and behavioral adaptations (e.g., Weaver et al. 1996; Nelson et al. 2002).

Plants and fungi synthesize steroids that are known to regulate development and reproduction, respectively. Plant steroid hormones signal via membrane receptor kinases. The nuclear hormone receptors that underlie many of the actions of steroid hormones in mammals and other vertebrates are absent from plants and are estimated to have arisen over 1 billion years ago after the divergence of the metazoans and fungi (Thornton et al., 2003).

From a single ancestral corticosteroid receptor, it is possible to construct a molecular phylogeny of steroid hormone receptors. Making note of the number of sequences in each clade shows the great variety of gene duplication and variation that can exist within a clade for a single receptor. For example, six teleost aldosterone-insensitive glucocorticoid receptor sequences and three teleost aldosterone-activated receptor sequences have been derived from a shared ancestral corticoid receptor.

A fundamental concept in evolution is that all species share a common ancestor. For any two species, if you trace back their descent far enough, you will arrive at a species that is a common ancestor. Phylogenies can be constructed that estimate both the relationships among species and the length of time since the different lineages shared a common ancestor. These phylogenies are based on anatomy, fossils, and molecules, from proteins to DNA. An implication of this concept is that the molecules involved in the processes of life also share common ancestor molecules. In other words, the metabolic and signaling pathways of two related species are likely to have similarities based on being derived from the same ancestral pathways.

The more distantly related the species are, the less likely that the molecules and pathways will be similar. This forms the underlying principle for the molecular evolution studies that have been used to estimate the time of divergence of the human lineage from the great apes (approximately 5–7 million years ago), and recently, to compare bone collagen proteins recovered from 65-million-year-old Tyrannosaurus fossils with bone collagen from modern animal species. Of those tested, the fearsome dinosaur is most closely related to the modern chicken (Schweitzer et al. 2007).

A kind of speciation (retaining or not retaining a common function) is realized through segregation (oxytocin, separated from vasopressin) outside and in different end organ systems. For instance, oxytocin is within auditory regions of the brain at all levels of the neural axis (Yu et al. 1996; Winslow et al. 1993). It could well be involved in animals that "see by hearing," in sensory processes that utilize vital ultrasonic vocalization coordination (Griffin 1958; Kanwal and Ehret 2006).

Evolution selects the diverse sensory systems that are linked to the organization of action (e.g., Hodos and Butler 1997). Oxytocin and vasopressin expressed in auditory nuclei, as in several species of bats, and in a wide array of brain regions (e.g., Kanwal and Rao 2002; Rao and Kanwal 2004), could well be linked to part of the systems in detecting auditory signals vital for approach (mother infant signals) or avoidance (predators; Insel 1992; Hofer and Sullivan 2001; Blass 1996).

Corticotrophin-releasing factor (CRF) systems are tied to signals that are perhaps less familiar, and perhaps dangerous. And depending upon the sensory system of the animal, olfactory, acoustic, and visual signals are tied to distress, in which CRF will be one information molecule among others (Kikusui et al. 2000).

A number of mechanisms, for instance, for peptides, include exon duplication and loss, excision, and gene duplication (Holland and Short 2008). Variation on the production of these information molecules, traced back hundreds and millions of years (e.g., neuropeptide Y, angiotensin; Conlon and Larhammar 2005; Holye 1998; Strand 1999), play diverse functions: for example, food ingestion and metabolism (Schulkin 1999; Herbert 1993; Fitzsimons 1979, 1998; Smith 1997a, b; Smedh et al. 2015).

Diverse peptides and steroid hormones have a long co-evolutionary history. One form of regulation in the brain is steroid induction of diverse information molecules (e.g., Herbert 1993; Kalra 1993; Fink 1997; Pfaff 1999). These interactions between steroid hormones and neuropeptides underlie behavioral responses; they range across vertebrates and invertebrates (Bentley 1982; Strand 1999). An exaggerated emphasis has been placed on restraint as a primary model for steroid–peptide interactions with regard to CRF and glucocorticoids. The continued evidence, both at a neural and a behavioral level, is less in conformity with this view.

Gene duplication is a major source of the genetic variation on which evolution acts. In other words, occasionally a segment of DNA containing a functional gene is duplicated in the genome. Then there are two (or more) identical genes, at least initially. The genes can have independent evolutionary paths after the duplication event; in fact, that is the expectation, though selective pressures may act similarly on both genes. An evolutionary perspective predicts that the molecules produced after a duplication event will have multiple and diverse functions and their regulation, function, and mode of action will vary among different taxa, tissues within taxa, and even with developmental stage. Every ligand can probably bind to multiple receptors; every receptor can probably bind to multiple ligands. Variation in function can come from changes in either the ligand or the receptor. Whenever scientists discover a function for a molecule, that function is likely merely one of many.

## CRF and Its Affiliates

CRF is an excellent example of the principles behind the evolution of molecules and signaling pathways. The CRF molecule itself is an ancient regulatory molecule found in fish, amphibians, and birds, as well as in mammals. It is remarkably conserved within mammals: primate, rodent, carnivore, and equid CRF are identical in amino acid sequence, although bovine CRF differs from the others in several amino acids (Seasholtz et al. 2002). This probably reflects the rapid adaptive radiation of the ruminants during the late Eocene and early Miocene, when grasses began to dominant many environments.

The evidence suggests that CRF derived from a molecule found in the common ancestor of vertebrates. The CRF superfamily throughout geological time shows sequence substitutions indicative of adaptation, evolution, and lineage divergence. Several events demarcate major moments in the evolution of CRF, such as peptide structural changes, innervation and fiber descension, and creation of distinct lineages.

Over the last three decades, a peptide research revolution has been in progress that continues to accelerate; new peptides and new functions for previously known peptides continue to be found. Consider CRF, which was first identified in extracts from sheep hypothalamus (Vale et al. 1981) and is probably best known for its initiating role in the hypothalamic pituitary adrenal (HPA) axis (Dallman et al 1995). We now know that CRF is part of a family consisting of at least four ligands (CRF, urocortin, urocortin II, and urocortin III, also known as stresscopin); at least two receptors, CRF type 1 receptor (CRF-Rl) with eight splice variants (Pisarchik and Slominski 2001) and three splice variants of the CRF type 2 receptor (CRF-R2: cx, p, and y); and a binding protein (CRF-BP) that is capable of binding all four ligands. Recently, a novel peptide, teleocortin (Tcn), was identified as a potential member of the CRF family in a teleost fish species, which activates both CRF type 1 and 2 receptors (Hosono et al. 2015). The chordate CRF peptide and receptor coevolved; therefore, the affinity between the peptide ligand and receptor has been maintained and evolved as well. These molecules serve not only as neurotransmitters and as neuromodulators but also as autocrine, paracrine, and endocrine hormones in peripheral tissues, and they are found throughout the body (Swanson et al. 1987; Power and Schulkin 2006).

CRF has the capacity for complex and subtle regulation of physiology and metabolism. The effects can vary with the ligands (and their up- and downregulation) and will depend on which receptor types are expressed (and their up- and downregulation). In addition, the effects of CRF and its related peptides will depend on the up- or downregulation of CRF-BP. In mammals, CRF-BP seems restricted to an inhibitory role with the ligands, presumably by binding them and making them unavailable to the receptors, and possibly increasing their clearance. There is no evidence of a receptor for CRF-BP or of any membrane effects for this peptide in mammals. Different tissues appear to express different ligands and receptors, or at least to have differing ratios of these molecules. Regulatory steroids, peptides, and cytokines appear to have tissue-specific regulatory effects on the CRF hormone family. For example, glucocorticoids can result in either up- or downregulation of ligands and/or receptors, depending on the tissue (Watts 2005; Schulkin et al. 2005).

Protein analysis has detected CRF-binding protein in the brain extracts of species from each vertebrate class, in varying quantities (Fellman et al. 1984). Urocortin and CRF have an affinity for CRF-R1 of approximately the same magnitude. However, the three urocortin ligands have a higher affinity for CRF-R2 than does CRF, by an order of magnitude (Lovejoy and Balment 1999). Indeed, urocortin II and urocortin III are thought to be selective ligands for CRF-R2, with low affinity for CRF-R1 (Lewis et al. 2001). Urocortin II may even play a role in the proinflammatory pathways of pregnancy, alongside CRF (Novembri et al. 2011, 2015).

This implies that the CRF system might predate the divergence of insects from vertebrates. It was noted that insect diuretic hormone-I and its receptors share similarities with the vertebrate CRF hormone system (Huising and Flik 2005). And indeed, CRF affects feeding and reproduction in the desert locust (Van Wielendaele et al. 2012).

The descendant molecules of that original molecule include CRF (found in all vertebrates), the urocortins (found in mammals), sauvagine (found in frogs), and urotensin-I (found in fish). The urocortins are related to sauvagine and urotensin-I (Seasholtz et al. 2002), which implies that the separation of CRF from the urocortins likely occurred before the separation of mammals from other vertebrates. And urotensin-1-like peptide was found in the nematode *Caenorhabditus elegans*, suggesting that the origin of this peptide hormone family occurred over 500 million years ago (Lovejoy and Balment 1999). The CRF-binding protein was found to be expressed in honeybees, and to be remarkably conserved (Huising and Flik 2005; Fekete and Zorrilla 2007; Lovejoy and Lannoy 2013; Even et al. 2012).

The CRF/DH family of peptides begins with an ancestral peptide that evolved during or before the late pre-Cambrian period (Figure 2.1). Its evolution may have occurred sometime after the appearance of the radiates, and before the pre-Cambrian explosion. Paralogous systems likely evolved during the early radiation of arthropods and fishes during the Cambrian and Ordovician periods. Novel functions appeared in the following sequence: osmoregulation/ion regulation, diuresis/feeding, rudimentary HPA function, and HPA control. The development of function spans from the pre-Cambrian to the Ordovician period. Genomic duplication often resulted in familial divergence.

In multicellular organisms, CRF family peptides have been found in both deuterostomes and protostomes. While there is currently no evidence of CRF-like peptides in the Metazoa lineage before this bifurcation, its presence in both lineages suggests the existence of a CRF-like precursor peptide before the divergence occurred. Gene duplication has resulted in variation and paralogous forms of genes within the CRF family of peptides, such as urotensin-1, urocortin II and III, and diuretic hormone.

In both mammalian and nonmammalian vertebrates, central CRF plays multiple roles in regulating and coordinating the responses to external and internal challenges to viability, both through its effects on the pituitary and as a neurotransmitter/neuromodulator (Seasholtz et al. 2002). Central CRF has direct and indirect effects on appetite, energy metabolism, blood pressure, arousal, and sexual behavior, among many other areas (Dallman et al. 1995; Schulkin 1999).

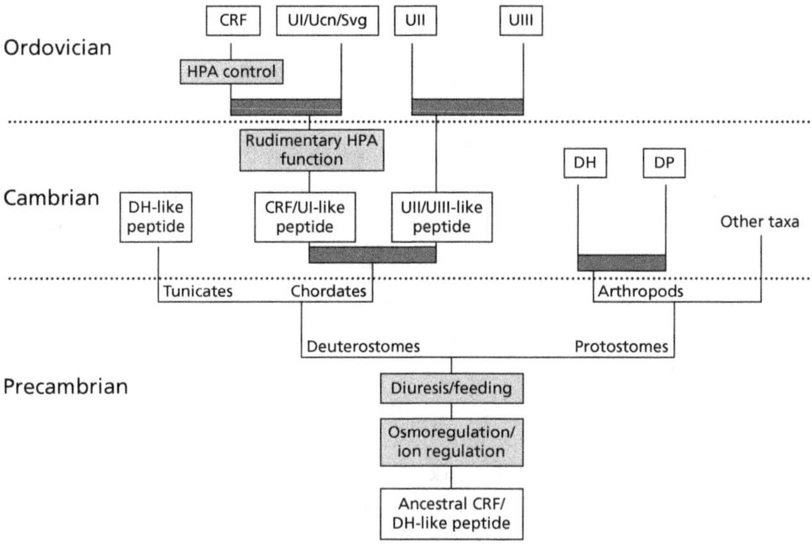


**Figure 2.1** Evolution of the corticotrophin-releasing factor (CRF)/DH family of peptides: a plausible scheme for the functional evolution of the CRF/DH family of peptides is depicted. The ancestral peptide may have evolved during or before the late pre-Cambrian period, sometime after the appearance of the radiates and before the pre-Cambrian explosion. Most of the known CRF/DH paralogous systems likely evolved during the Cambrian and Ordovician periods during the early radiation of arthropods and fishes. The appearance of novel functions of the CRF/DH peptides is shown as a light-gray box. Gene or genome duplications are shown as a dark-gray box. CRFI, urotensin-1; Ucn, urocortin; UII, urocortin II; UIII, urocortin III; DH, diuretic hormone; DP, diuretic peptide.



However, CRF is not solely a neuropeptide. It was found to be widely distributed in peripheral tissue, such as skin, liver, lung, and kidney (e.g., Emanuel et al. 2000; Slominski and Wortsman 2000; Ilias and Mastorakos 2003; Doztler et al. 2004). The function and regulation of CRF varies remarkably among tissues. In peripheral tissue, it generally acts in an autocrine or paracrine fashion, and it is often associated with inflammatory processes, where its actions can be both pro- and anti-inflammatory (Slominski and Wortsman 2000; Ilias and Mastornkos 2003). It also appears to serve a role in the periphery, regulating energy metabolism and energy substrate. Peripherally infused CRF increases energy expenditure and fat oxidation in humans (Smith et al. 2001), and CRF receptors are expressed in adipose tissue (Seres et al. 2004).

Many of the CRF family ligands and both receptors are expressed in different tissues of the alimentary tract (Tache and Perdue 2004). For example, stomach tissue was shown to express urocortin and the CRF-R2 receptor. Colonic epithelial cells express both urocortin (Ucn) and CRF, and both CRF receptors. Exogenous CRF and Ucn delivered to the stomach result in delayed gastric emptying, and this effect can be blocked with a CRF-R2 antagonist (Martinez et al. 2002). Exogenous CRF and Ucn delivered to colonic tissue results in increased colonic motility (Martinez et al. 2002; Zhou et al. 2013), and delivery of a CRF-R1 antagonist reduces colonic motility (Martinez et al. 2002; Sagami et al. 2004). CRF is also associated with patients with colonic ulceration (Kawahito et al. 1995; Yuan et al. 2010). CRF helps regulate intestinal angiogenesis through both type 1 and type 2 CRF receptors (Im et al. 2010).

Expression of CRF-R1 is widespread in the skin and subcutaneous fat cells of humans and, to a lesser extent, in mice. In contrast, CRF-R2 is widely expressed only in mouse skin. In human skin, only cells derived from the scalp (hair follicle keratocytes and hair follicle papilla fibroblasts) showed appreciable expression of CRF-R2. Human skin cells express CRF, Ucn and Ucn II mRNA, and peptide; mouse skin cells express Ucn and Ucn II mRNA and peptide, but only CRF peptide, implying that CRF is not synthesized in skin cells, but is transported to the skin from elsewhere (Slominski et al. 2004). Slominski proposed an analogy with the central CRF "stress system," suggesting that CRF has a function in human skin via CRF-R1 to respond to cellular damage, for example, from overexposure to ultraviolet light (Figure 2.2). In contrast, the role of CRF-R2 receptors in human skin may be restricted to hair growth.

Mice, being nocturnal and fur-covered, do not express this cutaneous "stress system" via CRF-R1 to the same extent but do share the CRF-R2 hair-growth system with humans. In most mammals, CRF-BP appears to be an exception to the widespread distribution of CRF hormone family peptides in peripheral tissue. The distribution of CRF-BP appears to be restricted to the brain in mammals other than anthropoids. In many, but apparently not all anthropoids, CRF-BP is expressed in the liver and is detectable in the circulation (Bowman et al. 2001; Paschos et al. 2013).

The CRF molecule is ubiquitous in vertebrates, with orthologs found in birds, amphibians (Stenzel-Poore et al. 1992), and fish (Okawara et al. 1988), where it plays multiple roles in regulating and coordinating the metabolic responses to external and internal challenges to an organism's viability (Denver 1999; Seasholtz et al. 2002). Residues 9-21 are identical among all CRF orthologs so far sequenced (Lovejoy and Balment 1999), and mammalian CRF compares with fish CRF in the range of 75% in tilapia and salmon, to 95% in the suckerfish (Seasholtz et al. 2002; Lovejoy and Balment 1999; Arai et al. 2001; Chang and Hsu 2004; Van Enckevort et al. 2000). Thus, CRF is an example of an

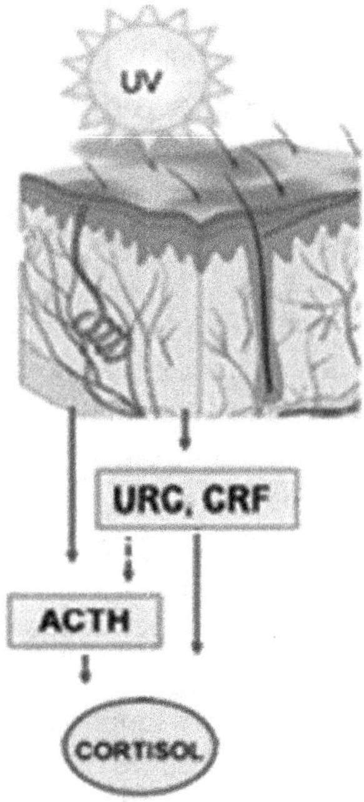


**Figure 2.2** (See color plate section) *The* skin stress-response system can activate the central HPA with its direct regulatory, metabolic, and phenotypic consequences. *We hypothesized that global responses (on the organism level) to UVR initiated in the skin involved simultaneous activation of sensory receptors and local production of humoral messages. The net effect of all of these processes is release of cortisol/ corticosterone and induction of steroidogenesis with subsequent metabolic and regulatory effects.*

Reproduced from Andrzej T. Slominski, Michal A. Zmijewski, Cezary Skobowiat, Blazej Zbytek, Radomir M. Slominski, and Jeffery D. Steketee, Sensing the Environment: Regulation of Local and Global Homeostasis by the Skin's Neuroendocrine System, *Advances in Anatomy, Embryology and Cell Biology*, Volume 212, pp. 1–115, Figure 17, Copyright © 2012, Springer-Verlag Berlin Heidelberg.

information molecule whose structure and function appear to be remarkably conserved over lineages separated by 500 million years or more.

The amino acid sequences of CRF vary little from species to species. Rat, human, and mouse CRF sequences are identical, and there is little variation from this sequence in ovine, *Xenopus*, and sucker CRF.

At the same time, this potent molecule has given rise to three other paralogous peptides in mammals, the urocortins, most probably through gene duplications and subsequent independent evolutionary change within lineages. In amphibians, a peptide found in skin, sauvagine (Montecucchi and Henshen 1981), derives from a common molecular ancestor to the urocortins.

In fish, the peptide urotensin-I also stems from this ancestor molecule common to all vertebrates (Seasholtz et al. 2002). Because sauvagine and urotensin-I are more closely related to the urocortins than to mammalian CRF, the implication is that urocortin, sauvagine, and urotensin-I are orthologs of each other, and paralogs of CRF. Thus, the original separation of the urocortins from CRF occurred before the separation of mammals from the other vertebrates (Seasholtz et al. 2002).

Mammalian CRF-like peptides show variable affinity for CRF-binding protein (CRF-BP) and both CRF-receptor types 1 and 2 (CRF-R1 and CRF-R2, respectively). CRF shows strong binding affinities for CRF-BP and CRF-R1, and a weaker affinity for CRF-R2. Urocortin-1 (UCN1) shows a strong affinity for both CRF-receptor types, as well as CRF-BP. Urocortin-2 (UCN2) shows a strong affinity for only CRF-R2, and a weaker affinity for CRF-BP. And urocortin-3 (UCN3) shows an affinity for CRF-R2. In X. laevis, similar patterns are found, except there is no known UCN2, and UCN1 only displays a weak affinity for CRF-BP.

A comparison of the CRF receptors supports this hypothesis. The type 1 receptor (CRF-R1) is generally associated with CRF, whereas the type 2 receptor (CRF-R2) appears to be primarily a high-affinity receptor for the urocortins (Lewis et al. 2001). Both CRF-receptor types are found in fish, amphibians, and birds, as well as in mammals (Arai et al. 2001). A phylogenetic analysis of the CRF receptors indicates that the two receptor types diverged before the divergence of fish from other vertebrates (Arai et al. 2001), thus an ancestral CRF hormone family existed in the common ancestor of all extant vertebrates. A third receptor type, more closely related to CRF-R1, has been identified in catfish (Arai et al. 2001). This receptor (CRF-R3) is sufficiently different from catfish CRF-R1 that the following possibilities cannot be distinguished: CRF-R3 is unique to fish and was present in the common ancestor of fish and tetrapods but was lost in the tetrapod radiation, or there is a third, undiscovered CRF-receptor type in mammals, birds, and amphibians.

A number of CRF receptors are ancient; genomic duplication of CRF receptors plays a most ancient role in osmoregulation and energy balance, GI regulation and digestion, and wake and sleep cycles (Romanowski et al. 2010). It is possible to depict phylogenetic relationships of the CRF-R1 (Arai et al. 2001). From this phylogenetic tree, clear relationships between CRF-R1 types are depicted through points of divergence. As with CRF, whereas mice and rats have the closest relationship owing to their more recent divergence, the relationship between mice or rats and *Xenopus* extends further back to a much older shared ancestor.

The CRF-binding protein, CRF-BP, is also found in these lineages and it appears remarkably conserved (Huising et al. 2004, 2007). All vertebrate CRF-BP so far sequenced have ten cysteine residues that form five consecutive disulfide bonds. Interestingly, CRF-BP has been found in insects, including honeybees, the malaria mosquito, and fruit flies (Huising and Flik 2005). The insect CRF-BP shares 23%–29% amino acid identity with vertebrate CRF-BP, and eight of the cysteine residues are conserved (Huising and Flik 2005; Westphal and Seasholtz 2006).

The existence of a honeybee CRF-BP ortholog is supporting evidence that an ancestral form of the vertebrate CRF signaling system was in existence before the split between the insect and vertebrate lineages. Insect diuretic hormone-I has been proposed to derive from a common ancestor with CRF, based on similarities in structure, anatomical location, and function (Huising and Flik 2005).

When looking at diuretic hormone-I of three specific insects (the honeybee, housefly, and cockroach), it is clear that most of the residues that conform to the CRF motif, as well as many other sites in the sequence, have been conserved among insect species.

Thus CRF reflects many aspects of evolutionary principles; it displays remarkable conservation of structure, related to the selective constraints owing to its functional signaling modalities. At the same time, it has given rise to substantial variation, in producing descendant molecules that diverged through gene duplication hundreds of millions of years ago. Over this vast time period, the CRF hormone family appears to have continued to add new functions to its signaling pathways. In mammals, the CRF hormone family signaling pathways are widespread, with virtually all tissues that have been examined (e.g., skin, heart, stomach, and intestine) being shown to express some member of the CRF family. Analysis of CRF family members in diverse peripheral tissues shows expression in various quantities and locations, while all members are at least present in the brain and pituitary. In the X. laevis specifically, CRF, urocortins, CRF receptors, and CRF binding proteins are present within most of the tissues studied (e.g., adipose, skin, and heart, Figure 2.3).

There are a number of related peptides and receptors in the fly brain that are tied to activation of or disruption of sleep (Sellami et al. 2011). CRF immunoreactivity was demonstrated in the rat retina (Skofitsch and Jacobowitz 1984), and the colocalization of CRF with other peptides in sensory ganglia has been demonstrated as well (Skofitsch et al. 1985). Photomicrographs of CRF-like immunoreactive neurons extending into various layers of the rat central retina suggest that CRF plays a role in retinal function. The regular rhythmical function of CRF and CRF-R1, associated with the sleep and waking cycle, has been explored in the literature (e.g., Ehlers et al. 1986; Romanowski et al. 2010; Chang and Opp 1998). Several studies have shown the effect of

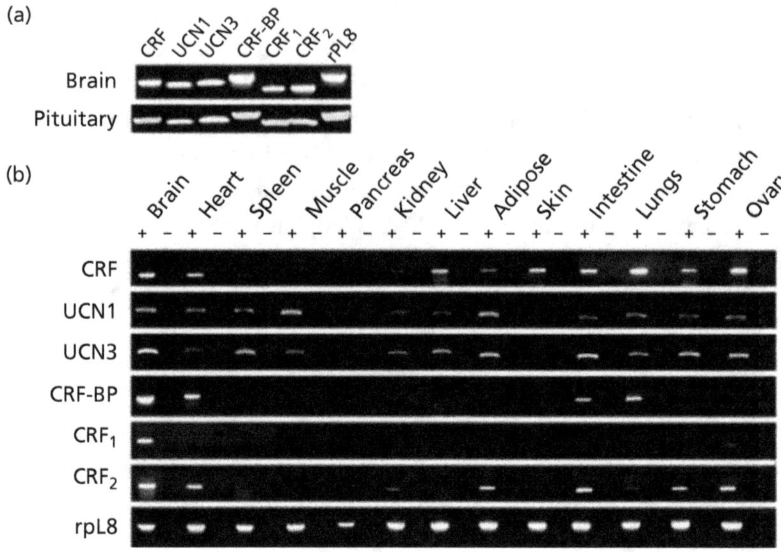


**Figure 2.3** Tissue distribution of mRNAs for CRF, urocortins, CRF receptors, and CRF binding protein in juvenile X. laevis tissues analyzed by RT-PCR. (a) Expression analysis in brain and pituitary. (b) Expression analysis in diverse peripheral tissues. Ribosomal L8 (rpL8) was used as a housekeeping gene to control for RNA quality and loading.

Reprinted from *General and Comparative Endocrinology*, 146 (1), Graham C. Boorse and Robert J. Denver, Widespread tissue distribution and diverse functions of corticotropin-releasing factor and related peptides, pp. 9–18, doi:10.1016/j.ygcen.2005.11.014, Copyright © 2005 Elsevier Inc., with permission from Elsevier.

intracerebroventricular administration of CRF in disrupting sleep regularity in rodents, or in diminishing capacity for sleep recovery after periods of sleep deprivation or after anxiety-inducing experiences (e.g., Wellman et al. 2015; Gvilia et al. 2015; Machado et al. 2010; Gonzalez and Valatx 1997; Kimura et al. 2010; Opp 1995). But the role of CRF in arousal should be considered not only as a dysregulation of sleep but also as an adaptive function, which emphasizes an alertness and awareness to novel stimuli, as well as a basic necessity for daily rhythmicity (Watts et al. 2004; Steiger 2002; Buckley and Schatzberg 2005).

The largest neurosecretory center of the fly brain expresses CRF. In drosophila, the pars intercerebralis secretes a CRF homolog, DH44 (diuretic hormone 44; Cardoso et al. 2014; Cavanaugh et al. 2014). This hormone apparently plays a requisite role in the rest: activity rhythms that dictate the individual's circadian clock (Cavanaugh et al. 2014).

Evidence suggests that CRF expression in flies is tied to anticipation and change, as seen in sleep; CRF expression has also been activated in anticipation of rewarding and aversive stimuli in flies (Figure 2.4; Nitabach [2014], unpublished observations).

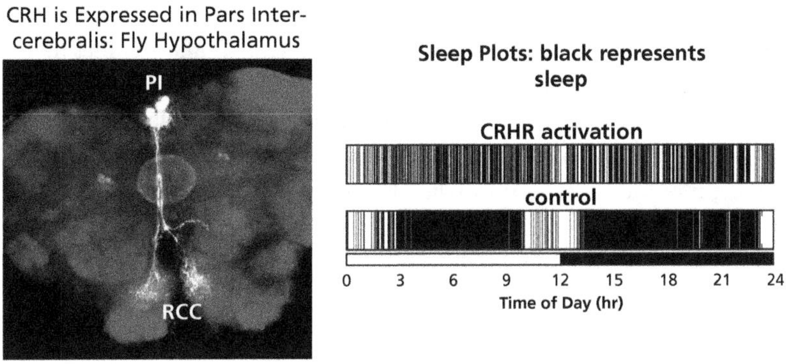


**Figure 2.4** (See color plate section) (Left) Corticotrophin-releasing hormone (CRH) peptide is secreted by large neurons of the fly pars intercerebralis, a brain region homologous to the vertebrate hypothalamus. (Right) Activation of central nervous system (CNS) CRH receptors using a genetically targeted membrane-tethered CRH ligand strongly suppresses and fragments sleep.

What was selected for in these brains was speed and a limited but very able set of behavioral options. But diverse forms of learning can nevertheless be revealed (e.g., food source location), as well as a rich and diverse assortment of information molecules (e.g., diverse amines, including serotonin; Neckameyer and Leal 2002) that underlie development. Additionally, the metamorphoses that underlie diverse forms of social interaction, the division of labor in insect societies, and the fundamental changes in response to others are all readily available for study as well (e.g., response to pheromones; Bloch and Meshi 2007; Dingle 2002; Fahrbach and Weeks 2002). In other words, invertebrates (e.g., insects or even plants) are also richly endowed with diverse hormonal systems that facilitate diverse events, including molting , other life transitional stages, and even parental behaviors (Ewer and Reynolds 2002; Trumbo 2002; Dingle 2002).

Amazingly, at least to this investigator, something like an HPA axis is also expressed in insects (Scharrer and Scharrer 1944). The hypothalamus-pituitary complex has a functional endocrinological equivalent in locusts, called the pars intercerebralis-corpus cardiacum complex.

The world is perceived by the brain one is endowed with; the fly brain, so radically different from our own, is still composed of diverse regulatory cephalic systems that underlie two fundamental events: feeding and reproduction (Dethier 1964; Dethier and Stellar 1961). The insect brain has no cortex and a massively different set of features for so competent and elegant a species (flies and bees; Dethier 1961, 1964; von Frisch 1994; Wilson 1962). A fly's head contains its central brain and neurons that support the central nervous system, as well as neurons that extend to the peripheral nervous system.

Vertebrate releasing hormones have an invertebrate counterpart and a shared potential ancestral purpose. CRF in vertebrates and CRF-like peptides in invertebrates originally served to maintain water homeostasis. Other vertebrate releasing hormones and the invertebrate counterparts display a similar pattern. One very interesting evolutionary depiction of a common ancestor of the CRF receptors and a structurally related anthropod diuretic receptor DH44R is shown in Figure 2.5 (Cardoso et al. 2014).

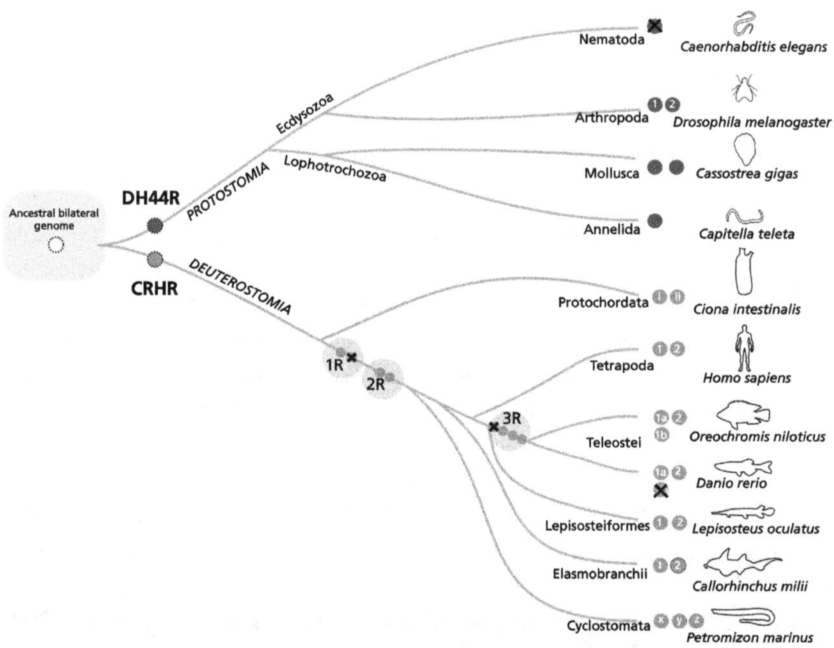


**Figure 2.5** (See color plate section) Outline of the evolution for the CRF/DH44 receptors in relation to several important evolutionary lineage divergences. The deuterostome CRF receptors and protostome DH44 receptors are represented by green and orange filled circles, respectively. The genome doubling events in the vertebrates are marked as 1R, 2R, and 3R, respectively. Lost genes are marked with black crosses. To facilitate interpretation, the Ciona duplicate CRFR-like receptors were designated by i and ii, the duplicate receptors in lamprey by x, y, and z, and the teleost CRFR1 duplicates by a and b. The timeline is not drawn to scale having in consideration the species evolutionary events.

Reprinted from *General and Comparative Endocrinology*, 209, João C.R. Cardoso, Rute C. Félix, Christina A. Bergqvist, and Dan Larhammar, New insights into the evolution of vertebrate CRH (corticotropin-releasing hormone) and invertebrate DH44 (diuretic hormone 44) receptors in metazoans, pp. 162–170, doi:10.1016/j.ygcen.2014.09.004, Copyright © 2014 Elsevier Inc., with permission from Elsevier.

## Conclusion

CRF and related molecular sequences are ancient. They appear in invertebrates' and chordates' ancestral genes and, therefore, appeared very early in animal physiology, even before the appearance of jawed fish (Lovejoy and Balment 1999; Lovejoy and Jahan 2006). Residue conservation is both easily measured and well conserved between CRF peptide sequences of diverse species. Human/rat CRF sequences compared with that of tilapia display a remarkably low number of residue substitutions. Urocortin is a related peptide to CRF, and duplication of genes can appear even before there is a known receptor site (Lovejoy and Jahan 2006; Lovejoy and Balment 1999). Analysis of the two CRF-receptor types (types 1 and 2) suggests that there was evolutionary divergence of fish from other types of vertebrates (e.g., Arai et al. 2001). Some clarity about the link across species between the CRF receptors has been outlined (Hillhouse and Grammatopoulos 2006; Chang and Hsu 2004). Both kinds of CRF-receptor types are found in all vertebrates studied to date, and there is a third subtype.

There are many examples of regulatory molecules that, over time, become adapted to serve multiple, diverse functions. This book is concerned with CRF, but similar reviews could be written about leptin, oxytocin, and most, if not all, other molecules that regulate physiology. Evolution is limited in that it can only work on existing variation. However, once a gene for a potent regulatory molecule exists, the potential for the differentiation of function, regulation, and mode of action exists as well.

An evolutionary perspective provides a powerful tool for understanding and predicting variation in the ways that regulatory molecules have been co-opted for different functions in different tissues and in different evolutionary lineages. An evolutionary perspective predicts that these molecules will have multiple and diverse functions, and that their regulation, function, and mode of action will vary among different taxa and among different tissues within taxa. The evolutionary perspective cautions that whenever scientists discover a function for a molecule, that function is likely merely one of many.

One fundamental form of nutritional and social attachment in mammals is lactation; the mammary glands, a cardinal feature of mammals, dates back to Triassic and Jurassic insectivores. In fact, spraying eggs with water may be an evolutionary precursor toward mammary gland development (Oftedal 2002). And hormones such as angiotensin, vasopressin, oxytocin, and prolactin (Fitzsimons 1998; Denton 1982), which are tied to fluid balance, are used in diverse functions (Manzon 2002; Kim et al. 2015), including the expansion of regulation that perhaps led to the formation of new glands. Indeed ancestral genes that precede CRF were tied to diuretics (Chang and Hsu 2004).

Information molecules, such as glucocorticoids and CRF, are ancient, and they are widely distributed across diverse organs and play various regulatory roles. The genes that underlie the production of these information molecules stretch back millions of years (Lovejoy and Balment 1999; Lovejoy and Jahan 2006; Strand 1999; Bentley 1982; Tan et al. 2012). One evolutionary mechanism is gene duplication, diversification, and variation on the production of these information molecules (e.g., Conlon and Larhammar 2005; Hoyle 1998; Hughes and Friedman 2013). In a number of species and organ systems, glucocorticoids and CRF underlie the physiology of change and adaptation.

Chapter 3

# Conservation of CRF in Brains and Its Regulation by Adrenal Steroids

## Introduction

Darwin (1881) was interested in everything biological, including vegetable mold and worms. In a short paper, he described the activity of worms in vegetable mold, as well as the behavior and biology of a worm he found interesting.

I suppose worms are interesting and perhaps even beautiful to some; personally, I am not one who finds them beautiful.

But worms, like the type Darwin was interested, in are rich in information molecules: one of which is CRF. As noted in the previous chapter, CRF is just one of many information molecules. One piece of a conserved evolutionary heritage is found on the genes that render CRF-like peptides and receptor-like features possible in *C. elegans* (Jee et al. 2013). In *C. elegans*, seb-3 is a CRF receptor-like G-protein coupled receptor involved in locomotion and other behaviors. There is notable conservation between sequences when seb-3 is aligned with rat, mouse, and human CRF-R1.

In this chapter, I begin with a comparative section, which links CRF to the biology of change, and then I turn to the more traditional way of understanding, that is, within the HPA axis, to the newer findings suggesting that glucocorticoids not only restrain CRF expression but also can enhance it in several extrahypothalamic regions. These alternative or diverse regulatory mechanisms are phylogenetically conserved. CRF can be regulated by both fast and slow genomic mechanisms.

Zebrafish (or teleost) development and reproduction is a robust and ancient process. Comparative genomics have traced zebrafish genes back 300 to 400 million years ago (Taylor et al. 2001). While vulnerable and easily accessible post fertilization, the short gestation period of zebrafish (2 days) results provides numerous viable embryos per cycle (Steenbergen, Richardson, and Champagne 2011). CRF cells are well conserved in teleosts (Matz and Hofeldt 1999; Pepels et al. 2002), and research has shown the coexpression of CRF and other neuropeptides in the hypothalamus (Herget and Ryu 2015). Zebrafish are just one example among others of the conservation and coexpression of CRF.

One easy feature of the consequences of CRF activation is motor-related behavior; of course, depending upon the animal and the behavioral range, that can encompass a lot. But locomotor activation is a common theme. There is genetic variation, and some zebrafish are more bold than others are (Oswald et al. 2012), an assertion that is apparent in experiments (Ghisleni et al. 2012). The locomotor capability reflects the effects of both glucocorticoids and CRF on bodily tissues (Yeh et al. 2013; Piato et al. 2011). Experiments conducted by Fosque et al. (2015) testing transgenic zebrafish have found CaMPARI (fluorescent activity sensor) expression under various experimental conditions.

Importantly and consistently, CRF is intimately tied to the brief developmental trajectory, or life span, of diverse teleosts (e.g., zebra fish and rainbow trout; see e.g., Alderman and Bernier 2007, 2009; Bernier et al. 2008). And both cortisol-related steroids and CRF promote development and activity. Foraging for food, for instance, is a prominent activity in most species including fish (Alderman et al. 2012). Glucocorticoids are well expressed in teleosts, and their regulation of CRF is ancient (Yeh 2015).

Moreover, CRF is differentially impacted by environmental events (hypoxia, food deprivation, and social adversity) in diverse teleosts. CRF expression and regulation is site specific; preoptic regions (HPA) versus caudal brainstem CRF sites (Bernier et al. 2008). Differential regulation and site-specific expression is a theme in the evolution of CRF, in addition to development of tissue, which is a core feature of CRF function. Interestingly isolation and subordination both have differential effects on CRF expression in forebrain versus hindbrain of CRF producing cells in rainbow trout. In experiments where rainbow trout were kept under isolation stress, the stressed conditions resulted in a three- to fourfold increase in CRF mRNA levels within the preoptic area at both measured time points (24 and 96 hours). However, within the caudal neurosecretory system, there was only a change in CRF mRNA levels at the 96-hour time point, and CRF was significantly decreased. Under subordination stress, there was a twofold increase in CRF levels in the preoptic area (after both 8 and 24 hours), but no change was found in CRF levels at measured time points in the caudal neurosecretory system (Bernier et al. 2008; Figure 3.1).

Moreover, significant differences have been found in the amount of CRF expressed in embryonic head/brain compared with tail/caudal neurosecretory

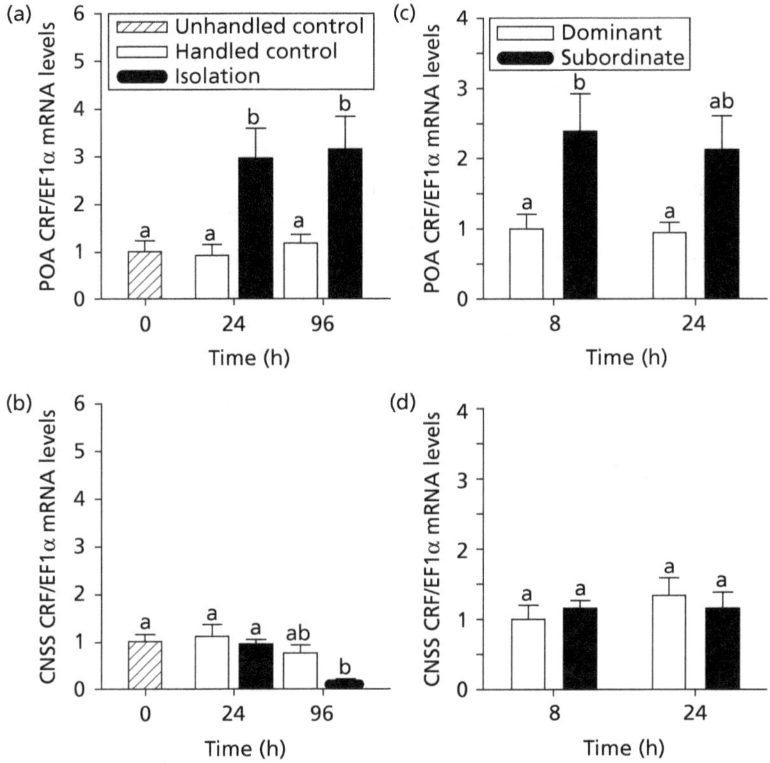


**Figure 3.1** The ratio of corticotrophin-releasing factor (CRF) mRNA levels to elongation factor 1a (EF1a) mRNA levels in the preoptic area (POA; a and c) and caudal neurosecretory system (CNSS; b and d) of rainbow trout. Fish left undisturbed (unhandled control), anesthetized, and returned to their original 125 l tank for either 24 or 96 hours(handled control), or anesthetized and individually confined to 9 l tanks for either 24 or 96 hours (isolation) are shown in a and b. Values are means CS.E.M. (nZ11). For comparative purposes, the expression ratios are presented relative to the respective 0-h unhandled control value. Bars that do not share a common letter are significantly different from each other as determined by one-way analysis of variance (ANOVA) and by pairwise Student–Newman–Keuls test (P < 0.05). Size-matched rainbow trout allowed to interact for either 8 or 24 hours and identified as either dominant or subordinate through behavioral observations are shown in c and d. Values are means CS.E.M. (nZ8). For comparative purposes, the expression ratios are presented relative to the respective 8-h dominant value. Bars that do not share a common letter are significantly different from each other as determined by one-way ANOVA and by pairwise Student–Newman–Keuls test (P < 0.05).

Adapted from Nicholas J Bernier, Sarah L Alderman, and Erin N Bristow, Heads or tails? Stressor-specific expression of corticotropin-releasing factor and urotensin I in the preoptic area and caudal neurosecretory system of rainbow trout, *Journal of Endocrinology*, 196 (3), pp. 637–648, Figure 5, doi: 10.1677/JOE-07-0568, Copyright © 2008, Society for Endocrinology.

system gene expression at several ages (Chandrasekar et al. 2007). CRF mRNA expression in zebrafish first occurs before 24 hours post fertilization (hpf) and the larvae display elevated cortisol at 3 days post fertilization (dpf) as a result of osmotic stress and at 4 dpf owing to handling stress, forming the first endocrine response (Alderman and Bernier 2008). CRF and its distribution are depicted in the rainbow trout in the next figure (Figure 3.2).

CRF is present in young zebrafish, as shown by immunostaining. CRF expression under experimental light–dark conditions indicates that it is necessary for camouflage behavior. Light facilitates camouflage and the induction of CRF in different regions of the brain (Figure 3.3; Wagle et al. 2011).

CRF is even a feature in the camouflage system that facilitates melanosome dispersal in zebrafish larvae (Wagle et al. 2011). Pigment cells in zebrafish have been linked to CRF gene expression, which is reminiscent of CRF found in human skin. CRF is mostly essential for diverse forms of inflammatory

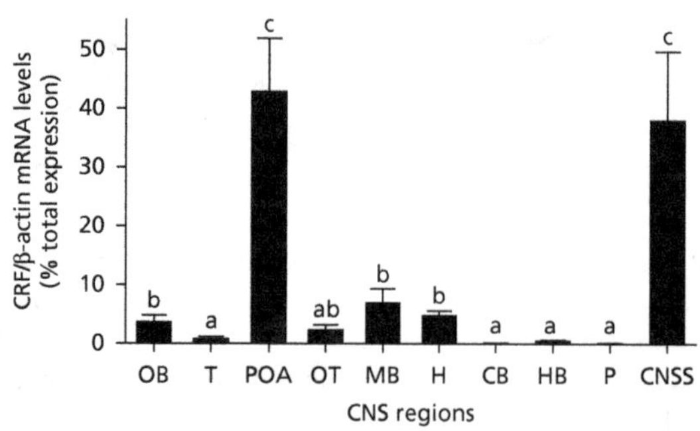


**Figure 3.2** Distribution of corticotrophin-releasing factor (CRF) mRNA in the central nervous system (CNS) of rainbow trout. The trout CNS had ten regions assayed by real-time RT-PCR for CRF and UI mRNA levels: the caudal neurosecretory system (CNSS), cerebellum (CB), hypothalamus (H), hindbrain (HB), midbrain (MB), olfactory bulbs (OB), optic tectum (OT), pituitary (P), preoptic area (POA), and telencephalon (T). The ratio of CRF to b-actin mRNA levels as a percentage of total expression. For a given gene, bars that do not share a common letter are significantly different from each other as determined by one-way analysis of variance (ANOVA) and by pairwise Student–Newman–Keuls test.

Adapted from Nicholas J Bernier, Sarah L Alderman, and Erin N Bristow, Heads or tails? Stressor-specific expression of corticotropin-releasing factor and urotensin I in the preoptic area and caudal neurosecretory system of rainbow trout, *Journal of Endocrinology*, 196 (3), pp. 637–648, Figure 1, doi: 10.1677/JOE-07-0568, Copyright © 2008, Society for Endocrinology.

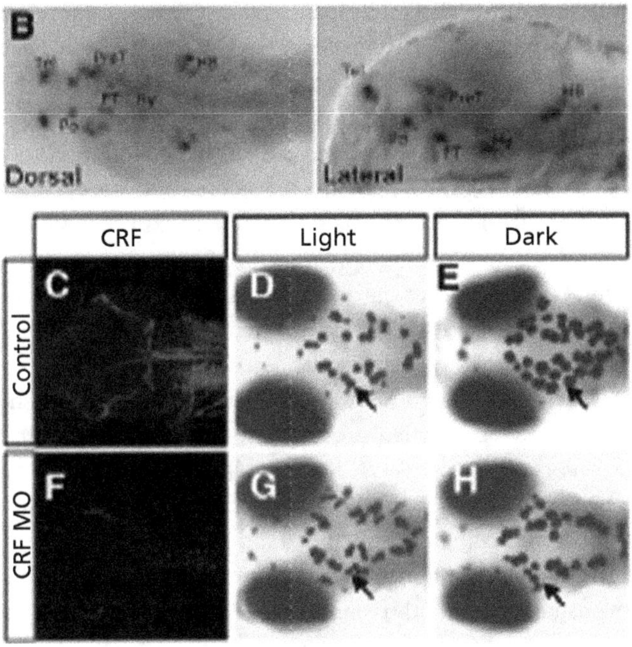


**Figure 3.3** (See color plate section) Corticotrophin-releasing factor (CRF) is necessary for the camouflage behavior. (b) Dorsal and lateral views of 5dpf zebrafish larvae showing CRF expression in the telencephalon (Tel), pretectum (PreT), preoptic (Po), posterior tubercle (PT), hypothalamus (H), and hindbrain (HB). (c-h) The CRF immunostaining is detected in axonal processes of 5-day old control morphant (c) and is greatly reduced in the crf morphant (F). Images of 5 dpf control (d, e) or the crf morphant (g, h) exposed to light (d, g) or darkness (e, h).

Adapted from Mahendra Wagle, Priya Mathur, and Su Guo, Corticotropin-Releasing Factor Critical for Zebrafish Camouflage Behavior Is Regulated by Light and Sensitive to Ethanol, *The Journal of Neuroscience*, 31 (1), pp. 214–224, Figure 1, doi: 10.1523/JNEUROSCI.3339-10.2011, Copyright © 2011 by The Society for Neuroscience.

expression, including arthritis, psoriatic skin, and systemic lupus erythematosus (O'Kane et al. 2006; Loite et al. 2013; Schmitz et al. 2015). Inflammation from lipopolysaccharides in adult epidermal keratinocytes (Zbytek et al. 2008) is mediated in part by CRF (Zbytek and Slominski 2008). Preliminary data show similar results in the ovine cotyledon (Frasch, personal communication).

As noted, and amazingly, indeed the whole skin is like a brain on the outside; there is something like an HPA axis with CRF, ACTH, and cortisol within our skin, where cortisol production is increased with the addition of progesterone in normal human melanocytes, and CRF activates POMC expression and corticosterone expression (Slominski et al. 2005, 2013). Moreover, diverse amines are also expressed in skin. In other words, skin is a massive endocrine and regulatory system for information molecules. The regulation and function of these ligands and receptors are not yet well understood. However, glucocorticoids were shown to both up- and downregulate ligands and receptors in different tissues. For example, in heart tissue, glucocorticoids upregulate urocortin but downregulate CRF-R2 (Kageyama et al. 2000). In skin cells, glucocorticoids downregulate CRF (Slominski and Wortsman 2000). The expression of CRF hormone family ligands and receptors in the alimentary tract and in skin provides useful examples of the variability and flexibility of function for this regulatory system, both within and among species.

CRF is exaggerated under conditions of adversity in the zebra fish (Ghisleni et al. 2012). And CRF and cortisol are linked to behavioral responses in avoiding duress (Steenbergen et al. 2011; Khor et al. 2011), and either facilitate or compromise the transition in the life stages of the zebra fish. The biological capability within the ecological milieu dictates the diverse roles that information molecules play in regulatory physiological adaptation. What is the niche? What are the capabilities? With regard to CRF as an information molecule, a common theme is the diversification of function.

## Corticosteroids

Corticosterone in the rat and cortisol in humans are well-known adrenal steroids. The ability to synthesize corticosteroids (e.g., aldosterone, the other adrenal steroid, and corticosterone) probably arose before the origin of the jawed vertebrates (hagfish and lamprey; Denver 1998, 2009). The ancestral glucocorticoids and its descendants were structurally preadapted to bind aldosterone, and thus were prepared to utilize the ligand, which appeared around 200 million years after the receptor (Bridgham et al. 2006).

Glucocorticoid receptor phylogeny (Bridgham et al. 2006) has an ancestral profile that dates back perhaps 450 million years ago (Ortlund et al. 2007; Denver 1998, 2009). Tissue capable of corticosteroidogenesis is found in extant chondrichthyan fishes such as the holocephalans (ratfish) and elasmobranchs (sharks, skates, and rays). In modern bony fishes, the teleosts, cortisol is the major steroid produced by adrenocortical tissue (Bentley 1982).

Glucocorticoid function is broad with regard to its impact on regulatory behavior and physiology. Moreover, glucocorticoids vary in their effects during seasonal variation, in droughts, and in predictability for the allocation of food resources and other metabolic requirements (Wingfield 2004; Landys et al. 2006). One idea here is that molecules may not change during evolution, but that their use does.

While glucocorticoids are linked to social status across many species (e.g., Sapolsky 1992), they are also involved in many behavioral functions (e.g., Goymann and Wingfield 2004; Schulkin 2003). The point here is that glucocorticoids respond to a variety of demanding situations; they are not specific to any one in particular.

Glucocorticoids are multidimensional in action and effect. They are fundamental to glucose and energy metabolism and are involved in every core end organ system and their regulation; in the short term, they mobilize and redistribute energy resources. In the long term, they deplete key resources that are essential for the organization of behavioral action (Dallman et al. 1994, 2003; Herbert 1993; Sapolsky 1992; McEwen and Stellar 1993; Lightman 2008; McEwen and Wingfield 2003). Depending upon context, corticosteroids mobilize and restrain diverse organ systems (e.g., hypothalamic/pituitary function).

Glucocorticoid and other steroid receptors are part of a major class of DNA binding factors that regulate gene transcription (Chen et al. 2003; De Kloet et al. 2005; Denver 2009). These corticosteroid-receptor complexes regulate transcription of numerous genes in most organs of the body and the brain (Colucci and Ortlund 2012) in which both genomic and nongenomic signaling systems are operative (Moussatche and Lyons 2012).

Progesterone shares a long evolutionary past with other steroids and competes for many of the same receptor sites as does, for instance, cortisol; it is the degree of specificity that is site specific and context dependent (Funder et al. 1988; McEwen 1995).

Human progesterone reveals evidence of adaptive changes at transcription levels (Chen et al. 2008). Progesterone's history stretches over a billion years and is closely tied to the other primary steroids through a molecular family of steroids and receptors sites (Karalis et al. 1996). For instance, the lamprey, a jawless fish at the lower end of the vertebrate evolution, contains both corticoids and progesterone receptor sites (Sower et al. 2009).

Glucocorticoids participate with other metabolic hormones (e.g., insulin) in the regulation of metabolism and in the generation of feeding behavior (Dallman et al. 2003, 2007; Laugero et al. 2002), in part by the activation of diverse neurotransmitters and neuropeptides (e.g., Herbert 1993; Dallman et al. 2003). Glucocorticoids are importantly linked to diverse forms of foraging behaviors, and anticipation of rewards (Pecoraro and Dallman 2005; Wingfield 2004).

It has been known for some time that adequate levels of adrenal steroids are essential for survival (Cannon 1916/1929, 1927, 1932/1963; Richter 1942–43), though with great variation in species and function of responses (Romero and Wikelski 2001; Sapolsky 1992; Wingfield 2004). An interesting experiment of nature occurred in the Galapagos Islands following El Nino (1997–1999) in marine iguanas. Levels of corticosterone (following mild restraint stress), though with great variation, predicted population health and viability and was linked to levels of the adrenal steroid following the weather changes (Romero and Wikelski 2001; Romero et al. 2009). The glucocorticoids are essential for physiological and behavioral adaptation (e.g., maintaining energy viability).

In general, the elevation of the adrenal steroids is adaptive in the short term and disruptive in the long term (McEwen 1998; McEwen and Wingfield 2003). The important point in understanding regulatory physiology is that promoting viability over time is the purpose of the adaptive nature of different hormones, and it is dependent on the circumstances both internal and relative to niche and circumstance (Wingfield and Romero 2001; Sapolsky 1995).

Glucocorticoids set the conditions for at least four states of physiological regulation of CRF (Sapolsky et al. 2000). They are depicted below:

**Four Contexts for Glucocorticoid Actions**

1. Permissive Actions
2. Suppressive Actions
3. Stimulating Actions
4. Preparative Actions

Most information molecules, particularly cortisol and neuropeptides, also express variation across the light–dark cycle. CRF expression in the brain is an example of one such variation; recall that cortisol is the "wakeup" hormone tied to activity and action. Both CRF and cortisol are wakeup-signaling systems; most of us engage with the world while there is light. Of course, that has all changed in the last several hundred years. But light–dark rhythmicity is ancient; the circadian clocks that still run through our biology, including the release of CRF in the morning in the anticipation of activity, are a fundamental fact of life. In the context of cortisol and CRF, both are under circadian regulation, which underlies the normal expression of information molecules such as cortisol and CRF (Dallman et al. 1978; Watts et al. 2004).

Dark-field photomicrographs can depict the range of CRF heteronuclear and messenger RNA levels in the medial parvocellular part of the hypothalamic paraventricular nucleus in intact animals throughout the day (Watts et al. 2004; Figure 3.4). CRF heteronuclear RNA levels in the medial parvocellular part of the hypothalamic paraventricular nucleus and plasma ACTH concentration in rats trends inversely across a given day. Peak CRF values occur during periods when lights are off (Watts 2005).

**Expression of CRF in the brain:** CRF mRNA is widely distributed in the central nervous system (CNS). For instance, in frogs, CRF is widely distributed

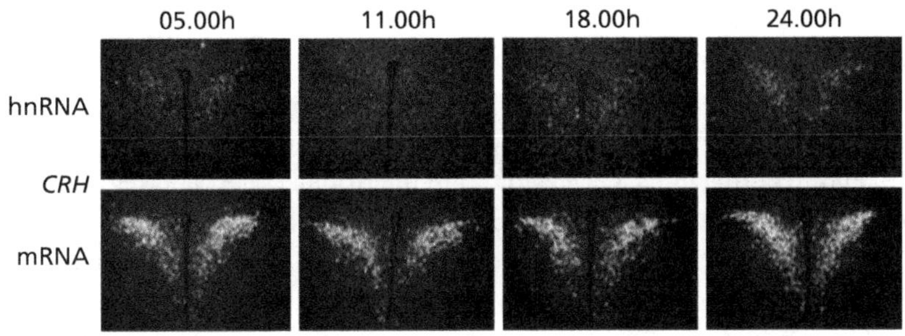


**Figure 3.4** Dark-field photomicrographs of CRF hnRNA and CRF mRNA hybridization in the PVHmp in intact animals at various times of the day to illustrate the range of RNA levels seen throughout the day. At any one time, serial sections are illustrated from the same animal.

from the forebrain to the brainstem: from the amygdala to the hypothalamus and cerebellum, down to the locus coeruleus and in the spinal cord (Kuhar 1984; Yao et al. 2004). Corticotrophin-releasing factor is found in several areas of the brain.

To a smaller extent, there are CRF cells in the lateral hypothalamus, prefrontal (Uribe-Mariño et al. 2016), and cingulate cortex. In brainstem regions, CRF cells are clustered near the locus coeruleus (Barrington nucleus) (Valentino et al. 1993, 1995; Valentino and Bockstaele 2008), the parabrachial region, and regions of the solitary nucleus. Tellingly, overexpression at the level of the Barrington nucleus influences bladder function (McFadden et al. 2012; Wood et al. 2013).

In addition to the hypothalamus, CRF and related peptides are widely expressed throughout the mammalian CNS in limbic structures, the cortex, and brainstem nuclei associated with autonomic functions (Rodaros et al. 2007; Swanson et al. 1983; Gray 1999; Makino et al. 1994a, b). The Laboratory of Clinical Sciences was one of the first to map the CRF cell bodies and receptors in the brain (Olschowka et al. 1982a, b; Skofitsch and Jacobowitz 1984). They were found to be widely distributed in the brain (Swanson et al. 1983). Jacobowitz's group and others found that the majority of CRF neurons within the PVN (paraventricular nucleus) are clustered in the parvocellular division. Other regions with predominant CRF-containing neurons are the lateral bed nucleus of the stria terminalis and the central division of the central nucleus of the amygdala, in addition to diverse nuclei from the brainstem to the forebrain. They included diverse regions such as the tegmental region, nucleus lateralis, septum, accumbens regions, diverse hypothalamic nuclei (preoptic and supraoptic nuclei), regions of the parabrachial and solitary nucleus and the dorsal horn of the sacral spinal cord, and cells that are clustered near the locus coeruleus (Olschowka et al. 1982a, b).

Corticotrophin-binding protein is also widely distributed in the CNS (see e.g., Swanson et al. 1983; Sawchenko et al. 1993; Behan et al. 1995; Gray et al. 2016) and hypothalamus. CRF is found in the frontal cortex and regions of the thalamus, as well as brainstem regions near the locus coeruleus (Valentino et al. 1993, 2012). Like CRF, expression of the binding protein is increased under conditions of adversity.

Reaching back in evolution, there are many diverse experiments across phyla demonstrating the release of CRF or CRF binding protein. For instance, CRF and urocortin in rainbow trout (Alderman et al. 2008; Bernier et al. 2008) has been studied under normal conditions and under adversity. In the rainbow trout, the preoptic region shows elevated CRF (and neuropeptide Y) under social conditions of adversity (Doyon et al. 2003). Secretion of CRF during activation of the HPA axis is prominent in diverse species, and peptides such as CRF or urotensin elicit diverse appropriate behavioral responses (e.g., aggression in rainbow trout; Backström et al. 2011).

In the CNS of the rainbow trout, there is diverse expression of CRF and urotensin-I (UI) throughout the ten regions of brain assayed by Bernier, Alderman, and Bristow (2008). Regions of interest included: the caudal neurosecretory system, cerebellum, hypothalamus, hindbrain, midbrain, olfactory bulbs, optic tectum, pituitary, preoptic area, and telencephalon.

In both rats and primates, CRF immuno-reactive fibers (Millan et al. 1987) are found in regions of the brain that included the prefrontal, orbital, cingulate and temporal cortex, the septum, amygdala, hippocampus, dorsomedial thalamus, basal ganglia, and brainstem sites such as the locus coeruleus, parabrachial nucleus, and solitary nucleus, in addition to regions of the spinal cord and cerebellum.

As discussed, CRF is an ancient molecule that has widespread expression throughout the body, and multiple and diverse functions in living organisms. Gene duplications have resulted in a small but potent hormone system consisting of multiple ligands and receptors that add to the complexity and ability of the CRF family to regulate physiology, metabolism, immune responses, and other adaptive features that protect and enhance an organism's viability. To interpret the function and regulation of CRF solely through its function and regulation in the HPA axis is more likely to confuse than to inform (Denver 1998 1999; Lovejoy et al. 2014). There is a profound change in the expansion and utilization of this information molecule beyond our conception of it in the HPA axis (Lovejoy 2009). CRF and diuretic perhaps were diversified into a plenitude of regulatory functions in which the CRF receptors spread across bodily functions and capabilities.

Neuropeptides ushered in another form of synaptic transmission (Swanson 1983; Kiss et al. 1984; Hökfelt et al. 1983). CRF was first identified in extracts from sheep hypothalamus (Vale et al. 1981; McDonald et al. 2000). It is probably best known for its initiating role in the HPA axis. In response to a threatening or challenging event, CRF release (and synthesis) is increased in the PVN. Hypothalamic CRF stimulates adrenocorticotrophin hormone (ACTH) production and release by the anterior pituitary gland, which in turn stimulates cortisol production and release by the adrenal glands.

CRF neurons in the medial parvocellular part of the hypothalamic paraventricular nucleus (PVHmpd) are poised to have intrahypothalamic synaptic, paracrine, and endocrine influences via neuroactive substances released from all of its axon terminals. Its paracrine effects (at the level of the median eminence) and endocrine influences (at the level of the anterior pituitary), in addition to its intrahypothalamic influences, illustrate the diverse nature of the CRF neuron in the PVHmpd (Figure 3.5).

**Glucocorticoid/CRF interactions:** Cortisol acts on diverse end organ systems to prepare the animal to respond to the initiating event (e.g., heart rate may be increased, gastric emptying may be inhibited, colonic motility may be increased, blood flow to major muscle groups may be increased, and blood glucose levels may rise). All of these actions prepare the animal for "flight or fight" responses. In addition to a marked inhibition of corticotroph responsivity to CRF, a negative-feedback mechanism was proposed, in which cortisol crosses the blood-brain barrier and inhibits CRF production by the PVN, restraining the system (Swanson 1991; Sawchenko 1987; Dallman et al. 1987).

Thus, CRF from the PVN serves to coordinate the peripheral response to threats to viability perceived by the CNS. The negative-feedback system serves to regulate the peripheral response so that it remains commensurate with the perceived threat (Sapolsky et al. 2000). As a neuropeptide, CRF plays key roles in integrating hormonal, autonomic, metabolic, and behavioral responses to challenges to viability. Many of these challenges are associated with circumstances that require high levels of alertness, attention, and possibly energy mobilization. For example, "stressful" circumstances that animals would be likely to attempt to avoid or to escape from (e.g., sensory evidence of the presence of a potential predator) are circumstances where CRF would be expected to be elevated in both the PVN of the hypothalamus and the CEA, among other brain regions.

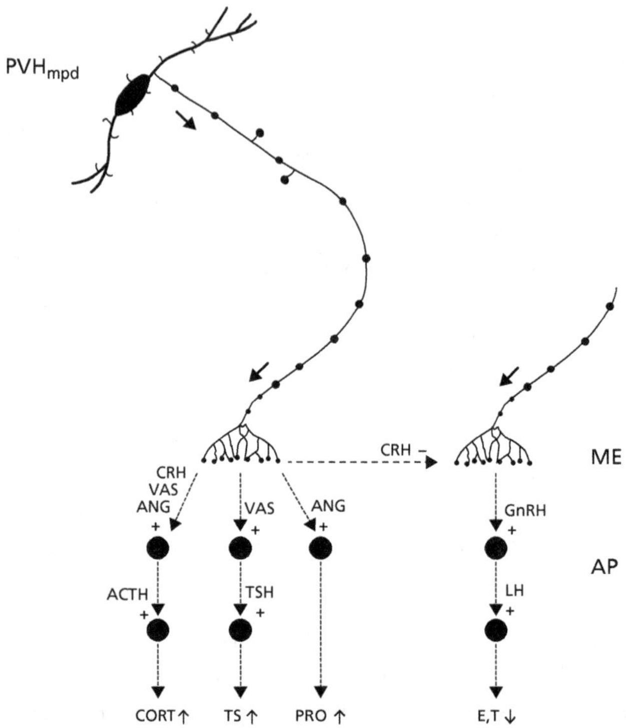


**Figure 3.5** An idealized CRF neuron in the PVHmpd. As discussed in the text, such neurons appear to be in a position to exert intrahypothalamic synaptic influences from axonal terminal boutons, paracrine influences at the level of the median eminence (ME), and endocrine influences at the level of the anterior pituitary (AP). It is important to point out that at this time it seems reasonable to assume that all of the neuroactive substances expressed in such a neuron are released from all of its axon terminals. Abbreviations: ACTH, adrenocorticotropic hormone; ANG, angiotensin; CORT, corticosterone; E, estrogen; GnRH, gonadotropin releasing hormone; LH, luteinizing hormone; PRO, prolactin; T, testosterone; TSH, thyroid-stimulating hormone; T3, triiodothyronine; VAS, vasopressin.



Beautiful depictions of the activation of the paraventricular nucleus using green fluorescent protein for the regulation and control of CRF have been conducted (Alon et al. 2009), showing activation of PVN CRF neurons. Using genetic tools (CRF-Irese CRe mutant mouse, e.g.; Cusulin et al. 2013;

Figure 3.6) the activation is revealed electrophysiologically and with neurosecretory signaling specificity (Cucuslin et al. 2013; Partridge et al. 2014, 2016).

In the brain, the CRF family has at least two receptors (and perhaps more); CRF-R1 and CRF-R2 are localized in rodent and primate brain. Activation of both the CRF-R1 and CRF-R2 receptors is linked to a G-protein, and it activates the adenylate cyclase cascade, resulting in an increase in intracellular cyclic adenosine monophosphate (cAMP) and calcium levels (Perrin and Vale 1999; Chen et al. 2005; Nicholson et al. 2004; Majzoub 2006). Calcium, sodium, and cation-conducting channels are all involved in CRF-induced excitability in mice (Zemkova et al. 2016). The distribution of CRF-R1 receptor sites includes regions of the hippocampus; the septum, amygdala (medial and lateral region), and neocortex; ventral thalamic and medial hypothalamic sites, and sparse numbers of receptors are located in the PVN and the pituitary gland. The distribution is widespread in the cerebellum, in addition to brainstem sites such as major sensory nerves and the solitary nucleus (e.g., De Souza et al. 1985; Potter et al. 1994; Pett et al. 2000; Makino et al. 1995; Lim et al. 2005; Walker and Lawrence 2016).

The distribution of CRF-R2 is more limited than that of CRF-R1 and is found primarily in subcortical regions including the amygdala, septum, bed nucleus of the stria terminalis, and PVN and ventral medial nucleus of the hypothalamus (Shemesh et al. 2016; Lovenberg et al. 1995a, b).

**Glucocorticoids and CRF in the PVN and amygdala:** The HPA axis is usually thought of in a CRF context; but this is not quite right. What appears dominant across phyla is the expression of the diverse forms of HPA or HPG or HPT regulation under negative restraint; for instance, negative restraint is expressed across vertebrates in and out of the water (Sower et al. 2009).

Negative restraint by glucocorticoids has been demonstrated across phyla (e.g., Vandenborne et al. 2005; Denver 2009). In bony fish, both the pituitary and the urophysis (caudal neurosecretory system) play a role in regulating interrenal glucocorticoids. Hypothalamic CRF, as well as urotensin-I in the brain, contributes to the release of pituitary ACTH. This ACTH in conjunction with urotensin-I from the urophysis acts on the interrenal glands, and regulates glucocorticoids (Lovejoy and Balmert 1999).

The restraint of HPA activation by glucocorticoids is rapid and profound (Dallman et al. 1987, 2005; Keller-Wood et al. 1983). It is also specific; mineralocorticoids do not produce these effects (Sawchenko 1987; Watts and Sanchez-Watts 1995).

Dogs infused with a combination of cortisol and corticosterone or saline displayed a remarkable difference in plasma ACTH levels after injection with

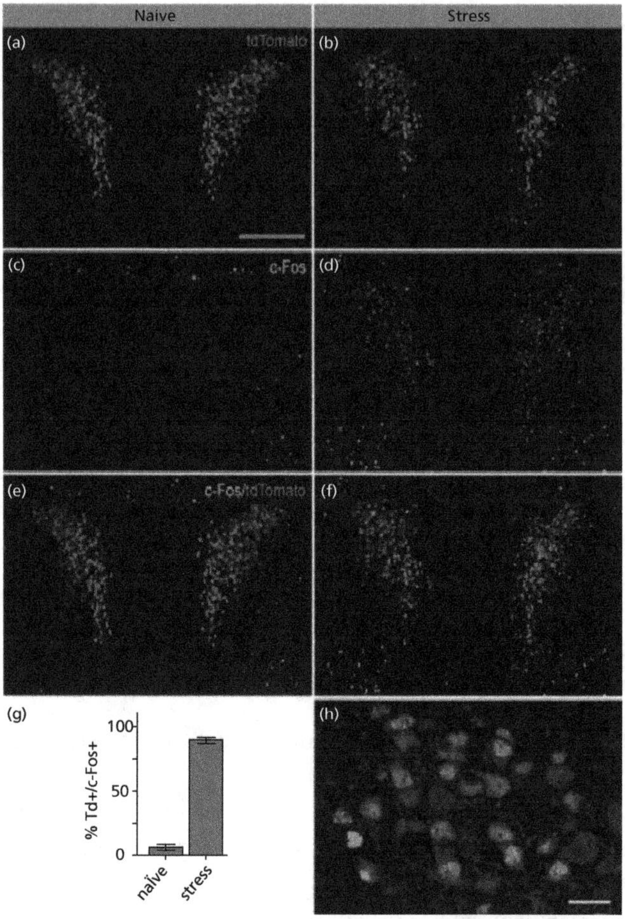


**Figure 3.6** (See color plate section) Induction of cFos protein in PVN tdTomato neurons following stress. (a and b) tdTomato expression in PVN of naïve CRF-IRES-Cre; Ai14, and one exposed to a 15 min forced swim stress, 90 min before sacrifice. Confocal image at 15x. (c and d) cFos immunoreactivity in these naïve or stress brain sections, showing an increased number of cFos containing nuclei following stress. (e and f) Merged image showing high neurons co-expressing cFos in naïve (n = 5) versus stressed (n = 6) mice. (h) Higher magnification (100x) of cFos immunoreactivity in tdTomato cells from a stressed mouse. Data summarized in (g) are mean + / - SEM. Scale bars are 200 um (a–f) and 20 um (h).

Reproduced from Wamsteeker Cusulin JI, Füzesi T, Watts AG, Bains JS, Characterization of Corticotropin-Releasing Hormone neurons in the Paraventricular Nucleus of the Hypothalamus of *Crh-IRES-Cre* Mutant Mice, *PLoS ONE*, 8(5): e64943. doi:10.1371/journal.pone.0064943, © 2013, Wamsteeker Cusulin et al.

insulin (at 120 minutes). Those dogs infused with saline showed a jump in their plasma ACTH levels in response to hypoglycemia, while those infused with cortisol and corticosterone experienced a decrease followed by a smaller increase in ACTH levels (Keller-Wood and Dallman 1984).

One should also note that evidence suggests that negative restraint of CRF may not be confined solely to the PVN. For example, it may also constrain CRF activity in the locus coeruleus (Pavcovich and Valentino 1997). While the profound effect of inhibition is indisputable, there are neuronal populations within the PVN that project to the brainstem that are not inhibited by glucocorticoids, and some regions which are actually enhanced (Swanson and Simmons 1989; Watts and Sanchez-Watts 1995; Makino et al. 1994a, b). That is, CRF neurons en route to the pituitary are restrained by glucocorticoids, but CRF en route to other regions of the brain appears not to be restrained (Palkovits et al. 1998; Swanson and Simmons 1989; Watts and Sanchez-Watts 1995; Cusulin et al. 2013). Thus, even within the PVN, CRF neurons that project to the brainstem are not restrained by glucocorticoids, and CRF neurons elsewhere in the brain are not restrained in production by glucocorticoids (e.g., Imaki et al. 1991; Swanson and Simmons 1989; Palkovits et al. 1998; Watts and Sanchez-Watts 1995; Valentino et al. 1993).

The degree of the HPA activation is coordinated by both humoral and neural mechanisms (e.g., Ma et al. 2001; Herman et al. 2003; Watts 2005; Sawchenko 1987). Efferent pathways from the hippocampus and amygdala regulate the expression of CRF in the PVN (Herman et al. 1998, 2003; Raper et al. 2014). However, elevated levels of glucocorticoids do not appear necessary for CRF at the level of the PVN to return to normal levels after a mild acute aversive stimulus suggesting other mechanisms in addition to glucocorticoids (Shepard et al. 2005). It has been shown in normal rats that serum corticosterone levels increase when the rats are under restraint, compared with adrenalectomized rats given constant corticosterone replacements. Of interest in this experiment was there was no difference in the pattern of CRF heteronuclear RNA synthesis between the two groups of mice undergoing restraint (Shepard et al. 2005).

High and low circulating levels impact neuropeptide context: this holds for probably the whole array of neuropeptide or information molecules in a very active functioning brain (Swanson 1991). Negative restraint by glucocorticoids holds for vasopressin as it does for CRF in this parvocellular region of the PVN (Sawchenko 1987; Kiss et al. 1984), and the same cellular region and perhaps cell bodies can have multiple peptide and transmitter releases (Swanson 1991). The important point is that the neuropeptide milieu varies as a function of activity and neural central state; cortisol is a prime regulator of such events; restraining on the one hand and enhancing on the other. For instance, neurons of the medial parvocellular part of the hypothalamic paraventricular nucleus after adrenalectomy and colchicine treatment display relative density that varies with CRF, vasopressin, and angiotensin II expression (Swanson 1991).

Clearly, CRF gene expression is differentially regulated in the brain by glucocorticoid hormones (Swanson and Simmons 1989; Imaki et al. 1991; Tanimura and Watts 1998). In the parvocellular region of the paraventricular nucleus, glucocorticoid activation decreases CRF gene expression, while in the central nucleus of the amygdala and lateral bed nucleus of the stria terminalis, CRF gene expression is enhanced by glucocorticoids (Figure 3.7; Swanson and Simmons 1989; Makino et al. 1994a, b; Watts and Sanchez-Watts 1995; Watts 1996; Thompson et al. 2004.) These effects on the upregulation of CRF gene expression may be mediated by both type I and type II glucocorticoid receptor sites (Watts and Sanchez-Watts 1995; Tanimura and Watts 1998).

Corticotrophin-releasing factor regulation by negative feedback is not axiomatic even at the level of the paraventricular nucleus (Swanson and Simmons 1989; Tanimura and Watts 1998). Interestingly, under various conditions, glucocorticoids in the paraventricular nucleus of the hypothalamus stop inhibiting

**Figure 3.7** Effects of corticosterone in adrenally intact rats. Corticosterone increases corticotrophin-releasing factor (CRF) in the CEA (a, b, and c), while it decreased it in the PVN (d, e, and f). Corticosterone effects on CRF mRNA in the central nucleus of the amygdala and the parvocellular region of the paraventricular nucleus of the hypothalamus.

Reprinted from *Brain Research*, 640 (1–2), Shinya Makino, Phillip W. Gold, and Jay Schulkin, Corticosterone effects on corticotropin-releasing hormone mRNA in the central nucleus of the amygdala and the parvocellular region of the paraventricular nucleus of the hypothalamus, pp. 105–112, doi:10.1016/0006-8993(94)91862-7, Copyright © 1994, Elsevier B.V., with permission from Elsevier.

CRF gene expression (Pich et al. 1993; Kalin et al. 1994; Albeck et al. 1997; Tanimura and Watts 1998; Vandenborne et al. 2005).

Interestingly, aldosterone also tended to increase CRF gene expression in the CEA (Watts and Sanchez-Watts 1995). The replacement of corticosterone in adrenalectomized rats restores the levels of CRF gene expression in the CEA (Palkovits et al. 1998).

Glucocorticoids are known to regulate CRF mRNA at transcription factors (e.g., Ma et al. 2001; Nicholson et al. 2004; Watts 2005). And while cAMP-induced CRF gene regulation has been linked to glucocorticoid mediated receptor function, there is no consensus on the regulatory component of the CRF promoter region and transcriptional related changes (Nicholson et al. 2004; King et al. 2002; Shepard et al. 2005; Dallman et al. 2003; Donley et al. 2005; van der Laan et al. 2009).

Behavior itself can regulate CRF expression; the ingestion of sucrose can regulate CRF expression in rats. The ingestion of sucrose directly modulates the expression of CRF gene expression in the PVN and the central nucleus of the amygdala (Dallman and Bhatnagar 2000; Dallman 2003; Figure 3.8). In bilaterally adrenalectomized rats, Dallman and colleagues showed an inverse relationship between the amount of CRF mRNA in the paraventricular nuclei and the total sucrose ingested during a 9-day period.

Experiments using microdialysis to measure peptides strengthen the argument, namely that corticosterone can facilitate CRF release (Merali et al. 2008; Cook 2002). For instance, in one study, we found that the release of CRF was enhanced from the amygdala (or prefrontal cortex). Rats were treated with corticosterone for several days systemically and had indwelling catheters for measuring CRF by microdialysis aimed at the CEA or prefrontal cortex. CRF release was enhanced by corticosterone in rats treated with a mild aversive air-puff stimulus (Merali et al. 2003, 2008).

Using sheep, Cook (2002) studied the release of CRF, measured by dialysis, in the CEA. Following acute exposure to a dog for 6 minutes, both venous and amygdala levels of cortisol increased after 10–30 minutes. With the ability to escape, CRF decreases. Importantly, amygdala CRF had a large increase during exposure to the dog and a second peak 10–30 minutes after, corresponding to the increase in cortisol. Similar dual peaks of CRF release were found with a foot shock. Administration of a glucocorticoid receptor antagonist blocked the second CRF peak in the amygdala without affecting the first peak (Figure 3.9).

These data indicate that the initial response of CRF in the amygdala to an acute aversive social event is independent of cortisol, but the second delayed peak is cortisol-dependent. In addition, the initial CRF response to a stressor following repeated inescapable exposure to the dog came under the control

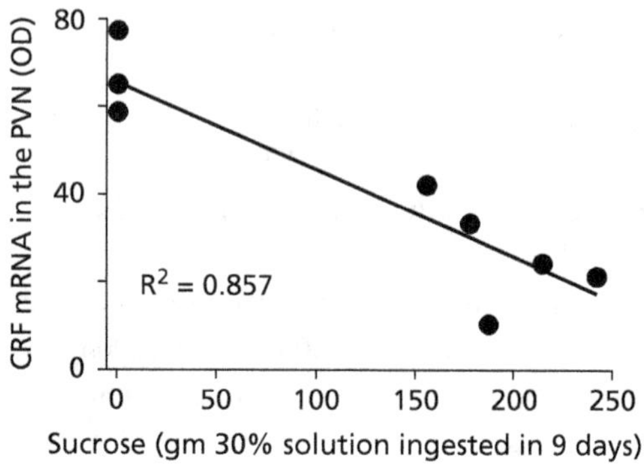


**Figure 3.8** Sucrose ingestion is inversely related to CRF mRNA in the paraventricular nuclei (PVN) in bilaterally adrenalectomized rats, the total sucrose intake during the 9 days it was available is tightly related to the expression of hypothalamic CRF. CRF mRNA was restored to values similar to those in sham-operated rats (5 points on the right) reducing the normally elevated CRF seen in adrenalectomized rats (4 points on the left) to values similar to rats with intact adrenals (data from Laugero et al. 2001).

Reproduced from Mary F. Dallman, James P. Warne, Michelle T. Foster, and Norman C. Pecoraro, Glucocorticoids and insulin both modulate caloric intake through actions on the brain, *The Journal of Physiology*, 583 (2), pp. 431–436, doi:10.1113/jphysiol.2007.136051 © 2007, The Physiological Society.

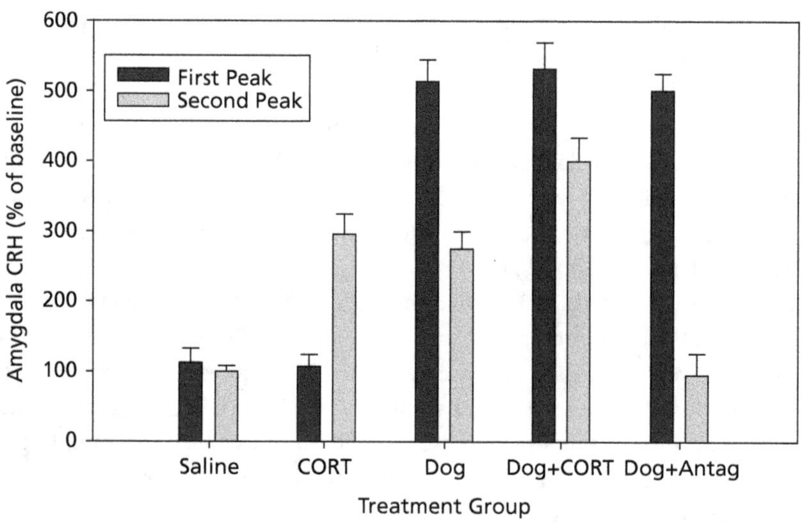


**Figure 3.9** The effects of drug treatments, with and without the presence of a dog, on measured values of cortisol (CORT; ng/ml) and a glucocorticoid receptor antagonist (Antag; ng/ml); percent of baseline CRF (%) shown as mean +/– S.E. and mean area under peaks.

Reprinted from *Physiology & Behavior*, 75 (4), Christian J Cook, Glucocorticoid feedback increases the sensitivity of the limbic system to stress, pp. 455–464, Copyright © 2002, Elsevier Science Inc., with permission from Elsevier.

of cortisol. Sheep were given 7 days of repeated exposure to the dog, either with the ability to escape or not to escape from the dog. On the eighth day, the sheep were given a foot shock. While venous and amygdala cortisol levels in response to the foot shock were identical in escape and nonescape groups, both peaks of CRF release in the amygdala were higher in the repeated nonescape group compared with the escape group and became regulated by cortisol (Cook 2002, 2004).

A fearful event (the sight of a dangerous predator) can induce CRF within seconds from the CEA (e.g., Cook 2002; Merali et al. 1998, 2008) and thus is a site-specific action of a peptide (Herbert 1993; Herbert and Schulkin 2002; Chen et al. 2012; Muglia 2000; Radley 2012). To sustain the fear, the adrenal gland is activated in the mobilization of coherent action, and in this regard, the CRF release is sustained (Cook 2002; Merali et al. 2001, 2008)—hence, a role of central CRF for the perception of danger. Thus, the initial release of CRF is independent of the adrenal gland hormone. The second release of CRF is glucocorticoid dependent; blocking its release or action reduces CRF expression in the amygdala.

**Evolutionarily conserved glucocorticoid regulation of corticotrophin-releasing factor expression:** The distribution of CRF and glucocorticoid immunoreactivity in the frog brain is evolutionarily conserved with mammals. Glucocorticoids are co-localized with CRF and response components in the proximal promoters of the frog CRF genes (Yao et al. 2008a, 2008b).

In many species (primates, rodents, reptiles, and amphibians), CRF is expressed in the amygdala, bed nucleus of the stria terminalis, and medial pallium, and its expression, and that of the immediate early gene c-fos, are increased following exposure to a stressor, as well as to elevated levels of corticosterone (Yao et al. 2008a, 2008b; Denver and Crespi 2006). CRF and corticosterone are well known across phyla to promote locomotor activity and affect diverse forms of behaviors (Moore and Rose 2002; Rosen et al. 1994; Wingfield 2004; Giardino and Ryabinin, 2013) including sexual, feeding, and fear-related behaviors among others.

In the frog, just as in mammals, CRF is increased by treatment with corticosterone (given in their bathing water over several days). The increase of CRF was noted in the central/medial amygdala and medial BNST, whereas it was decreased in the PVN. The regulation of CRF by glucocorticoids could be—at least in part—directly related to CRF expressing neurons because CRF and glucocorticoid receptors (GR) were found to co-localize in these cells in the rat brain (e.g., Yao et al. 2008a, 2008b; Sawchenko 1987). Thus, as in mammals, glucocorticoids decreased CRF expression in the major site of hypophysiotropic CRF neurons in the frog that are homologous to the mammalian PVN.

By contrast, glucocorticoids elevated CRF in the frog central/medial amygdala and BNST.

The amygdala is phylogenetically ancient and expressed in all vertebrates (Johnston 1923; Herrick 1905, 1926/1963; Martínez-García et al. 2002; Laberge et al. 2006; Swanson and Petrovich 1998; Swanson and Mogenson 1981). Taken together, previous findings suggest that a functionally homologous limbic neurocircuitry was present in the earliest tetrapods, (Laberge et al. 2006; Martínez-García et al. 2002) and that the nature of the feedback regulation might have occurred before the divergence of the amphibian and amniote lineages, and may be a common feature in tetrapods. Importantly, within nonmammalian species, CRF expression is also differentially regulated by glucocorticoids in discrete regions of the CNS. Both glucocorticoids and CRF, in addition to diverse other information molecules, are influenced by environmental events.

For instance, adverse conditions in tadpole water resources and environments affect CRF expression in the central brainstem (Boorse and Denver 2003; Boorse et al. 2005). Juvenile X. laevis undergoing shaking and confinement for various durations display variable levels of plasma corticosterone concentration, CRF immunoreactivity, and phosphorylated cAMP response element binding protein (CREB) immunoreactivity. In experiments conducted by Denver and colleagues, there were significant differences in all of these levels between unstressed and stressed conditions in the anterior preoptic area, at almost all time points. Hindbrain regions (rhombencephalon and brain structures caudal to the pituitary) collected from S. hammondii tadpoles subjected to high and low water treatments displayed a significant increase in CRF content after 24 hours subjected to normal and adverse conditions (high and low water volumes respectively; Boorse and Denver 2003.

The regulation of CRF expression in mammals is cell-type specific, being inhibited by glucocorticoids in the hypothalamus/POA, but stimulated in the amygdala. Researchers have found that CRF expression is negatively regulated by glucocorticoids in the frog POA. Inhibition of corticosteroid synthesis increased the levels of CRF primary transcript (mRNA) and peptide in the POA. The level of CRF mRNA was decreased by 6-h glucocorticoid treatment but was not altered by metyrapone (MTP; Yao et al. 2008b). Negative regulation of CRF expression has been reported in several teleost fishes (Alsop and Vijayan 2009).

An inhibitory action of glucocorticoid on hypophysiotropic CRF neurons is supported by the finding that CRF and glucocorticoid receptors (GR) co-localize in cells of the frog POA. In the rat, CRF and GR were found to co-localize in the parvocellular neurons of the PVN (Ceccatelli et al. 1989; Cintra et al. 1987, 1991). There is also evidence for indirect glucocorticoids providing negative feedback on PVN neurons in mammals via a descending pathway originating in the hippocampus (Makino et al. 2002; Schulkin et al. 1998; De Kloet et al. 1998). High expression of GR is observed in cells of the mammalian hippocampus, which send inhibitory projections to the PVN (Yao et al. 2008a; Makino et al. 2002; Sapolsky et al. 1991; Herman et al. 2003, 2005). Note that the frog medial pallium, which is a brain region homologous to the mammalian hippocampus, expresses high levels of GR (Yao et al. 2008a). Like the mammalian hippocampus, the frog medial pallium expresses CRF, and GR are expressed in these cells (Denver 1998).

As with the human CRF promoter, dexamethasone (DEX) did not alter basal activity of the frog CRF promoters, but it significantly reduced Fsk-induced activity. Glucocorticoids decreased mammalian CRF gene activation caused by cAMP analogs, phorbolesters, or cytokines (Guardiola-Diaz et al. 1996; Van et al. 1990; Maloski et al. 1997; King et al. 2002; Rosen et al. 1992; Vamvakopoulos and Chrousos 1993b; Wagner et al. 2006). Together, these findings show that glucocorticoids may directly influence, via the GR, transcription of the CRF gene, and that such regulation is phylogenetically ancient (Yao et al. 2008b).

Although glucocorticoids exert negative feedback on PVN CRF neurons in mammals, and similar regulation in amphibians, several groups, as already discussed, have shown that glucocorticoids have opposite effects on CRF expression in the amygdala and BNST (Makino et al. 1994a, b; Shepard et al. 2000; Swanson and Simmons 1989, Watts and Sanchez-Watts 1995; Thompson et al. 2004). Similar relationships occur in the frog, in which there are significant effects on CRF immunoreactivity by way of corticosteroid treatment. Exposure to corticosteroid and metyrapone (corticosteroid synthesis inhibitor) treatment (both individually and together) resulted in significant differences in X. laevis brain regions. In the BNST and medial amygdala, CRF levels significantly increased after corticosteroid treatment, and decreased in the medial amygdala after metyrapone treatment compared with the control, (metyrapone treatment in the BNST had minimal effect). In the POA, there was a significant decrease in CRF levels after corticosteroid treatment and an increase after metyrapone treatment. When both corticosteroid and metyrapone treatment was administered simultaneously, there was no apparent difference between CRF immunoreactivity in a control X. laevis and those treated simultaneously by both in all three brain regions

(Yao et al. 2008a). Thus, positive in addition to negative regulation of CRF is ancient.

## Conclusion

Glucocorticoids increase both CEA and BNST CRF gene expression in mammals (Makino et al. 1994a, b; Watts and Sanchez-Watts 1995). Several lines of evidence exist that suggest evolutionary conservation in the function of the limbic system between amphibians and mammals. CRF is expressed in the amygdala, BNST, and medial pallium of the frog, and its expression and that of the immediate early gene c-fos are increased after exposure to a stressor (Yao et al. 2004). The glucocorticoid receptor sites are expressed in the frog amygdala, BNST, and medial pallium, and GCs increased CRF in the amygdala and BNST (Yao et al. 2008a).

An interesting finding by Fenoglio is that early life experiences (e.g., handling) have differential effects on CRF expression in the CEA and the PVN (consistent with the main theme of this book). In the CEA (and BNST) by postnatal day 9, CRF gene expression is enhanced. Rats having underdone handling for 7 days show significantly increased CRF mRNA expression in the central amygdala and anterodorsal BNST, and significantly decreased expression in the PVN. Photomicrographs of coronal brain sections suggest typical regulation of CRF mRNA signals.

Differential regulation of CRF gene expression is a feature of adaptive design: Perhaps not surprisingly regulatory systems of CRF are diverse and multiple. I have emphasized the endocrine side and neural side of this regulation; the design is flexible, adaptive, and contextual. Most investigators outside of the field only consider the regulatory system of CRF with regard to the HPA axis.

The system is active, and clock-related endocrine expression is operative. Cortisol, the hormone of energy balance, and CRF, the neuropeptide tied to the biology of change and danger, are going to be involved in most active events that matter. We will see in later chapters the diverse ways in which this occurs. For now, note multiple regulation of CRF beyond the HPA negative restraint conception. Neural afferents affect CRF neurons through two main, basic processes—slow-acting peptide synthesis and immediate peptide release. The pathways of peptide synthesis and peptide release merge at the CRF neuron terminal for release and coupling (Watts 2005).

As detailed in later chapters, other regions of the brain are not constrained by glucocorticoids with regard to CRF expression (e.g., regions of the frontal cortex; Merali et al. 2008; Frim et al. 1990; Weninger et al. 1999), while glucocorticoids enhance CRF expression in the placenta, setting the stage for maturation of the fetus and development in general and eventual parturition (Majzoub et al. 1999). The next chapter turns to dramatic forms in which CRF participates in the regulation of the internal milieu.

Chapter 4

# Radical Change: Metamorphosis and Parturition

## Introduction

Hardly anything in nature is as dramatic as watching the transformation or transition of one physical form into another. In fact, metamorphosis has intrigued humans throughout our history and has even been the subject of great works of literature, such as in the novel *Metamorphosis* by Franz Kafka (1915), one of the great writers of the twentieth century. His novel, though, was a fable about loss of control and human existential angst. Our story of metamorphosis, however, is about change and adaptation or preadaptation and the expansion of use and capability.

Change and adaptation are conceptually linked in biology and certainly, they persist in our considerations of CRF. CRF is fundamentally tied to the world in which animals are adapting to changes including the seasons, times of day, droughts, and breeding cycles. Cycles of stability and change set the conditions for adaptive viability; underlying such events is predictive coherence, where diverse information molecules drive physiological and behavioral adaptations (Wingfield 2004; Power 2004; Gluckman et al. 2007). CRF and glucocorticoids, the focus of this book, are just two of the diverse information molecules. CRF can have both rapid and slower effects and it can be regulated by both membrane-related fast mechanisms and slower genomic mechanisms.

Life stage changes underlie many of the events in animal adaptation; certainly metamorphosis exists in the stages from egg to larva and finally to adult form as a function of some type of adaptation. Early, irreversible stages before transitioning to an adult cycle make up the vertebrate life cycle. Transitions between ontogenetic stages are examples of metamorphosis that continue throughout the life cycle. One feature of CRF among others is that it responds to life-cycle events; both endogenously generated and accelerated in a threatening context.

This chapter directly explores the role of CRF and other information molecules in the act of parturition (birth) in mammals, particularly in primates. At the heart of what it is to be a mammal is the placenta. The placenta is an endocrine gland containing nearly every information molecule that we know of; it is essentially all organs in one gland. Of course, there are always exceptional counter examples (e.g., egg-laying mammals in the order monotremata: the platypus and echidna).

**Metamorphosis**: One of the most remarkable events is the biology of radical change in metamorphosis; many information molecules underlie and may accelerate this change. One species among others that has been studied in terms of its metamorphic life cycles is the tadpole (Figure 4.1). The development of *Xenopus laevis* from tadpole to adult amphibian is an iconic and predictable example of metamorphosis. Tadpoles of several amphibian species have been shown to accelerate metamorphosis when their ponds dry. Ecological studies suggest that tadpoles use special senses to detect deleterious changes in their larval habitat. Amphibian tadpoles show strong responses to environmental signals, both by expressing multiple, discontinuous phenotypes (i.e., polyphenism) and by exhibiting continuous variation in the timing of developmental events. The most important environmental variable for a tadpole is water availability (Crespi and Denver 2005; Yao et al. 2007, 2008).

**Figure 4.1** Metamorphosis of Xenopus laevis.
Photo Credit: © David Bay, 2016.

Thyrotropin-releasing hormone (TRH) in a number of species can affect metamorphosis (Kühn et al. 2005; De Groef et al. 2003, 2006) perhaps via the CRF-R2 receptor (Okada et al. 2007). Hypothalamic-derived CRF injected into tadpoles facilitates the rate of metamorphosis (Denver and Crespi 2006). One functional role of CRF is to escalate the rate of metamorphosis when the tadpole is subjected to hostile environmental conditions (Denver 1997). Interestingly, in studies examining food or habitat adversity and the presence of these information molecules, particularly the levels of CRF in the brain, it is brainstem CRF that is elevated 24 hours after deprivation or other conditions of adversity (Boorse and Denver 2003. Indeed diverse information molecules participate in this transformation and can accelerate its expression including thyroxine, corticosterone, and TRH. CRF receptor blockers inhibit the acceleration.

CRF infusions facilitate the development of frog morphology (*Eleutherodactylus coqui*) and when compared with TRH infusions, they facilitate the morphological development largely. Young *E. coqui* injected with thyrotropin-releasing hormone (TRH), CRF, and saline displayed morphological differences. Those animals injected with CRF have only a short stub remaining of a tail, thick adult skin, and pointed snouts, suggesting accelerated development. In contrast, those injected with saline and TRH seem to be less developed morphologically. TRH infusions trigger CRF expression (Kuhn et al. 2004). *E. coqui* injected with varying amounts of CRF displayed increased expression of TRb (TH-direct response gene) mRNA compared with controls, displayed by reverse transcriptase-PCR (Kulkarni et al. 2010). In another experiment, body weight, gill length, and tail height were slightly decreased with thyroxine and CRF injections, but both were significantly decreased compared with measurements taken after saline injection (Kuhn et al. 2004).

Information molecules play diverse roles depending on which end organ tissue expresses them, what sites are within that tissue, and where there are receptors. A beautiful depiction of the diverse information molecules is shown in the next figure, showing the interdependent effects of thyroid-stimulating hormone (TSH), corticoids (Cort), and adrenocorticotropic hormone (ACTH) on metamorphosis (Fig. 4.2).

Several endocrine systems play a role in controlling tadpole metamorphosis. From the pituitary gland, ACTH has a stimulatory effect on the interrenal gland, and TSH stimulates the thyroid gland. Corticoids from the interrenal glands both stimulate and inhibit the brain. Thyroid hormone from the thyroid gland also has stimulatory and inhibitory effects on the brain. Habitat desiccation results in the precocious activation of the thyroid and the interrenal axes, the hormones that drive metamorphosis (Denver 1998). Clearly, links between several hormonal signaling systems induce dramatic change (e.g., thyroxine and TRH; adrenal

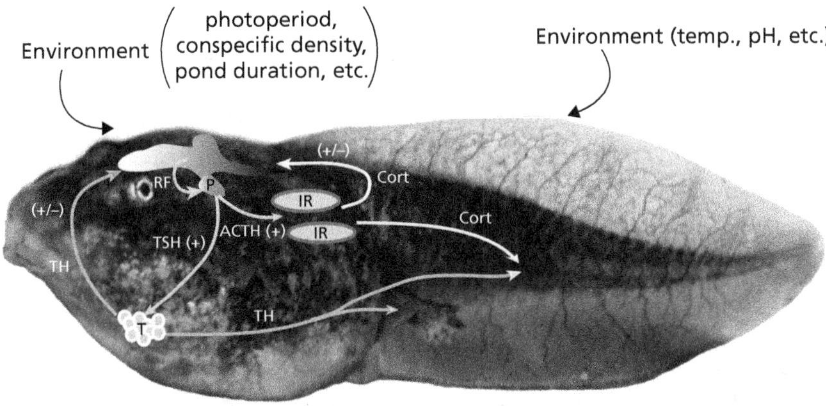


**Figure 4.2** (See color plate section) Endocrine systems controlling tadpole metamorphosis. P, pituitary gland; RF, releasing factor (CRF regulates both TSH and ACTH secretion in tadpoles); IR, interrenal gland; ACTH, adrenocorticotropic hormone; TSH, thyroid-stimulating hormone; T, thyroid gland; TH, thyroid hormone; CORT, corticoids. Plus signs indicate a stimulatory effect and minus signs a negative feedback. In the case of TH and Cort effects on the brain, (+/–) indicates that these hormones promote differentiation of neurosecretory centers (and other brain regions) in addition to their negative feedback effects on neurohormone and pituitary hormone secretion.

Reproduced from Robert J. Denver, Karen A. Glennemeier, and Graham C. Boorse, "Endocrinology of Complex Life Cycles: Amphibians," in Donald W. Pfaff, Arthur P. Arnold, Susan E. Fahrbach, Anne M. Etgen and Robert T. Rubin, ed., *Hormones, Brain and Behavior*, pp. 469–513, doi:10.1016/B978-012532104-4/50030-5, Copyright © 2002, Elsevier Inc.

steroids and CRF). With regard to CRF, the CRF signal is transduced by at least two receptor subtypes and may be modulated by a secreted binding protein (Challis 2001; Denver and Crespi 2006; Smith 2007).

## CRF and the Reproductive Tract

Throughout the Cenozoic, Mesozoic, and Paleozoic eras, there were major developments in the vertebrate subphylum that lead to the great diversity that exists today. This is easily visualized through a basic phylogeny depiction, of which there are many.

Many information molecules are fundamental in the varied forms of adaptive behaviors in reptiles and birds. These adaptations range from live births to egg laying. CRF, like other information molecules, is tied to the change and adaptation in the life, birth, and death of fish, reptiles, and birds. The range of information molecules that underlie egg laying and cultivation are familiar and ancient (progesterone, estrogen, cortisol, prolactin, oxytocin, and CRF). Egg laying is perhaps the most familiar action in which diverse information molecules are critically involved. Of course, there is the egg-laying platypus, a great biological wonder in terms of it being an egg-laying mammal.

The placenta is an endocrine organ intimately involved in regulating metabolism, signaling to both maternal and fetal tissue, and regulating the metabolism, growth, and development of both. The placenta serves as a central regulator of maternal-placental-fetal metabolism and produces perhaps the broadest array of information molecules (hormones, cytokines, and all other classes of signaling molecules) of any other organ except for the brain. Placentas from diverse mammals may appear slightly different, but they play the same role and have the same capacities (Power and Schulkin 2012).

Implantation sites of an early pregnant rat uterus have high concentrations of CRF (Makrigiannakis et al. 2004); CRF is thought to play an important paracrine role in decidualization and implantation. Rats given a CRF-R1 antagonist in early gestation had a dose-dependent reduction of endometrial implantation sites and live embryos (Makrigiannakis et al. 2004). Progesterone reduces CRF-related inflammation in peritoneal cells (Tariverdian et al. 2010).

The developing fetal brain is incapable of performing all of the necessary regulatory functions. Instead, signals from maternal organs pass through the placenta, signaling the placenta, which regulatory roles it needs to fulfill for the developing fetus. The placenta also sends signals back to the maternal organs, including her brain, in order to coordinate maternal physiology to benefit the growing fetus.

The placenta is thus a transformative organ; ancient and vital for the transition from fetus to adult mammal (e.g., Murphy et al. 2007; Moffett and Goeders 2007; Brawand et al. 2008; Wildman et al. 2006). In marsupials, the evolution of the placenta is linked to brain size (Weisbecker and Goswami 2010), but in most species it is tied to gestational length (Capellini et al. 2011).

We now know that the placenta performs more than the transfer of nutrients as a passive organ: it is a vital 9-month engine of physiological regulation and long-term changes on brain, behavior, and physiology (Barker 1997, 2004; Power and Schulkin 2012). In other words, it is a central regulator (mini brain; Yen 1994) of maternal–fetal physiological interactions necessary for reproduction and has long-term impact. There is nothing quite like this organ anywhere else in the body, so rich it is in information molecules (Petraglia et al. 1990; Power and Schulkin 2012), although nonplacental animals have other routes to facilitate maternal/fetal interactions and fetal development (e.g., Reznick et al. 2002).

At the onset of pregnancy, a new organ (placenta) develops and matures, participating in many of the regulatory functions from maternal tissue, including the maternal brain. For example, growth-hormone secretion from the maternal pituitary essentially shuts down by the second trimester, with a placental form of growth hormone taking its place (Frankenne et al. 1988; Alsat et al. 1997). A key bodily function is to allocate resources within the body in order to defend its own viability (Peters et al. 2004). Oxygen and glucose delivery to the fetal brain is tightly regulated compared with the flow to other organs (Gao et al. 2012; Smith et al. 2012).

The placenta also acts to allocate resources from the maternal to the fetal compartment. A well-regulated flow of oxygen and glucose is vital to healthy placental function and fetal growth and development as well. Regulatory control is fundamentally altered during pregnancy, in part because metabolism and physiology must account for two organisms instead of one, but also because regulatory signaling has been allocated to the maternal brain and placenta. Many physiological set points are adjusted, often many times, during gestation. For example, circulating levels of glucose, insulin, cortisol, and leptin all increase. Pregnant women are thus often termed hyperglycemic, hyperinsulemic, hypercortisolemic, and hyperleptinemic.

## CRF and Female Reproduction

The HPA axis and its associated products have significant effects on female reproduction. Hypothalamic CRF inhibits hypothalamic gonadotropin-releasing hormone, and glucocorticoids induced by the release of CRF from the paraventricular nucleus of the hypothalamus (PVN) inhibit pituitary luteinizing hormone and ovarian estrogen and progesterone secretion (Kalantaridou et al. 2004, 2007). Findings in CRF knockout (CRF-KO) mice, however, have complicated this issue, as reproductive suppression owing to restraint stress in female CRF-KO mice is indistinguishable from that of CRF replete controls (Jeong et al. 1999).

Peripheral expression of CRF was found to be widespread throughout the female reproductive tract, where it appears to exert paracrine actions related to inflammatory components of reproduction such as ovulation (Mastorakos et al. 1994; Kalantaridou et al. 2004), luteolysis, implantation (Athanassakis et al. 1999; Karteris et al. 2004), decidualization (Zoumakis et al. 2000), and early maternal tolerance (Kalantaridou et al. 2004). In addition to the placenta, CRF was found expressed in most female reproductive tissue, including the ovary, and in the uterus of both pregnant and nonpregnant women (Clifton et al. 1998, 2012; Muramatsu et al. 2001; Zoumakis et al. 2001; Wypior et al. 2011). Intracellular levels of CRF detected by immunofluorescence in cultured uterine cells reach the highest levels of cells positive for CRF during proestrus. Levels of secreted CRF in uterine cells reach their highest peak during anestrus (Athanassakis et al. 1999).

In contrast to the placental expression of CRF, ovarian and uterine expression appears not to be a solely anthropoid primate adaptation; both the rat and mouse ovary and uterus express CRF (Mastorakos et al. 1993; Athanassakis et al. 1999; Power and Schulkin 2006). There is differential expression of CRF receptors in uterine tissue, depending on whether or not the female is pregnant (Grammatopoulos et al. 1998).

## CRF and the Placenta

The activation of the fetal hypothalamic-pituitary-adrenal (HPA) axis promotes the maturation of the fetus and parturition (Challis et al. 1995). (As discussed in previous chapters, CRF is a peptide originally discovered in the paraventricular nucleus of the hypothalamus that is fundamental to the regulation of pituitary secretion of ACTH [Vale et al. 1981]). CRF was subsequently isolated in the placenta, and found to be identical to CRF in the hypothalamus (Kovo et al. 2013; Kropach et al. 2016; Shibasaki et al. 1982), and in extrahypothalamic sites in the brain (Swanson et al. 1983). In placental tissue, CRF is largely localized in syncytiotrophoblast and intermediate trophoblast cell bodies (see e.g., Frim et al. 1988; Petraglia et al. 2005).

Pregnancy is in part an inflammatory state. Proinflammatory cytokines are expressed by uterine epithelium in the pregnant state and by the placenta. The underlying cause of the inflammation of pregnancy is not well understood. There are probably multiple effectors giving rise to increased monocyte activation during pregnancy. Possible triggers include a variety of molecules released into maternal tissue by the placenta (e.g., cytokines and anti-angiogenic factors). Inflammation-related genes are present in varying quantities in human white adipose tissue and term placenta. These genes include TNF-a, IL-1Ra, IL-8, MCP-1, PAI-1, and VEGF.

It was thought to be paradoxical that although glucocorticoids are known to inhibit the HPA axis, including production of hypothalamic CRF, cortisol stimulates the CRF mRNA and peptide levels in some areas outside the hypothalamus, including the placenta and extrahypothalamic regions of the brain. In addition to human beings, chimpanzees, gorillas, and rhesus monkeys also have an exponential rise in placental trophoblast CRF levels during the second half of normal gestation (Frim et al. 1988; Riley et al. 1991; Robinson et al. 1989; Wu et al. 1995). Other primates studied, including baboons, have CRF in their placentas, but in contrast to the situation in human beings and chimpanzees, CRF levels peak during the first half of pregnancy and decline thereafter (Goland et al. 1992; Smith et al. 1993; Power and Schulkin 2006). The nonprimate mammals studied so far have not had CRF in their placentas (Robinson et al. 1989; Power and Schulkin 2012).

In gorillas and humans, placental CRF messenger RNA and CRF peptide secretion in maternal plasma (Shibasaki et al. 1982; Sasaki et al. 1984, 1987;

Goland et al. 1986; Campbell et al. 1987; Emanuel et al. 1994) have been shown to rise as much as fifty-fold during the third trimester of normal pregnancy. A parallel rise in fetal cortisol production occurs during the same period, as measured by fetal cortisol sulfated metabolites in maternal blood and urine (Challis et al. 2000). There is large variation, as much as fifty-fold, in maternal CRF concentrations at term (200–10,000 pg/mL) among normal pregnancies (Goland et al. 1988; Wolfe et al. 1988; Majzoub et al. 1999; Erickson et al. 2001). Pulsatile secretion of placental CRF has been ruled out by frequent blood sampling and it cannot account for this wide range of CRF concentrations at the end of normal pregnancy. The elevation in placental CRF secretion is associated with a surge of fetal glucocorticoids and the production of estriol during the several weeks before normal parturition. This observation, as well as the wide variability in the level of CRF expression seen in different women, has led investigators to speculate that CRF may not be playing an important role in the mother throughout most of gestation but may play a role in initiating parturition (Wolfe et al. 1988; Goland et al. 1988; Challis et al. 1995).

CRF in humans is not typically detectable in plasma except during pregnancy (Sasaki et al. 1984; Campbell et al. 1987; Goland et al. 1988; Wolfe et al. 1988). In the second and third trimesters of normal pregnancy, CRF is elevated in maternal plasma that is derived from the placenta (Figure 4.3). At the same time, both fetal and maternal ACTH and cortisol levels are elevated (Challis et al. 1995; Goland et al. 1995). Following parturition, CRF levels in the plasma decrease to nondetectable levels (Laatikainen et al. 1987).

During normal parturition in primates, CRF expression is elevated in placental tissue by 15 weeks, and by weeks 35–40 it is quite high (Frim et al. 1988; Goland et al. 1986); one role of cortisol is to facilitate CRF in placental tissue (Majzoub et al. 1999).

Placental CRF dramatically increases after 37 weeks of gestation. Whereas progesterone, estradiol, and cortisol continue to increase throughout gestation, but not as remarkably as do placental CRF levels.

**Role of CRF in gestation**: In the human placenta, CRF is produced exclusively in syncytiotrophoblast; it is only after cytotrophoblasts fuse into syncytiotrophoblasts that CRF is expressed and secreted into both maternal and fetal circulations. Levels climb much higher in the maternal circulation than in the fetal circulation, in which the upper limit of concentration is in the 300 pg/mL range. One model that explains the simultaneous rise in CRF and cortisol suggests that within the placenta the exponential rate of increase in CRF is positively related to the concentration of cortisol (Figure 4.4, Majzoub et al. 1999; Smith et al. 1993).

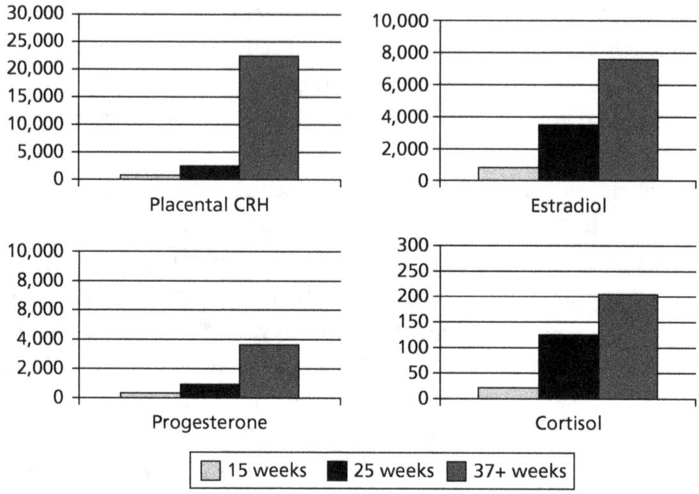


**Figure 4.3** Levels of diverse hormones during pregnancy.

Reprinted from *Trends in Endocrinology & Metabolism*, 23 (10), Roger Smith, Jonathan Paul, Kaushik Maiti, Jorge Tolosa, and Gemma Madsen, Recent advances in understanding the endocrinology of human birth, pp. 516–23, doi:10.1016/j.tem.2012.05.002, Copyright © 2012, Elsevier Inc., with permission from Elsevier.

The primate placenta is handicapped, in an endocrine sense, by its lack of 17-hydroxylase/lyase. Without these critical enzymes, produced from the same polypeptide, the placenta cannot directly synthesize estradiol from progesterone, the only steroid hormone that the primate placenta can synthesize de novo. The fetal adrenal zone, therefore, serves instead as the predominant source of that steroid's precursor, dehydro-epiandrosterone (DHEA; Ryan 1959; Siiteri and MacDonald 1963). Thus placental CRF and its ability to stimulate the fetal adrenal system may have evolved in primates to stimulate fetal ACTH, and ultimately the fetal adrenal system, to satisfy the high demand for DHEA synthesis (Figure 4.4; Karalis et al. 1996). This proposed feedback mechanism would function to stimulate a fetal adrenal "factory," contributing to placental estradiol production and to the endocrine sequelae necessary for parturition, including the stimulation of prostaglandin synthesis, oxytocin receptors, and gap junctions.

Consistent with this hypothesis, circulating levels of CRF in fetal blood are similar to those measured in the hypothalamic-pituitary portal circulation. CRF infusion into mid-gestational fetal baboons activates the fetal HPA axis; and androstenedione infusion into pregnant rhesus females results in preterm delivery (Goland et al. 1988; Berghorn et al. 1991; Mecenas et al. 1996).

Furthermore, fetal cortisol secretion is tied to maturation of fetal lungs and other systems before birth (Muglia et al. 1995). Thus the concomitant

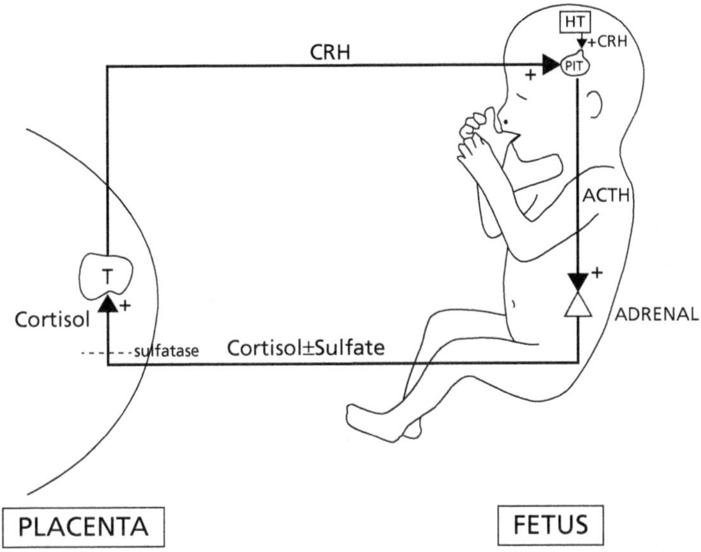


**Figure 4.4** The fetoplacental CRH-glucocorticoid positive-feedback hypothesis. CRH, secreted by the placental trophoblast (T), enters the fetal circulation via the umbilical vein and stimulates (+) fetal ACTH release from the fetal pituitary (PIT). Fetal ACTH stimulates secretion of fetal adrenal cortisol sulfate, which enters the placental circulation via the umbilical artery. Cortisol sulfate may be converted to cortisol by placental sulfatase, and cortisol stimulates further placental CRH secretion, thereby completing the positive-feedback loop. Fetal CRH, secreted from the fetal hypothalamus (HT), may independently stimulate fetal ACTH release, and placental and fetal hypothalamic CRH may be directly stimulated by environmental stresses. In addition, placental ACTH may stimulate the fetal adrenal directly.

Reproduced from Bruce G. Robinson, Rodica L. Emanuel, David M. Frim, and Joseph A. Majzoub, Glucocorticoid stimulates expression of corticotropin-releasing hormone gene in human placenta, *Proceedings of the National Academy of Sciences of the United States of America*, 85 (14), p. 5248, Figure 5, Copyright © 1988, National Academy of Sciences.

stimulation of fetal cortisol and DHEA by placental CRF is of obvious benefit for postnatal survival in that it synchronizes the glucocorticoid effects on fetal-organ maturation with the fetal DHEA production essential for the timing of parturition (Karalis et al. 1996).

**CRF and the timing of parturition:** The human fetal pituitary system develops early in gestation and begins to monitor levels of fetal cortisol at 7 to 8 weeks gestation, responding to low cortisol levels by secreting ACTH. This is best demonstrated in fetuses with enzymatic blocks in cortisol synthesis (congenital adrenal hyperplasia); adrenal systems compensate by increasing the synthesis of all steroids, including androgens, proximal to the enzymatic block, resulting in the virilization of female fetuses by 8 to 10 weeks' gestation (Pang 1997). CRF mRNA also is present in placental syncytiotrophoblast cells by 8 weeks' gestation, and CRF levels clearly rise exponentially, as much as 20 times, during the last 6 to 8 weeks of gestation (Frim et al. 1988). Similarly, CRF peptide levels in maternal blood are quite low until the final 8 to 10 weeks of gestation.

It is possible that peptide levels, although low in early gestation, increase at an exponential rate throughout gestation and are detected as rapidly rising only near the end of pregnancy (McLean et al. 1995). However, other data suggest that CRF levels are low until the beginning of the third trimester, and only after that do they begin to rise at an accelerating rate (Goland et al. 1986, 1988). If the latter case is true, what might restrain the system at earlier points?

There is some evidence that progesterone may play a role in this differential regulation of CRF in the hypothalamus and placenta, in essence acting as a brake on the latter system of positive regulation (Robinson et al. 1988; Majzoub and Karalis 1999). In vitro studies have shown that when progesterone is added back to placental trophoblast cultures during cortisol stimulation, there is inhibition of the increases in CRF mRNA and peptide.

Inhibition of CRF would be reversed only if the rise in fetal adrenal cortisol secretion in late gestation was great enough to compete with progesterone at the level of the glucocorticoid receptors, essentially causing a functional withdrawal of progesterone at the level of the glucocorticoid receptors. In this way, progesterone might serve as an early brake on the system, a brake that is overridden only after cortisol rises to an appropriately high level.

The molecular basis for the differing regulation of the CRF promoter gene in the PVN and placenta has begun to be elucidated (Cheng et al. 2000; King et al. 2001, 2002; Nicholson et al. 2004). The CRF promoter gene does not contain a consensus glucocorticoid regulatory element (GRE), but a number of regions where glucocorticoid receptors (GR) could possibly bind were identified (Guardiola-Diaz et al. 1996). A negative glucocorticoid regulatory element (nGRE) was identified that is involved in the inhibition of CRF promoter activity by glucose adenosine monophosphate response element (CRE) and a caudal-type homeobox response element (CDXRE) were also identified (Nicholson et al. 2004). Both CRE and CDXRE are involved with stimulating CRF promoter activity, but the response depends on nuclear factors. The GR does not bind to either site; this suggests that glucocorticoids stimulate the CRE indirectly through protein/protein interactions or by affecting nuclear factors (King et al. 2011; Nicholson et al. 2004; Cheng et al. 2000).

In cell types where glucocorticoids suppress CRF synthesis, three nuclear factors that bind to the CRF promoter gene were identified (Nicholson et al. 2004). Two were identified as CREB and cFos, and the third was suggested to be a novel CRE-binding protein termed CRE-associated protein (CREAPI). Inducible cAMP early repressor (ICER) is elevated in rat PVN following restraint stress, and was suggested to be involved in the inhibition of CRF following its elevation resulting from restraint stress by interacting with the CRF promoter gene at the CRE (Shepard et al. 2005). In placental cells, where glucocorticoids upregulate CRF synthesis, only two nuclear binding proteins were identified: CREB and cJun (Nicholson et al. 2004). Thus the tissue-specific regulation of CRF promoter activity by glucocorticoids is suggested to be mediated through differential stimulation of nuclear factors by glucocorticoids (Nicholson et al. 2004; Donley et al. unpublished).

## Placental CRF Timing in Human Gestation

A timing mechanism appears to exist in the human placenta that determines the length of gestation. A series of coupled endocrine and paracrine feed-forward loops serves to coordinate the timing of parturition with fetal maturation. Placental synthesis of CRF was suggested to reflect the setting of this fetal placental "clock," and to be a key component in the coordination of fetal maturation with the progression of human gestation (McLean et al. 1995). Prospective longitudinal studies (McLean et al. 1995; Leung et al. 2001, 2010) revealed that women destined to deliver preterm had more rapid exponential rises of serum CRF, whereas women who would deliver late had slower rates of rise of serum CRF. But there are many other molecules (see below) potentially involved in timing of delivery that need to be considered as well. A study that looked at the levels of vitamin D (a steroid) and CRF in preterm delivery found that low maternal vitamin D was associated with poorer birth outcomes, and that mothers who deliver early preterm had low vitamin D and high maternal CRF levels (Power et al. unpublished).

Several lines of evidence support the hypothesis that events early in pregnancy may set the timing of birth. As mentioned, women who subsequently enter preterm labor not only have elevated serum CRF at mid-pregnancy, but their rate of increase of CRF is accelerated from early on (McLean et al. 1995; Hobel et al. 1999; Leung et al. 2001). In a prospective study, forty pregnant women were identified who had experienced a 6.8 magnitude earthquake during pregnancy or shortly afterward. Women who experienced the earthquake earlier in their pregnancy had a significantly shorter gestational length than those who experienced it later in gestation (Glynn et al. 2001).

Therefore, events early in pregnancy that affect maternal expression of glucocorticoids through the stimulation of her HPA axis appear to be able to affect the pattern of placental CRF expression through the rest of gestation. Interestingly, the most successful treatment that has been found to delay parturition after the onset of preterm labor is administration of progesterone (Meis et al. 2003).

There is a positive feedback loop responsible for differential regulation of CRF in the fetal placental axis (Majzoub and Karalis 1999). In this model corticotrophin-releasing factor produced in the placenta is secreted into the fetal circulation at a concentration of about 200 to 300 pg/mL, consistent with CRF levels found in the hypothalamic-pituitary portal circulation and sufficient to stimulate corticotrophin secretion from the anterior pituitary. Placental CRF, through fetal ACTH or possibly through direct interaction with fetal adrenal CRF receptors, is proposed to stimulate the fetal adrenal to produce cortisol, which binds to placental glucocorticoid receptors to block the inhibitory effect of progesterone, further stimulating CRF production in a positive fashion (Majzoub and Karalis 1999).

The expression of CRF and its related peptides is widespread in peripheral tissue, with important functions in the immune system, energy metabolism, and female reproduction. For example, CRF is involved in the implantation of fertilized ova and in maternal tolerance to the fetus. In humans, abnormally high placental expression of CRF is associated with pregnancy complications (e.g., preterm labor, intrauterine growth restriction (IUGR), and preeclampsia). According to studies in humans, pregnant women who are between 26 and 31 weeks of gestation and have a greater rate of change in CRF levels and a higher level of CRF overall after 31 weeks, are more likely to deliver preterm babies (Sandman et al. 2006). Within anthropoid primates, there are at least two patterns of placental CRF expression over gestation: monkeys differ from great apes (and humans) by having a mid-gestational peak in CRF expression. The functional significance of these differences between monkeys and apes is not yet understood, but it supports the hypothesis that placental CRF performs multiple roles during gestation.

## Cortisol Induction of CRF and Profound Birth Changes

During the course of pregnancy in humans, cortisol acts to elevate CRF to promote the normal course of pregnancy (Challis 2001; Majzoub et al. 1999; Smith 2007). Placental CRF, transported through the umbilical vein to the fetus, could stimulate the fetal pituitary-adrenal axis to produce cortisol and cortisol sulfate, which would then be capable of further stimulating placental CRF production, creating a positive feedback loop. This model is supported by both in vitro and in vivo data, in that betamethasone administration to pregnant women results in either no fall or clear stimulation of placental CRF secretion into the maternal circulation (Challis 2001). This placental model is quite different from the regulation of CRF expression in the HPA axis, which responds to cortisol with downward regulation or negative restraint.

Diverse in vivo and in vitro studies have demonstrated that glucocorticoids agonists can enhance CRF gene expression in the placenta (Challis 2001; Smith 2007; Majzoub et al. 1999; Wang et al. 2012; Rosen et al. 2015). Placental tissue treated with glucocorticoid agonists or antagonists either increased or decreased CRF gene expression. Moreover, in several studies, women treated with glucocorticoid agonists, for instance, developed elevated levels of CRF in plasma and in placenta (Frim et al. 1988; Chan et al. 1988; Figure 4.5). When the glucocorticoid agonist dexamethasone is used to treat pregnant women, their levels of CRF peptide and mRNA significantly increase.

Placental CRF acts in an endocrine fashion, and thus has the capacity to exert profound developmental and metabolic effects on both maternal and fetal physiology (Sanderson et al. 2000). For example, CRF is a potent vasodilator in both

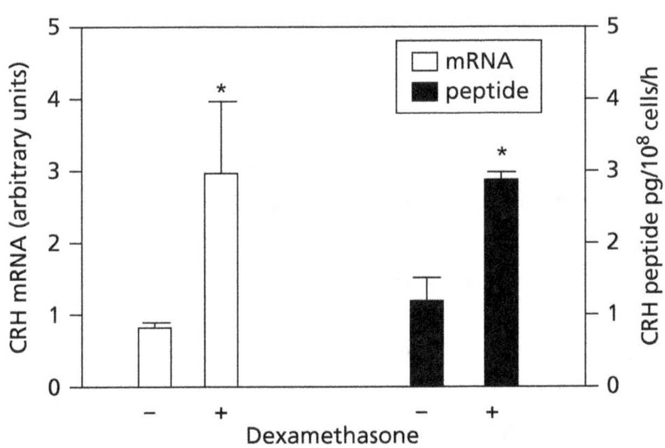


**Figure 4.5** Effect of dexamethasone on cultured trophoblast CRF and GR mRNA. Quantitation of CRF mRNA displayed in c together with CRF peptide levels in media in the presence (+) and absence (–) of dexamethasone. *, $P < 0.001$ compared with dexamethasone.

Reproduced from Bruce G. Robinson, Rodica L. Emanuel, David M. Frim, and Joseph A. Majzoub, Glucocorticoid stimulates expression of corticotropin-releasing hormone gene in human placenta, *Proceedings of the National Academy of Sciences of the United States of America*, 85 (14), p. 5247, Figure 3d, Copyright © 1988, National Academy of Sciences.

the maternal and fetal placental vascular trees (Clifton et al. 1995; Crompton et al. 2003), and the CRF-R2 receptor is tied to the reduction of vascularization (Bale et al. 2003). Another study has shown that increased exposure to placental CRF affects the development of the fetal CNS and the child's future mental health (in the form of self-reported internalizing symptoms; Howland et al. 2016). Hypercortisolemia is characteristic of human pregnancy, and evidence supports the hypothesis that placental CRF is partly, though not solely, a regulator of maternal cortisol levels (Challis et al. 2000; Florio et al. 2002; Petraglia et al. 2005).

This positive induction of CRF in the placenta by cortisol has been linked to cAMP mediated mechanisms; cAMP activates CRF promoter sites (Smith 2007; Nicholson et al. 2004) and non-canonical pathways such as NF-kb (Wang et al. 2012). Mechanisms such as cAMP are preserved across diverse taxa and are phylogenetically ancient (Denver and Crespi 2006).

There are a number of studies demonstrating glucocorticoid upregulation of CRF in the placenta. In one early study, it was reported that dexamethasone treatment did not suppress levels of CRF in the plasma of pregnant women (Tropper et al. 1987). It was then demonstrated that glucocorticoids do not inhibit the production of CRF in the placenta as expected, rather they increase CRF gene expression in the placenta (Robinson et al. 1988; Jones et al. 1989). Glucocorticoids increase CRF gene expression in primary cultures of human placental syncytiotrophoblasts. These effects are dose related, and may be greater in response to dexamethasone than for cortisol, suggesting that these effects are dependent upon type II glucocorticoid receptor sites (Jones et al. 1989; Challis et al. 1995). Studies have demonstrated that pregnant women treated with betamethasone after 30-weeks gestation had increased CRF levels in both plasma and placental tissue (Marinoni et al. 1998). An additional study also revealed that pregnant patients at 24 weeks also have increased levels of plasma CRF following betamethasone treatment (Korebrits et al. 1998).

Thus, in marked contrast to glucocorticoids' well-known inhibition of CRF via type II glucocorticoid receptor sites at the level of the paraventricular nucleus of the hypothalamus, and the negative restraint of the HPA axis (Munck et al. 1984; Sawchenko 1987; Swanson and Simmons 1989), glucocorticoids increase CRF gene expression in the placenta.

## A Role for CRF in Fetal Tissue Maturation

Fetal cortisol production can stimulate further CRF production from the placenta. In humans (Goland et al. 1994), chimpanzees, and gorillas (Smith et al. 1999), serum cortisol and CRF are correlated. This is consistent with the hypothesis that glucocorticoids drive placental CRF production via a feed-forward system linking the placenta with the fetal pituitary and adrenal glands. This links the timing of parturition with fetal maturation (Majzoub et al. 1999).

It is tempting to hypothesize that the organ-maturation ability of glucocorticoids might be mediated by the up- or down regulation of CRF (or other members of the CRF hormone family) in fetal tissue. This temptation is tempered by findings from homozygous and heterozygous CRF knockout mice matings. The mouse pups from homozygous matings died on the first day of life, mainly because of improper maturation of lung tissue, consistent with the hypothesis that glucocorticoids are necessary for appropriate lung maturation. However, homozygous pups from heterozygous dams and from homozygous dams provided with glucocorticoids during gestation had normal lung development at birth, and survived (Muglia 2000).

Thus, although glucocorticoids appear to be necessary for fetal lung development, CRF does not. This result does not prohibit a role for other members of the CRF family in fetal tissue maturation, nor does it refute the hypothesis that there are multiple mechanisms (all glucocorticoid sensitive) toward achieving lung maturation.

## Progesterone

Gestation involves both the maternal and fetal HPA axes, as well as many hormones, such as progesterone, estrogen, cortisol, oxytocin, and CRF (Smith et al. 2009). Through both positive- and negative-feedback loops, these two axes interact and promote growth and development.

Progesterone competes with glucocorticoids for the glucocorticoid receptor sites, and thus might suppress the feed-forward placental CRF-cortisol system (Karalis et al. 1996 Figure 4.6).

This hypothesis was further tested by culturing trophoblast cells in a cholesterol-free media, thus depriving them of the raw materials to produce progesterone on their own. Under these progesterone-poor circumstances, CRF mRNA levels were greatly elevated. When progesterone was added back in significant amounts (10 mol/L), CRF levels fell in a dose-dependent fashion. It may be the case that decreases in progesterone lead to elevated levels of CRF, perhaps by the greater efficacy of cortisol at the receptor sites for this steroid hormone (Robinson et al. 1988; Majzoub and Karalis 1999; Karalis et al. 1996; Majzoub et al. 1999; Smith 2007; Vrachnis et al. 2012). Therefore, progesterone can lead to decreases in CRF.

To delay parturition is to block cortisol actions. Progesterone is one such mechanism. Progesterone provokes quiescence in both the brain, perhaps through GABA receptors, and in the periphery in diverse end organ systems

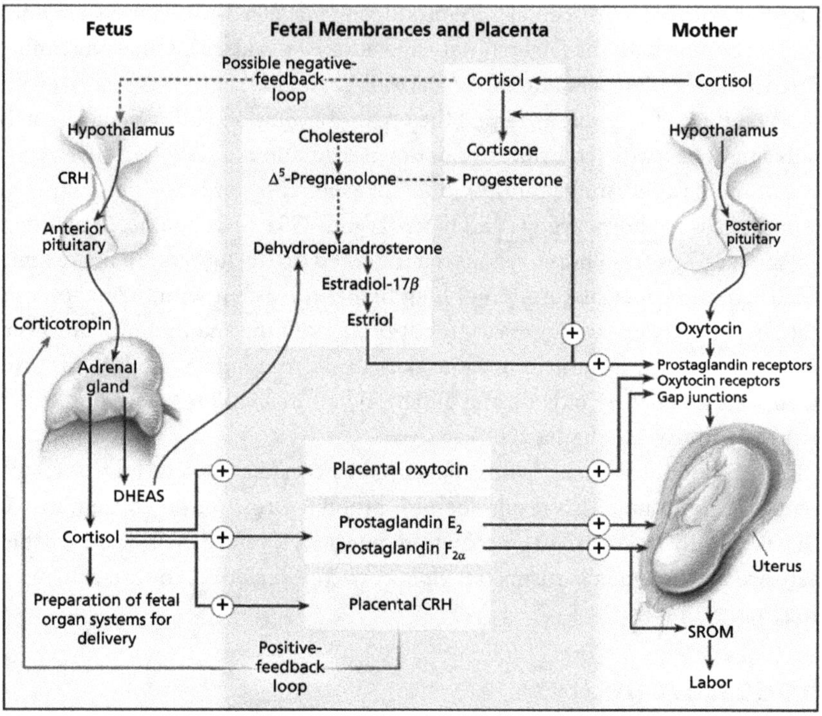


**Figure 4.6** (See color plate section) The maternal and fetal HPA axes are interconnected via the placenta, fetal membranes, and circulating information molecules. Some of the key events involved in gestation include progesterone, estrogen, cortisol, oxytocin, and CRF.



(Majzoub et al. 1999; Karalis et al. 1996; Orchinik et al. 1992). That is, the same hormonal mechanism that competes for receptor sites to promote or prolong gestation also acts in the brain to promote action or quiescence (McEwen 1995). In the brain, GABA is inhibitory and regulated by diverse steroids, one of which is progesterone. In other words, progesterone, a natural antagonist, does the converse and can delay parturition (Majzoub et al. 1999; Challis 2001; Smith 2007); interestingly, progesterone, via GABAergic mechanisms, facilitates a quiescence effect.

Progesterone directly affects both genomic and non-genomic mechanisms that regulate contractions at the level of myometrial cells (Mesiano et al. 2011) Progesterone interacts with both nuclear and membrane progesterone receptors to modulate myometrial contractility. Nuclear receptors work through genomic and extranuclear nontranscriptional methods to affect contractility, while membrane receptors work through intracellular signaling pathways. For the majority of pregnancy, progesterone maintains the myometrium in a relaxed state. Signaling systems also include progesterone-mediated CRF signaling systems (Wang et al. 2014).

Progesterone receptors in the uterus are tied to human birth, timing, and pregnancy (Mesiano et al. 2011). There are several progesterone receptors. The underlying progesterone is linked to progesterone receptor A expression in the myometrium (Mesiano et al. 2011). Like cortisol and other steroids, progesterone has both genomic and non-genomic actions (Moussatche and Lyons 2012). The opposite is true of cortisol. In other words, positive regulation of the CRF gene by glucocorticoids has been described in placental syncytiotrophoblast cells, where rising glucocorticoids of fetal origin upregulate placental CRF, which underlies parturition (Smith 2007; Majzoub et al. 1999; Challis 2001). The positive regulation may be mediated via the CRE, and a region located between -213 and -99 base pairs, which may mediate cAMP-dependent activation (King et al. 2001; King and Nicholson 2007).

Furthermore, RU-486 (a progesterone/glucocorticoid antagonist) stimulates CRF mRNA (Bluthgen et al. 2013), also suggesting that a withdrawal of progesterone stimulates CRF's effect in the placenta. Functional withdrawal contributes to parturition, but it has been hard to demonstrate the therapeutic action of progesterone to delay parturition (Iams 2014; Romero et al. 2013). One progesterone receptor tends to be anti-inflammatory and the other proinflammatory (Mesiano et al. 2002).

Progesterone receptors are present in trophoblast cells; it is therefore speculated that high levels of placental progesterone early in gestation could compete with cortisol for the glucocorticoid receptors, inhibiting cortisol and its effect on CRF synthesis (Karalis et al. 1996; Jeschke et al. 2005). Because progesterone appears to bind to glucocorticoid receptors about 25% to 50%, as does cortisol (Ojasoo et al. 1988; Chrousos et al. 1984), only when levels of cortisol become high enough to block effectively progesterone's inhibitory effect would the CRF elevation occur. Progesterone thus might inhibit the effects of cortisol on placental CRF in a dose-dependent fashion.

## Placental CRF and Anthropoid Evolution

Placental expression of CRF apparently may be a unique anthropoid primate adaptation (Robinson et al. 1989; Bowman et al. 2001; Power and Schulkin 2012), in which CRF acts as an endocrine, paracrine, and autocrine hormone to regulate and coordinate maternal, placental, and fetal development and metabolism (Petraglia et al. 2005). Placental CRF plays key roles in the progression of gestation to parturition (Majzoub et al. 1999; Challis et al. 2000; Huang et al. 2011). It is thought to be important in the transition from uterine quiescence to coordinated contractions, and in fetal development and maturation necessary for extrauterine life (Majzoub et al. 1999; Lockwood 2004; Smith et al. 2012).

The available evidence supports the hypothesis that anthropoid primates are the only species whose placentas produce CRF (Robinson et al. 1989; Bowman et al. 2001). Late pregnancy plasma samples from three lemur species contained no detectable CRF, and CRF peptide could not be extracted from placental tissue of the black lemur, Eulemur macaco (Bowman et al. 2001). This sets anthropoid primates apart from other mammals, and suggests that understanding this apparently unique anthropoid primate adaptation may be key to understanding both the normal course of human pregnancy, and various metabolic disruptions and complications of pregnancy.

All anthropoid primates studied produce placental CRF. However, there are at least two patterns of placental CRF production and secretion. The pattern seen in humans (e.g., Goland et al. 1986; Campbell et al. 1987; Laatikainen et al. 1987; Sasaki et al. 1987; McLean et al. 1995) and great apes (Smith et al. 1999). In humans, there is an exponential increase in CRF from the end of the first trimester until parturition. This is likely a derived pattern, as it differs from the pattern found in two quite divergent monkey species. The pattern seen in both the baboon (Goland et al. 1992; Smith et al. 1993), an Old World monkey, and the common marmoset (Power et al. 2011), a New World monkey, consists of a rapid increase in CRF early in pregnancy, followed by a decline to a plateau value, with a possible rise at the end of pregnancy.

Considering the evolutionary separation between these two species, this pattern may represent the ancestral anthropoid pattern. However, results from two species do not constitute overwhelming evidence for a consistent pattern of CRF expression during gestation in monkeys (Power and Schulkin 2006). Interpretation of the two known different patterns of anthropoid placental CRF expression is complicated by the fact that most, though possibly not all, anthropoid primates exhibit detectable levels of circulating CRF-BP of either hepatic or placental origin (Bowman et al. 2001). Circulating CRF-BP was detected in pregnant and nonpregnant humans and chimpanzees, pregnant gorillas, and nonpregnant orangutans, gibbons, lion-tailed macaques, common marmosets, and squirrel monkeys, but not in ruffed lemurs, horses, and sheep. Surprisingly, however, CRF-BP was also not detected in nonpregnant mandrills, baboons, and spider monkeys (Bowman et al. 2001; Robinson et al. 1989).

The human placenta expresses CRF-BP (Petraglia et al. 2005), and thus studies of pregnant monkeys are required to determine whether circulating CRF-BP is a common feature of anthropoid pregnancy. The pattern of free (active) CRF will depend on the pattern of CRF-BP. This pattern is known for humans (Linton et al. 1993), gorillas (Smith et al. 1999) (stable until the third trimester, and then a decrease in maternal serum CRF-BP concentration), and chimpanzees (Smith et al. 1999) (stable throughout gestation), but remains to be determined in other anthropoids.

In humans, concentrations of maternal serum CRF-BP in early pregnancy are probably sufficient such that most placental CRF secreted into the maternal compartment is bound, and thus inactive. Placental CRF is thought to have its major effects on maternal physiology in the third trimester, thus the concentration of CRF begins to equal and then exceed that of CRF-BP. It is unknown whether the early-mid-gestational peak in placental CRF expression in baboons and common marmosets results in substantial free CRF in the maternal circulation.

Placental CRF produced during gestation might have multiple effects on many maternal and fetal organs besides the pituitary and adrenals. These effects are likely tissue-specific during anthropoid gestation, depending on the period of gestation and the target tissue (maternal, placental, or fetal). For example, CRF may perform an autocrine or paracrine function in the human chorion, regulating prostaglandin concentration via production of 15-hydroxy prostaglandin dehydrogenase, and thus may contribute to myometrial quiescence (not stimulation) during most of pregnancy (McKeown and Challis 2003; Challis et al. 2000).

The functions of the early gestational rise in placental CRF production in all anthropoids and the significance of the differences among monkeys, apes, and humans are currently not known. Diverse data as indicated (Choy et al. 2004) suggest that CRF in early human pregnancy is preferentially expressed by cytotrophoblasts, and it plays an important role in early placental growth and development. This is consistent with data from the common marmoset, in which the rapid increase in placental CRF expression coincides with the period of rapid growth and development of the placenta (Power and Schulkin 2012; Power et al. 2006). In the common marmoset, low levels of CRF during the time of expected peak concentration were associated with fetal death, although this finding may have merely reflected placental failure (Power and Schulkin 2012). In humans, elevated CRF in the second or third trimester is associated with serious pregnancy complications that affect fetal health. Deviations from the normal pattern of placental CRF expression signal pregnancy complications and poor fetal health.

**Danger signals, parturition, and CRF:** A more complete understanding of the variation in function and expression of placental CRF among anthropoid species could provide important insight into the functions of placental CRF in normal and pathological pregnancies, and better inform our clinical approaches to preterm labor, intrauterine growth restriction (IUGR), and other pregnancy complications.

As I have indicated, most infectious diseases create an adverse climate for preterm babies, with elevated levels of CRF or metabolic strains (HIV, preeclamsia, metabolic disorder, and Group B strep). For instance, in placental tissue, chorioamnionitis is tied to preterm delivery with elevated levels of CRF, CRF receptors, and some but not all urocortin receptors (Torricelli et al. 2011a, b; Voltolini et al. 2014). Human vaginal epithelium with lacatobacillus, gardnerella, prevotella, and streptococcus, to name a few (Hyman et al. 2005; Gravett et al. 1986; Goldenberg 2002), set the conditions for prematurity. These diseases can exist throughout the uterus, making it a major potential site for bacterial infection. Indeed vaginal bacteria, or microbial communities, can induce immunological responses that induce CRF expression; in turn, CRF and fetal placental inflammation help to facilitate prematurity (Trivedi et al. 2012). Premature rupture of membranes (Lannon et al. 2014) and membrane weakening, intrauterine inflammation (Bastek et al. 2011), cervical length (Bastek and Elovitz 2013), and perhaps the induction of CRF in diverse tissue (e.g., intrauterine wall) can result in a premature birth.

There is a precedent of CRF being induced in diverse regions of the bodily tissue that does not normally express it; infections such as bacterial vaginosis or group B streptococcal, which can result in preterm delivery, may elicit CRF and cytokines as an inflammatory warning in target tissue before it becomes dangerous (Gravett et al. 1986; Eschenbach et al. 1988; Adams Waldorf et al. 2011). Lung-associated disorders are associated with CRF expression in the lungs (Gonzales et al. 2008), including asthma (Silverman et al. 2004), and in the brain (Carlson et al. 2007).

Bacterial vaginosis, for instance, has a much greater frequency in African Americans, and they are at least twice as likely to have low birth-weight babies (Fettweis et al. 2014). CRF measured in plasma is lower in African American women across trimesters as compared with other groups (Chen et al. 2010: Glynn et al. 2007). CRF levels may serve as an additional biomarker of risk, specifically of preterm birth, in pregnant minority women (Ruiz et al. 2015; Majzoub et al. 1999).

Women with preterm premature rupture of membranes (pPROM) exhibited significantly greater levels of CRF mRNA compared with women with early preterm delivery (PTD) and preterm labor (PTL) with intact membranes in the absence of histological chorioamnionitis, and women with pPROM and no choriamnionitis. Women with pPROM and chorioamnionitis had significantly higher levels of CRF and CRF-R1, but lower levels of CRF-R2.

**Danger:** In addition to preterm birth, maternal serum CRF is elevated for a number of pregnancy complications: preeclampsia, (Laatikainen et al. 1991; Perkins et al. 1993; Goland et al. 1995; Florio et al. 2004) and IUGR (Goland et al. 1993), and in metabolic challenges of pregnancy (e.g., twins; Warren et al. 1990).

Increased placental expression of CRF appears to signal a metabolic disruption or complication of pregnancy in humans, with the effects of accelerated birth (preterm) and IUGR being approximately equal and independent, and thus additive. High levels of CRF were detected in the myometrium and fetal membranes of women who experienced habitual spontaneous abortions. Women with at least two spontaneous abortions displayed significantly higher levels of CRF than did both women with elective abortions and women with only one spontaneous abortion (Madhappan et al. 2003).

The core hypothesis is that cortisol, by pumping up the levels of CRF before term, signals danger (Goland et al. 1993; Erickson et al. 2003; Wolfe et al. 1988; Hobel et al. 2008). The normal change in which cortisol and CRF would be facilitated is now exaggerated by the induction and feed-forward regulation at the level of the placenta (Frim et al. 988; Robinson et al. 1988; Majzoub et al. 1999). But when there is potential danger to both the mother and the fetus, CRF may contribute to the production of a low birth-weight baby (Schulkin et al. 1998; Shannon et al. 2007). One mechanism is via CRF regulating proteins such as RelB-NFkB2 (Wang et al. 2014, 2015). Interestingly, cortisol may also serve as a predictor of infant birth-weight from a mother's saliva collected between pregnancies (Guardino et al. 2016).

Importantly, the greater the carry load is, the greater the expression of CRF (Power et al. 2006; Figure 4.7). With a multiple birth comes an increase in the peak maternal circulating CRF. A greater number of offspring carried per pregnancy leads to a greater metabolic demand on the mother. The greater the metabolic strain, the greater the likelihood of preterm delivery. The activation of CRF in addition to its putative clock-like 9-month mechanism in our species (Smith 2007) promotes fetal movement and maturation by cortisol. In other words, a normal endocrine-related system linked to

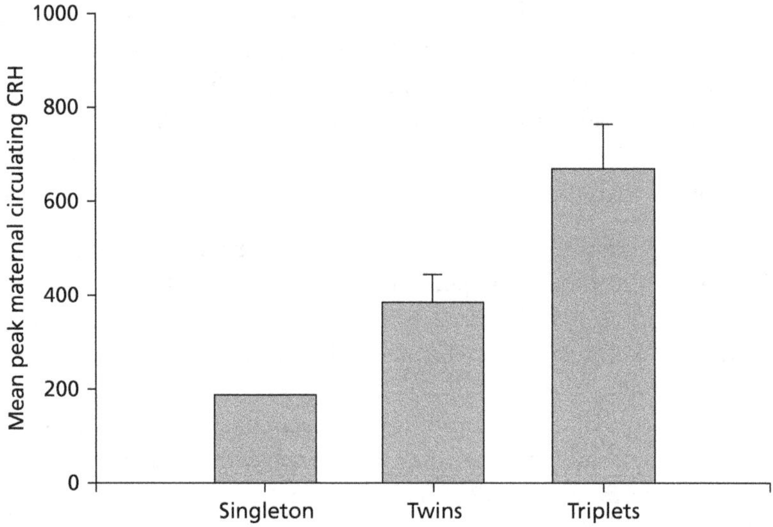


**Figure 4.7** Peak levels of maternal circulating CRF in the common marmoset pregnant with singleton, twins, or triplets.

Reproduced from Michael L. Power and Jay Schulkin, Functions of corticotropin-releasing hormone in anthropoid primates: From brain to placenta, *American Journal of Human Biology*, 18 (4), pp. 431–447, DOI: 10.1002/ajhb.20521, Copyright © 2006 Wiley-Liss, Inc.

dramatic change is pushed beyond what it can tolerate, lending itself to vulnerability in both the mother and the offspring (Schulkin 1999, 2003; Shannon et al. 2007).

## Conclusion

Evidence from work on CRF and diverse information molecules underlies the life-cycle changes surrounding metamorphosis and birth, with the onset of the placenta and mammalian trajectory much changed. The phylogenetic history for the induction of such change is quite ancient. Indeed, CRF and corticosterone facilitate metamorphosis in tadpoles and diverse forms of behavior including fear-related behaviors via CRF-R1 receptors (Denver and Crespi 2006; Denver 2009). Information molecules play diverse roles depending on which end organ tissue they are expressed in, what sites within that tissue, and where they have receptor sites. In mammals, diverse organs, including the placenta, contain CRF-like peptides, CRF receptors, and CRF-binding protein (Challis 2001; Majzoub 2006).

Evolutionary conservation of the role of CRF and its positive induction by glucocorticoids is nicely demonstrated in developmental changes in mammals where CRF of fetal and/or placental origin determines the length of the gestational period (Smith 2007; Challis 2001). Placental CRF is elevated by glucocorticoids (e.g., Frim et al. 1988; Robinson et al. 1988; Jones et al. 1989; Borders et al. 2016). During pregnancy, glucocorticoids act to elevate CRF in the placenta and to promote the normal course of pregnancy (Challis 2001; Majzoub 2006; Smith 2007). These events facilitate parturition both during normal pregnancy and under duress (e.g., Goland et al. 1993; Erickson et al. 2001; Wadhwa et al. 2001, Huang et al. 2011). Placental tissue treated with glucocorticoid agonists or antagonists either increased or decreased CRF gene expression. Moreover, in several studies, women treated with glucocorticoid agonists, for instance, developed elevated levels of CRF in plasma and in the placenta (Frim et al. 1988; Chan et al. 1988).

There are tempting parallels in the patterns of neural and placental CRF expression. Central neural CRF also is elevated under conditions that represent a challenge to the viability of the organism. Placental CRF is elevated under conditions that threaten the viability of the pregnancy. Cortisol stimulates and maintains CRF expression in several neural areas, sustaining central motive states appropriate for the external challenge (e.g., fear). Similarly, cortisol upregulates CRF expression in the placenta. Placental CRF and fetal cortisol interact in a feed-forward system to sustain the progression of the fetus toward maturation for extrauterine life, and the uterus for progression from quiescence to coordinated myometrial contractions leading to parturition. Challenges to a pregnancy frequently result in a hastening of this cascade, and thus an early delivery of the fetus(es). Preterm birth may serve to "rescue" the fetus from an adverse environment (McLean et al. 1995). Alternatively, it may represent a maternal adaptation to reduce the allocation of resources to a pregnancy occurring during stressful circumstances where the outcome is uncertain (Power and Schulkin 2008).

The most consistent fact about CRF detected in plasma of pregnant women is its link to potential maternal–fetal distress and greater metabolic and physiological demands in women who go on to experience preterm labor (Wolfe et al. 1988; Warren et al. 1992; Berkowitz et al. 1996). Bacterial infectious diseases (Petraglia et al. 1995), preeclampsia (Goland et al. 1995), diabetes (Wolfe et al. 1988), growth-retarded fetal development (Goland et al. 1993), multiple gestation (Wolfe et al. 1988), and psychosocial stress (Hobel et al. 1999) all constitute contributors to distress and adverse conditions. In one study, treatment for depression with serotonin reuptake inhibitors (SSRI) in pregnant women resulted in increased levels of serum CRF; both the SSRI treatment and elevated CRF levels were associated with a shorter gestational length, independently and as a combined effect (Hannerfors et al. 2015). In other words, for conditions of adversity, or great metabolic demand, CRF may play an important role in parturition, and in some instances, may predict vulnerability to preterm delivery (c.f. McLean et al. 1995; Berkowitz et al. 1996; Bisits et al. 1998).

Adversity before pregnancy can affect pregnancy-related events, including body weight and longer term effects of fear-related events (Leshem and Schulkin 2012; Shachar-Dadon et al. 2009). Preconception adversity and neuronal structure and function can ultimately affect CRF expression during pregnancy (Gillman et al. 2006; Leshem and Schulkin 2012).

A recent study associated both low and high levels of placental CRF with brain-damage disorders in infants born before 28 weeks of gestation (Leviton et al. 2016). But CRF is just one signaling system, and there are other systems at play.

Oxytocin and prostaglandin, two other information molecules, are also linked to parturition and they enhance CRF gene expression in placental tissue (Petraglia et al. 1989). Bacterial microbial induced cytokine production may facilitate parturition via changes in CRF. There is a well-established link between cytokine production, the release of CRF (Chrousos 1996), and a reduction in reproductive fitness (Ferin 1995).

Decreases in progesterone levels lead to the up regulation of CRF in the placenta, perhaps by enhanced binding of glucocorticoids (Karalis et al. 1996). In addition, decreases in CRF-binding protein are linked to the upregulation of CRF (Behan et al. 1995, 1996). One study found that a CRF-R1 antagonist could delay parturition in sheep (Chan et al. 1998).

Glucocorticoids serve an important fetal-organ maturation function in late gestation. Glucocorticoids are crucial for maturation of the fetal lung, and administering a single dose of exogenous glucocorticoid is the single most effective treatment for infants destined to undergo preterm birth (Goldenberg 2002). The prepartum increase in cortisol production by the fetal adrenal gland is important for fetal-organ maturation, especially of the lungs and kidneys, and it has effects on the fetal HPA axis and on extrahypothalamic brain regions. The mechanisms by which glucocorticoids mature fetal organs have not been elucidated.

During pregnancy, chronic adversity may be hormonally linked to early and abnormal elevation in CRF and to such sequelae as fetal growth restriction and preterm delivery. Conversely, consequences of increased placental CRF secretion during pregnancy could explain the development of a suppressed HPA axis after delivery in some women with postpartum depression (Chrousos 1995, Meltzer-Brody 2011; Meltzer-Brody et al. 2011; Yim et al. 2009). For instance, higher levels of CRF increase the vulnerability for postpartum depression (Meltzer-Brody et al. 2011), for which the CRF-R1 receptor has been implicated

(Engineer et al. 2013). But the data are inconsistent and unclear. Another study found no association between cerebrospinal fluid CRF levels in late pregnancy and the onset of postpartum depressive syndromes (Zaconeta et al. 2015).

Indeed, there is evidence, for instance, that serotonin receptors in the brainstem are dependent on serotonin secretion in utero. Of course, we know, for example, that glucocorticoids given to rat mothers have an effect on CRF expression in the brain (e.g., enhanced expression in the central nucleus of the amygdala (Welberg and Seckl 2001).

Chapter 5

# CRF: Contributor to Fear and Anxiety

## Introduction

There is no one essential property of fear, but rather it is a cluster of properties that function in diverse ways. These behaviors are understood in light of biology and evolutionary significance and various cognitive/motor and memory capabilities. A biological understanding of fear reaches from genes, to temperament, to social and ecological context, to cephalic capabilities, and individual and social experiences. Each factor contributes to understanding that fear is not simple or located in a single place.

We have come to realize that there are diverse neural systems that underlie different forms of behavioral adaptation and can be linked to fear-related behaviors. There are both general forms of expression, such as startle and freezing, and forms of association that are more specific that elicit fear-related flight or cover and freeze.

Context matters in understanding the behavioral adaptation and the neural circuits that cut across the cephalic tapestry from brainstem to cortex. Just as there are more than one kind of learning and memory system, there are diverse forms of fear-related systems for which neural circuits are both common and unique, depending upon the fear-related behavior.

The senses of fear encompass both the specific (e.g., odors and sights of predators) and more general learning about events, including fear of the unfamiliar (a rather basic fact) and the unlearning of fear or preventing the expression of fear. There are specific kinds of fear that are both learned and unlearned and highly semantic (Kagan 2002). Temperament, the way in which we will react to a given event, colors or biases the cephalic systems (Hebb 1946; Pavlov 1928). Temperament colors the way we engage others; it is a broad feature that is displayed across at least mammals (Davis et al. 2005; Kalin et al. 2016; Schmidt et al. 1997). Temperament is a piece of biology (Pavlov 1928); one well-studied feature is social wariness, or social presentation (Kagan 2002).

Predator adversity is the everyday occurrence of prey animals; one well-worked measure of this adversity is reproductive capabilities. One example is snowshoe hares; in the wild, snowshoe hares' reproductive rate declines when exposed to predation and glucocorticoids are increased (Boonstra et al. 1998). In the laboratory, CRF is elevated in these conditions.

In what follows, I briefly cover some of the information molecules associated with fear and then provide a discussion of the behavioral and biological basis of fear, endocrine regulation (namely the roles of CRF and glucocorticoids), and its aberration into pathology. CRF can be released quickly and can be regulated by both fast membrane-related mechanisms and slower genomic-related mechanisms.

**Information molecules and the brain**: Action does not necessarily mean movement. Freezing when fearful, for instance, is an action, albeit an inhibitory one, in which GABA is a primary neurotransmitter. GABA affects a diverse circuitry following a fear-related event, going from cortical neurons to the brainstem and spinal neurons; the result is freezing, for instance, at the sight of a predator or in expectation of an aversive/fearful event.

GABA is an information molecule, a core biochemical signature that plays diverse roles in the organization of behavior and physiology (Herbert and Schulkin 2002), of which there are many. GABA is a very general-function neurotransmitter, as are indolamines: serotonin, which sets the tone of experience; dopamine, which underlies the organization of action and the prediction of events; and norepinephrine, which is tied to attention. They serve just about everything we do. The classic neurotransmitters in the central nervous system fall into four major categories, each of which contains some noteworthy players: catecholamines (e.g., dopamine, norepinephrine, and epinephrine), indoles (e.g., serotonin and melatonin), cholinergic (e.g., acetylcholine), and amino acids (e.g., GABA, glutamate, and aspartate). CRF is coexpressed with many of the classical transmitters and peptides. Thus, the treatments for serotonin for example, affect CRF expression (Nemeroff and Owens 2004; Ohmura et al. 2010; Marcinkiewcz et al. 2016).

**Amygdala and bed nucleus and fear-related events:** Fear is basic, but is not easily attributed to a single behavior. Perhaps we are better off referring to diverse survival systems, as LeDoux (2012) recently suggests, and the diverse roles of CRF underlying fear of specific objects and angst about conditions more generally, rather than trying to define fear exclusively as one thing (Davis et al. 1997). Fear can turn to anxiety, as Freud and others noted, when it moves from specific objects to fearfulness in general, under diverse conditions (Davis et al. 2010). There are many potential roles for emotional stimuli. They can act as a trigger stimulus for survival circuits, an incentive, or an enforcer. Through both conditioned and unconditioned modes of action, these stimuli activate and modulate responses to certain situations. This can also occur when the fear of a specific object is overwhelming or interferes with one's functioning, thereby becoming a phobia. All survival circuits become hyperexcitable toward events that are perceived and reacted to as dangerous or harmful. Similar yet distinct neural systems underlie fear and anxiety (Rosen and Schulkin 1998; LeDoux 1996; Davis et al. 2010).

The amygdala is activated under diverse conditions of adaptation. Another region of the brain, the bed nucleus of the stria terminalis (BNST), is linked in origins to the amygdala and has been shown in experiments to be tied to more nonspecific anxious states (Walker and Davis 1997). Visual cues can bring on fearful responses by stimulating cortical and subcortical pathways in the brain. The thalamus is responsible for first processing these images. It sends cursory information to the amygdala quickly, prompting the onset of a fear response. The thalamus also sends sophisticated information to the visual cortex, which conducts a more definitive assessment of the visual cue. The visual cortex sends its assessment to the amygdala, which either incurs a heightened fear response or suppresses the existing one. A classic example is that of seeing a snake on a trail, or a snake-shaped stick. Both visual cues will incur the same immediate fear response from the amygdala. The visual cortex's final analysis of the threat determines whether heart rate and blood pressure will either be increased or returned to rest.

Activation or deactivation of regions of the amygdala disrupts normal fear conditioning (Regev et al. 2011, 2012; Sherrin et al. 2009) without disrupting general anxiety or phobias; the opposite holds for the bed nucleus of the stria terminalis. The underlying hypothesis of Mike Davis and his colleagues, in response to the larger clinical human literature (Freud 1959), is whether there is a difference between specific fear-related responses and more general anxiety (see Davis 1998; Walker and Davis 2008). According to Davis's hypothesis, CRF acts on receptors in the bed nucleus of the stria terminalis, rather than the central amygdala nucleus, resulting in an increase in the startle reflex and anxiety. A recent study focused on the role of nitric oxide in inducing effects within the BNST when an animal is exposed to aversive stimuli that affected both CRF and NMDA receptor action (Faria et al. 2016). Diffuse cues and long-term unconditioned activation play a greater role in this anxiety complex. In comparison, explicit cues and short-term conditioned activation result in a fear response. Both anxiety and fear are involved with similar hypothalamic and brain stem targets, but are differentially associated when it comes to their major target receptors. Other regions not included are the globus pallidus (Sztainberg et al. 2011), accumbens (Young et al. 2001), lateral septum (Henry et al. 2006), and raphe nucleus (Spiga et al. 2006).

Darwin (1872) in his great book on emotions depicted in some detail the sense that when fear turns from a normal adaptation toward pathology and devolution of function, horror or abject fear is born. Fear is, primarily, a signal of danger, an important piece of biological adaptation. Therefore, fear systems are ultimately diverse survival systems (LeDoux 2012).

One thing to note is the mixed nature of fear; fear is rarely a pure event. The stuff of adaptation is not about purity. It is what the primatologist Michael Power calls "viability": short and long (2004). An emotion like fear is easily linked to an aversion; aversion is the desire to remove ourselves from an event or remove an object from contact with us. A good example is an aversion to a specific food that may have made us ill: we are disgusted by it, we avoid it (Garcia et al. 1966; Rozin 1998), and it can provoke fear as a contaminant or imagined contaminant. Getting sick is not fun, and we normally avoid it if we can. It, therefore, can easily become a fear. We easily learn and avoid core fears (scorpions, or, for some, snakes; Shettleworth 1971). In other words, we come prepared to learn quite easily how to avoid something that is potentially harmful to us (Rozin 1976).

Fear is linked to a number of events and mechanisms, ranging from classical conditioning to biological prepotent stimuli (e.g., predators, heights, and unfamiliar events). "Original conditions" with regard to fear might include the fear of being eaten, the fear of being abandoned, and the fear of starving. Various forms of social competence and technical competence evolved to cope with fear-related events, yet the state of fear cuts into everything about our species (Kagan 2002).

The origins of fear are rooted in adaptation, in survival systems that are motivational (Lashley 1938; Stellar 1954; LeDoux 1996). This includes being prepared to assess danger, having the perceptual and motor capabilities to avoid what is dangerous, and having the ability to cope with it when it occurs (Frijda 1986; Rosen and Schulkin 1998).

Fear-like expressions appear early in ontogeny; babies look toward what is familiar and reliable, and they begin an early fundamental orientation to events in the world. We are anchored to objects that are reliable and that can be counted upon. Of course, we are also rooted in what can facilitate comfort and be soothing in those early years and, indeed, throughout life. Different kinds of people play diverse roles toward those ends (e.g., parents, teachers, friends, and colleagues). Fear is developed easily for those things that are unfamiliar or that cause discomfort (Kagan 2002). It is not simply the parental figures that ameliorate those fears, but also others who play diverse roles in the developing child's life and the evolving individual over the lifespan.

As a social species, fear is palliated by concurrent senses of social contact and social amelioration. We make contact to solidify our placement in a world in which there is coherence, so that we can use our neonatal capacity to predict the behaviors of those that matter around us. The detection of discrepancy from the expected is a feature in fear-related behaviors, particularly during development (Kagan 2002) and perhaps during aging. The detection of discrepancy also underlies learning; we are attuned to breakdowns in regularity, resulting in diverse forms of learning (Rescorla and Wagner 1972). It is learning not by simple contiguity (association of events in time) as was suggested in classical empiricism or early learning theory; rather, it is infused with cognitive competence. Prediction of events, rather than contiguity of stimuli, and what is at the center of detecting discrepancies is what matters (Kagan 2002).

Thus we are primed to behave in fear-like ways, which, depending upon the species, the context and capability vary in expression. These responses can vary, for example, by age, by culture, and by gender. Neuropeptides, like oxytocin, can elicit agitation, anxiety, and aggression, rather than attachment and social approach to an unfamiliar conspecific. Further, research has shown that oxytocin plays a role in regulation of CRF gene transcription through interference of normal processing (Jurek et al. 2015). Again, information molecules interact with each other to modulate and manage physiological/behavioral function. Fear is not a simple drive state, and so drive reduction is not an adequate explanation, although there is relief when the experience of fear fades from the purview of experience.

Context matters for the expression of fear (Valenstein 1976). The state of fear does not exist in the brain in a vacuum, but rather, it exists in a central state in which foraging for coherence in the expression of behavioral options is an adaptive response. The concept of central states, for instance the central state of fear, may not have one neural signature. For example, a glutamate circuit that is instantiated in the nucleus accumbens can elicit "dread" or fear or appetitive behaviors, depending upon the context and the site of stimulation (Berridge 2000). The caudal end of this region is associated with dread and the anterior end with appetitive response, while sites in-between elicit both (Faure et al. 2010). In other words, neuropeptides such as CRF and neurotransmitters such as GABA can elicit both appetitive and fear- or dread-related behaviors, depending upon where in the accumbens there is an injection. It is therefore apparent that we have unique neurological structures that code for diverse emotions within the same distinct brain regions.

Both dopamine and CRF are linked to the breakdown of expectation with regard to events (Schultz 2002; Merali et al. 2008); unexpected events result in further activation of both this peptide and neurotransmitter (Zweifel et al. 2011). Dopamine projections from the brainstem innervate forebrain sites in which CRF is expressed (e.g., parts of the accumbens), thereby potentially facilitating many behavioral options given the context and the species. Experiments with rats conditioned to anticipate snacks compared with rats with an "extinct conditioned" response (conditioning was discontinued) or no conditioning studied the release of CRF in the left and right medial prefrontal cortex (mPFC) by microdialysis. During the anticipatory phase, both conditioned rats and those with "extinct conditioning" experienced similarly high levels of CRF release. From the left mPFC, conditioned rats experienced a higher level of CRF release after snack presentation, while "extinct conditioned" rats had a higher level of CRF from the right mPFC. Rats with no conditioning, experienced overall a minimal (if any) increase in CRF release during anticipation and a decrease in CRF to below their baseline levels (Merali et al. 2004).

Importantly, and as I indicated at the onset of this book, CRF and CRF receptors subtypes are co-localized with a number of neurotransmitters; for instance, the CRF-R1 is co-localized with GABA in some but not all brain regions and with dopamine in others; both are fundamental neurotransmitters. Research has shown that CRF and other receptor systems, like the kappa-opioid receptor system, are intertwined and involved in activation of GABAergic neurotransmission in the central amygdala (Kang-Park et al. 2015). Shown by double in situ hybridization and double immunohistochemistry, CRF-R1 is expressed in many neurons of different neurotransmitter systems. In a model of traumatic stress, through a CRF-dependent mechanism, there was a long-lasting sensitization of noradrenergic receptors in the basolateral amygdala (Rajbhandari et al. 2015). In the cortex and hippocampus, CRF-R1 is expressed in glutamatergic (VGlut1, glutamate) neurons, as well as in GABAergic neurons of the reticular thalamic nucleus (RTN) and the globus pallidus (GP; within the basal ganglia). CRF-R1 is also expressed in dopaminergic (Dat, TH) neurons of the ventral tegmental area and the substantia nigra pars compacta. CRF-R1 is barely expressed in serotonergic (Sert, TPH) neurons of the dorsal and median raphe nuclei, and was not found in GABAergic neurons of the cortex or of the hippocampus (Refojo et al. 2011; Figure 5.1).

Dopamine, like CRF, is knotted to the organization of action (Schultz 2007; Berridge 2004), of which social contact is one; the fact that this transmitter and neuropeptide is both tied to organization and may be co-localized is quite interesting (see Koob and Bloom 1985). Indeed CRF even provokes the release of dopamine that is vital in the organization of action and social contact. In response to electrical stimulation, dopamine release is increased in those models that underwent CRF application. Dopamine levels increased in response

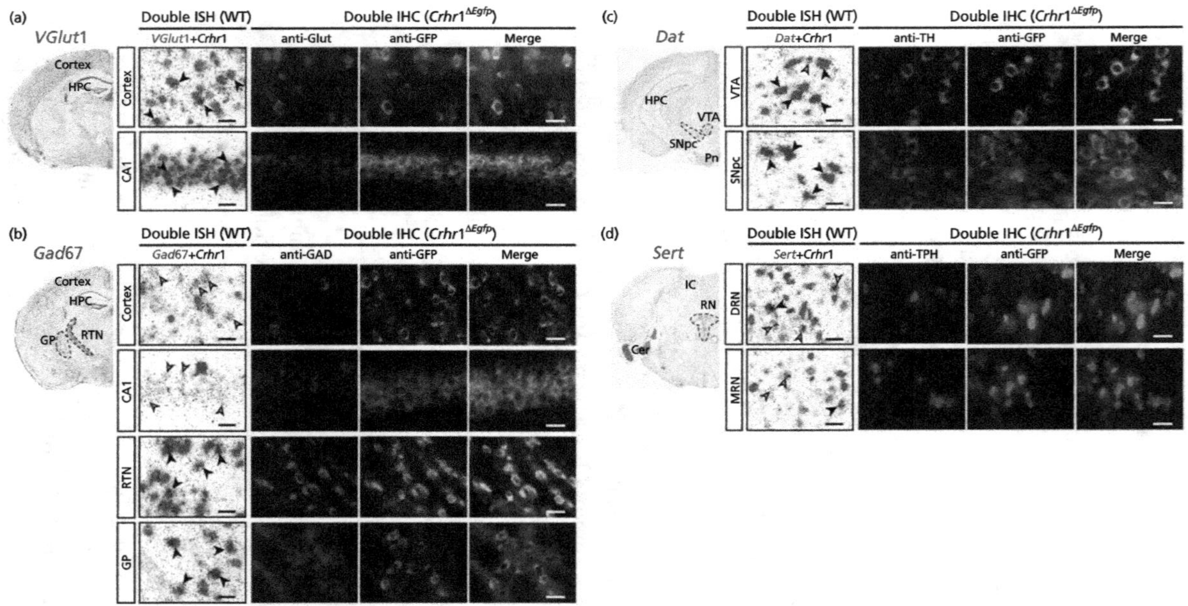


**Figure 5.1** (See color plate section) CRFR1 is expressed in neurons of different neurotransmitter systems. CRFR1 was co-localized with neurotransmitter-specific markers by double ISH and double immunohistochemistry (IHC), using wild-type and CRFr1DEGFP mice, respectively. (a) In the cortex and hippocampus (HPC), CRFR1 is expressed in glutamatergic (VGlut1, glutamate) neurons. (b) CRFR1 is not expressed in GABAergic (GAD67) neurons of the cortex or of the hippocampus but in GABAergic neurons of the RTN and the GP. (c) CRFR1 is expressed in dopaminergic (Dat, TH) neurons of the ventral tegmental area (VTA) and the SNpc. (d) CRFR1 is scarcely expressed in serotonergic (Sert, TPH) neurons of the dorsal (DRN) and median raphe (MRN). Gray arrowheads indicate cells expressing only CRFR1 (silver grains). Black arrowheads indicate cells co-expressing CRFR1 and the respective markers (red staining).
Reproduced from Damian Refojo, Martin Schweizer, Claudia Kuehne, Stefanie Ehrenberg, Christoph Thoeringer, Annette M. Vogl, Nina Dedic, Marion Schumacher, Gregor von Wolff, Charilaos Avrabos, Chadi Touma, David Engblom, Günther Schütz, Klaus-Armin Nave, Matthias Eder, Carsten T. Wotjak, Inge Sillaber, Florian Holsboer, Wolfgang Wurst, and Jan M. Deussing, Glutamatergic and Dopaminergic Neurons Mediate Anxiogenic and Anxiolytic Effects of CRHR1, *Science*, 333 (6051), DOI: 10.1126/science.1202107, Copyright © 2011, American Association for the Advancement of Science. Reprinted with permission from AAAS.

to CRF, and decreased in response to CRF-receptor antagonists. In mice lacking the gene for either CRF receptor, there was no significant difference in the amount of dopamine released in response to CRF application. CRF-receptor agonists had a variable effect on the amount of dopamine released (Lemos et al. 2012; Figure 5.2).

**Social contact, cortisol, and CRF:** Social contact is essential for our species and a variety of mammals and it plays a role in the regulation of information molecules, such as cortisol and CRF (Backström and Winberg 2013). The secretion of cortisol and CRF are adaptive responses serving physiology and behavior. The continued secretion of cortisol or CRF without relief has consequences though; bone demineralization, compromised immune system, and shifts in metabolism are the result of persistent oversecretion of cortisol (Gold et al. 1988). Social withdrawal and fearfulness in macaques can also lead to increases in plasma cortisol and CRF in cerebrospinal fluid during more fearful resting conditions than in less fearful resting conditions (Kalin et al. 2000).

There is an important social adaptation designed to restore elevated levels of cortisol back to normal. Social attachment, the formation of core safe or relatively safe attachments, provides feedback on the inhibition of steroid levels such as cortisol and the regulation of the internal milieu of diverse neuropeptides and neurotransmitters (Herbert and Schulkin 2002).

For example, cortisol tends be elevated in troops of macaques under conditions of uncertainty, including uncertainty of social control (Sapolsky 1992). Perception of where an individual baboon is in the hierarchy of its social group influences the internal milieu, particularly cortisol levels. In addition, social perception, facilitating social approach or avoidance, results in the behavioral regulation of cortisol and CRF (Sapolsky 1990; Cheney and Sefarth 2007). In the laboratory, there are a number of conditions in which CRF is elevated during states of uncertainty, such as variable and less predictive access to food resources (Coplan et al. 2001). In experiments conducted by Coplan, macaques given variable access to food (uncertainty of food availability) exhibited higher levels of CRF in their cerebrospinal fluid than did macaques who always had food at their disposal.

But it is not axiomatic that cortisol is elevated in the downtrodden, insecure members of primate groups (c.f. Creel 2001; Sapolsky 1992; Saltzman et al. 1994, 1998). Behavioral correlations can result in elevated levels of cortisol in many contexts (Abbott et al. 2003). It is not the case that elevated levels of cortisol equals adversity. The important point about cortisol is not whether it is elevated in dominance or the socially less fortunate group member, but whether there are behavioral strategies for regulating it, turning it off, decreasing and

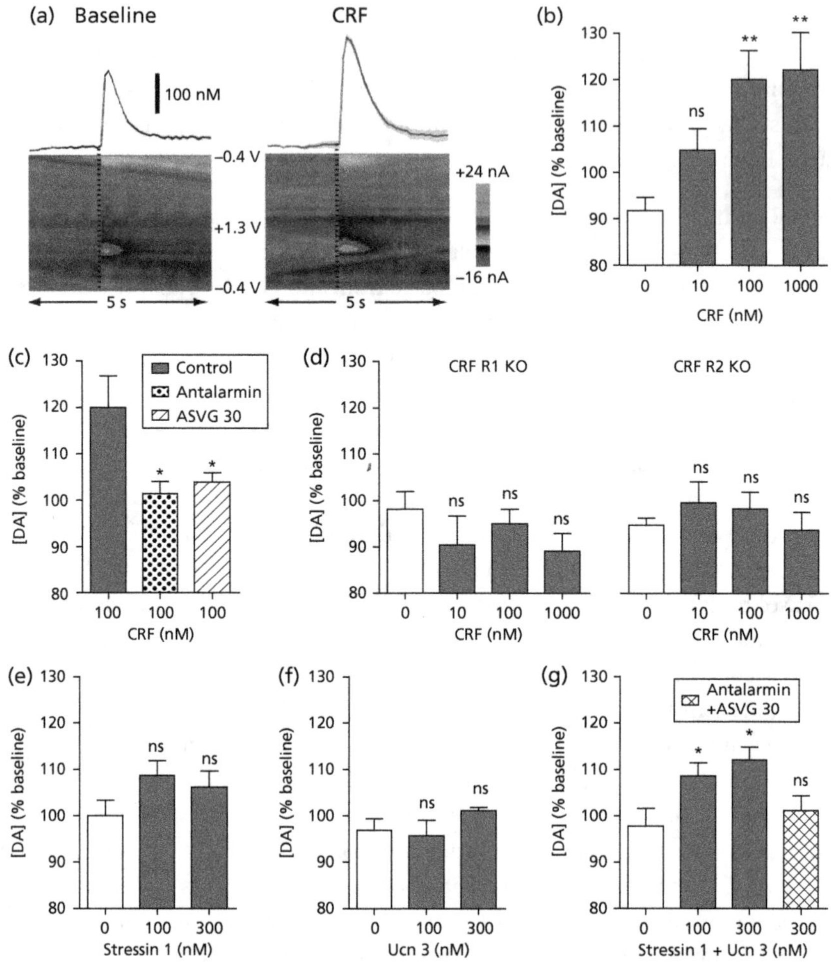


**Figure 5.2** (See color plate section) CRF increases dopamine release in the nucleus accumbens through coactivation of CRFR1 and CRFR2. (a) Representative dopamine release evoked by electrical stimulation (dashed lines) before (left) and after (right) application of 100 nM CRF (mean ± s.e.m. for 5 consecutive stimulations, top) and corresponding two-dimensional plots depicting changes in faradaic current (pseudocolor) with time as the abscissa and applied potential as the ordinate (bottom). (b) Concentration response to CRF, n = 11–18. (c) Effect of antagonists for CRFR1 (antalarmin, 1000 nM) or CRFR2 (anti-sauvagine 30 (ASVG 30), 250 nM), n = 18–20. (d) CRF in mice lacking the gene encoding the CRFR1 (left) or CRFR2 (right) receptors, n = 7–13. (e–g) Effect of the CRFR1 agonist, stressin 1, n = 9–15 (e), the CRFR2 agonist urocortin 3 (100 or 300 nM), n = 5–8 (f) or their co-application, n = 8–15 (g). Error bars, s.e.m. DA, dopamine; NS, not significant (with $P > 0.05$); $*P < 0.05$; $**P < 0.01$ versus vehicle.

increasing the level when necessary, and its costs when not (Sapolsky 1995; Wingfield et al. 1999; Eberhart et al. 1980, 1983).

So social contact is one way of managing the secretion of cortisol. More social contact, more grooming, and less cortisol circulating. Forming coalitions is essential for the regulation of the internal milieu, cortisol, the molecule of energy metabolism, along with central CRF and other molecules (Sapolsky; 1992; Schulkin 2003). In a similar way, young children who make social contact and seek out human warmth and security tend to have lower levels of cortisol under diverse conditions (Gunnar 1998; Gunnar and Davis 2005). The social context then feeds back upon the secretion of cortisol. And steroid hormones such as cortisol have diverse effects on the brain; they regulate the production of genes that produce neuropeptides (e.g., CRF), which are a common final pathway in promoting the behaviors that underlie approach and avoidance, sustain and persevere, in the myriad of social and complex behaviors (Palombit et al. 1997). With collections or friendships, cortisol is a bit lower in studies of primates and other animals in the wild (Engh 2006).

Disruption of social status like most events affects cortisol levels. Levels of cortisol for instance alone are not a measure of dominance or lack thereof, but the way in which dominance is achieved (e.g., Cavigelli et al. 2003; Saltzman et al. 1998; Backström and Winberg 2013). One recurrent theme is that social support is predictive of cortisol and CRF levels, mitigating diverse events that would elevate it in combating adversity (Abbott et al. 2003).

**CRF and basic survival:** CRF is activated and essential under diverse conditions including fear-related events (e.g., Risbrough et al. 2009; Kusnecov et al. 1999). Many of the hormonal signals linked to fear and other forms of behaviors have receptors in the amygdala; the degree of fear in this region has been shown to be associated with Fos activation and CRF (Lehner et al. 2008; Liebsch et al. 1995).

Indeed greater CRF release is tied to less familiar events and to food ingestion. In rats housed in their home cages, interstitial CRF levels remained stable, whether injected with a vehicle or diazepam. Rats housed in novel cages displayed increasingly higher levels of interstitial CRF, across all five sample collections, whether injected with a vehicle or with diazepam (Merali et al. 2004; Figure 5.3).

Importantly, psychological adversity and behavioral adaptation are preferentially regulated by the CEA rather than the PVN (Davis et al. 1997; Makino et al. 1999; Schulkin et al. 2005; Kolber et al. 2008; Lachize et al. 2009; Walker et al. 2008; Koob and Bloom 1985; Fenoglio et al. 2004). There appears to be differential regulation of the CRF promoter and transcription factors binding to CRF in these two regions of the brain by glucocorticoid activation (Donley et al. 2005).

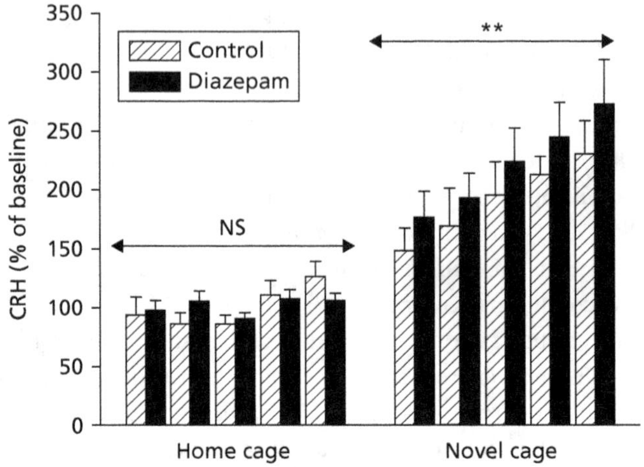


**Figure 5.3** Effect of novel versus home cage environment on the in vivo release of CRF at the CEA. The ordinate represents CRF levels, expressed as a percentage of baseline. Dialysate samples were collected every 20 min in the home cage and the novel cage. Following collection of the fourth sample in the home cage, rats were injected with either vehicle (hatched columns) or diazepam (2 mg/kg) (solid columns). In the home cage, interstitial CRF levels remained stable (first five samples). In the novel cage, levels of interstitial CRF increased progressively in both vehicle- and diazepam-treated rats over the course of five sample.

Reproduced from Z. Merali, S. Khan, D. S. Michaud, S. A. Shippy, and H. Anisman, Does amygdaloid corticotropin-releasing hormone (CRH) mediate anxiety-like behaviors? Dissociation of anxiogenic effects and CRH release, *European Journal of Neuroscience*, 20 (1), pp. 229–239, Figure 5, DOI: 10.1111/j.1460-9568.2004.03468.x Copyright © 2004, John Wiley and Sons.

However, the mechanisms for this differential regulation remain unclear (e.g., Kasckow et al. 2003; Ma et al. 2001; Shepard et al. 2005; Kolber et al. 2008; Lachize et al. 2009).

A plausible view of this differential regulation is the following: in the regulation of the PVN, restraint of the HPA is tied to vegetative physiological functions. But that does not mean that the CRF neurons in the PVN are not tied to behavioral orchestration; indeed, they are (Füzesi et al. 2016; Zhang et al. 2016). The behavioral coding in which CRF participates is tightly linked to regions such as the CEA and BNST in mobilizing behavioral adaptations. Thus, while restraining the HPA is a fundamental feature of regulatory physiology, the facilitation of CRF by glucocorticoids underlies the physiology of change and adaptation. Glucocorticoids can enhance CRF gene expression, conserved by evolution, and that underlies diverse forms of behavioral and physiological functions. Additionally, increasing CRF in the limbic system can override the PVN, putting a break on the HPA CRF effects.

**Behavioral effects of elevated CRF:** Elevated levels of CRF (Koob and Bloom 1985) in specific regions of the brain, including the central nucleus of the amygdala and the lateral bed nucleus of the stria terminalis, increase the likelihood that an event will be perceived as dangerous (see e.g., Koob et al. 1993, Kalin et al. 1994, Lee and Davis 1997). These two neural sites underlie the behavioral response to fear and anxiety (LeDoux 1987, 1996; Rosen and Schulkin 1998).

Transgenic mice with overexpression of CRF in the brain demonstrate exaggerated cautious behaviors (Stenzel- Poore et al. 1994; Kubler et al. 2008). In the elevated plus maze, mice with enhanced CRF exhibited exaggerated cautious avoidance of the open arm compared with control mice, suggesting an increase in trepidation owing to the CRF (Stenzel-Poore et al. 1994). Fear responses are attenuated by central pharmacological blockade of CRF expression (Swerdlow et al. 1989; Koob et al. 1993), but not peripheral blockade of CRF expression (Pich et al. 1993).

Differential levels of CRF activation in the central nucleus of the amygdala, as well as differential behavioral responses, can be attributed to the duration of time spent under various kinds of adverse conditions (Gallagher et al. 2008). For instance, CRF neurons in the central nucleus of the amygdala are responsive to fear predictive events; indeed, the CRF neurons are important for normal fear-related behaviors (Sanford et al. 2016).

Importantly, differential regulation in the CNS has been shown by genetically altering CRF CEA expression, with differing results; altering CRF under basal conditions tended to increase anxiety, but not under more stress related perhaps by enhancement of these adverse effects on behavior. During fearful situations, CRF neurons are selectively activated in the central amygdala. For instance, one study showed that in foot shock experiments, rats with a high responder status (i.e., rats with enhanced fear to novel tone 24 hours after foot shock and long-lasting avoidance in elevated maze) displayed high levels of precursor CRF mRNA in the CEA (Wang et al. 2015). Additionally, silencing CRF neurons in the central amygdala impairs the ability to undergo fear conditioning.

In particular, antagonists of CRF directed at the central nucleus of the amygdala reduce a number of fear-induced behavioral responses (Swerdlow et al. 1989; Koob 2009; Davis 1997). Infusion of CRF directly into the lateral or third ventricle (e.g., Koob et al. 1993), or the bed nucleus of the stria terminalis (extended amygdala), facilitates fear responses (Lee and Davis 1997). CRF is playing perhaps both tonic and phasic roles in the regulation of behavior

(Gallagher et al. 2008; Holly et al. 2016), potentially because of the differential location of the CRF release in the BNST and CEA, respectively.

**Glucocorticoids, CRF, conditioned and unconditioned fear:** Rat studies have demonstrated that chronic (i.e., from 4 days to 2 weeks) high levels induced CRF mRNA in the amygdala, and corticosterone was found to upregulate CRF mRNA transcription in the left BNST (Makino et al. 1994a, b). The studies already described herein and others, suggest that the amygdala is a site for glucocorticoid-mediated facilitation of CRF-related fear behavior (Myers and Greenwood-Van Meerveld 2007, 2010), and other studies indicate that the left BNST is also a site of behavioral action of CRF (Erb and Stewart 1999; Figueiredo et al. 2003; Lee and Davis 1997; Makino et al. 1999).

Moreover, enhancement or depletion of central glucocorticoid receptors is also known to influence diverse behaviors (e.g., De Kloet et al. 2005; Harris et al. 2013; see the review by Kolber et al. [2008]) and a balance of glucocorticoid and mineralocorticoid receptor sites may underlie a number of behavioral adaptations (Brinks et al. 2009; De Kloet 1991; De Kloet et al. 2005) and CRF gene expression (Watts 2005; Myers et al. 2005).

For instance, glucocorticoid receptor overexpression can increase fear-related behaviors (Wei et al. 2004). Manipulations in which the mineralocorticoid receptors are silenced and in which there was a 40% increase in corticosterone levels in mice resulted in greater fear-related responses (e.g., Brinks et al. 2009) and perhaps greater CRF expression in regions such as the central nucleus of the amygdala. In addition, mineralocorticoid receptor overexpression reduces corticosterone secretion and anxious behaviors (e.g., Mitra and Sapolsky 2008).

Corticosterone-mediated transcriptional upregulation of CRF in the CEA could have a significant effect on heightened fear, and on maintaining heightened fear over long periods. Lee and colleagues (1994) showed that repeated systemic corticosterone administration in rats over several days enhanced intracerebroventricular-CRF facilitated acoustic startle. This enhancement was found at doses of CRF and corticosterone that, when tested alone, did not affect startle. A related study found that repeated intracerebroventricular administration of dexamethasone, a synthetic glucocorticoid, decreased the threshold for CRF-induced anxious behavior and seizures (Rosen et al. 1994). In rats, chronic administration of corticosterone (daily injections for 4 days) increases CRF expression in extrahypothalamic regions and can enhance the learning and memory of contextual fear conditioning (Pitts et al. 2009, Thompson et al. 2004). The ability of glucocorticoids to sustain conditioned fear has also been demonstrated in several studies (Cook 2002, 2004; Stiedl et al. 2005); the type 1 receptor is critically involved in startle behavior (Flandreau et al. 2015).

As has been pointed out, corticosterone administration increased CRF mRNA expression in the CEA and decreased it in the parvocellular region of the PVN. These results indicate that repeated corticosterone injections increased the retention of contextual fear learning, perhaps through CRF induction. Moreover, injections of a glucocorticoid receptor antagonist or CRF-receptor antagonist (Donley et al. 2005; Roozendaal et al. 2002) into the basolateral amygdala influence fear-related behaviors, perhaps again by altering CRF gene expression in the CEA (Roozendaal et al. 2002). Moreover, they found that peripheral administration of antalarmin, a CRF type 1 receptor antagonist, reduced the behavioral and endocrine responses and the behavioral effects of both corticosterone implanted in either the CEA or BNST (Myers et al. 2005; Shepard et al. 2000; Lastein et al. 2008).

Shepard and colleagues (2000) demonstrated that corticosterone implanted directly into the CEA induced CRF expression in the CEA and reduced exploration of an elevated plus maze, a popular paradigm for studying cautious avoidance behaviors. Myers and colleagues (2005, 2007) also showed that implants of corticosterone into the CEA reduced exploration of an elevated plus maze and increased physiological distress (e.g., colonic distention).

In other sets of experiments, Cook (2002) found a significant relationship between cortisol and CRF in the amygdala of sheep in response to acute and repeated predator stress. With a single exposure to a dog, sheep demonstrated a biphasic CRF response in the amygdala, as measured by microdialysis; there was an initial rapid increase in CRF levels that decreased quickly and was a direct response to the dog. This was followed by a slower rising cortisol response that was paralleled by a second CRF peak, smaller and more prolonged than the first (Cook 2002). The first CRF response is cortisol-independent and is part of the initial fear response to a stressor, whereas the second response is a cortisol-dependent elevation of CRF that perhaps sustains the fear state and related behavior through and beyond the presence of the threat. The second response was mimicked by administration of cortisol in nonstressed animals. Further, following repeated exposure to a dog, sensitization of the CRF system (increased CRF release) in the amygdala was found by giving a novel foot-shock stressor to the sheep.

This study further demonstrated that enhanced release of CRF in the amygdala following repeated adversity was cortisol dependent, by blocking the increased CRF release with the glucocorticoid antagonist mifepristone (Byrne et al. 2004; Cook 2002; Ceccaldi et al. 2012); this is the same progesterone blocker that decreases CRF in placental tissue (Wang et al. 2013) and that can delay, under some conditions, parturition. These studies are crucial in demonstrating both that the prolonged release of CRF is cortisol- dependent and that this CRF response is sensitized by repeated stress in a cortisol-dependent manner. Moreover, the Cook (2002, 2004) study demonstrates that CRF is actually released in the amygdala during fear-related events (see also Merali et al. 2008).

The amygdala and left BNST might underlie different types of fear behavior. Infusion of CRF into the left BNST enhances unconditioned acoustic startle, but infusions into the CEA or the hippocampus do not (Lee and Davis 1997; Eraslan et al. 2015). Interestingly, lesions of the amygdala block enhancement of acoustic startle by intracerebroventricular CRF (Liang et al. 1992a), suggesting that amygdala efferents to the left BNST regulate the CRF-induced enhancement. Furthermore, lesions of the amygdala but not the left BNST block cue-conditioned fear; conversely, lesions of the left BNST, but not the amygdala, block unconditioned fear (Fendt et al. 2003; Walker and Davis 1997; Wallace and Rosen 2001). These data have led some to suggest that the amygdala regulates conditioned fear and the left BNST regulates unconditioned fear (Fendt et al. 2003), and others to suggest that CRF in the left BNST, rather than in the amygdala, mediates slow-onset, long-lasting responses typical of sustained threats, perhaps persisting beyond removal of the threatening stimuli (Walker et al. 2003).

The medial region of the amygdala, strongly innervated by the olfactory bulb, sexually dimorphic in morphology, and the repository of a number of gonadal and adrenal steroid receptors, is linked to a number of behaviors, including sexual dimorphic sexual behaviors (Newman 2002), mineral appetites (De Castro et al. 2006; Nitabach et al. 1989; Zhang et al. 1993), and fear-related behaviors knotted to olfactory cues. An olfactory cue like predator odor, for instance, induces freezing behavior (Takahashi et al. 2005).

Thus fear-related predator odors activate diverse gene products in the medial amygdala (Rosen et al. 2008) in addition to other brain regions; one gene product is CRF (Kalin et al. 1998). There is a prepotent sensory system designed to detect predators that are innately organized and somewhat separate from other fear-related neural circuits. For instance, diverse cat and ferret odors induce freezing behaviors in rats (Takahashi et al. 2005; Rosen et al. 2008). The fear-related behaviors are dose dependent; the greater the odor the greater the fear-induced freezing (e.g., trimethylthiazoline; Rosen et al. 2008). In the freezing behavior, diverse gene products are induced, including CRF, vasopressin, and various neurotransmitters.

This is innate prepotent activation by the ecological context; the odor results in the activation of corticosterone, which serves to maintain the fear-related response. Corticosterone-treated rats can increase conditioned fear-related behaviors; they also can decrease behavioral responses to unfamiliar events (Shepard et al. 2000; Myers et al. 2005). They do so via the activation of central CRF. Nonetheless, CRF type 1 receptor treatment does not seem to affect this prepotent innate behavior toward potential predators, though CRF is clearly induced by predator odor in regions of the brain such as the amygdala (Roseboom et al. 2007; Kondoh et al. 2016). It is also possible that predator odor will affect a preyed-upon animal through another avenue, potentially in concert with (though separate from) the HPA axis; the predatory signal may potentially stimulate CRF production in the vomeronasal organ or olfactory bulb. These induced levels of CRF may be maintained by other information molecules, as the CRF signal of a fearful event ultimately travels to and induces CRF in other brain regions (e.g., cortical and medial nucleus of the amygdala and the BNST). This avenue of signaling offers an extrahypothalamic mechanism for the odor-induced behavioral response to predator fear (e.g., the relationship between the peripheral representation of the odor and the central organization of the behavioral responses [e.g., CEA and BNST]).

Greater avoidance is one result of an innate predatory odor that relates to fear; one function of CRF—though this is speculative—is perhaps to act as a transduction mechanism from an external event to an internal via the olfactory sensibility. One scenario might be the following: the smell of the odor activates this primary sensory pathway, analogous to the sight of the dog in vision; the effects are immediate via the induction of CRF systems from the periphery to the CNS, namely in this case the medial amygdala. The result is cautious avoidance behaviors.

The sight of a frightening predator or competitor, within seconds, can induce CRF release from the central nucleus of the amygdala (Cook 2002; Merali et al. 2003, 2008). In experiments with sheep, using a dialysis probe to sample the chemical milieu bathing the amygdala, CRF is released at the sight or smell of an unfamiliar dog. CRF is clearly a signal about danger, and, among other things in the brain, it mobilizes action; if at a water spout or a feeding place the animal would flee; to sustain the fear the adrenal gland is activated in the mobilization of coherent action, and in response, the CRF release is sustained (Cook 2002). CRF in the amygdala after a forelimb shock is representative of sighting a predator. Experimentation with animals in an escape group (i.e., animals provided with an escape route) experienced a smaller increase in amygdala CRF than the non-escape group (Cook 2002; Figure 5.4).

The BNST is richly interconnected to the amygdala (Krettek and Price 1977a, b; Dong and Swanson 2006; Dong et al. 2001). The amygdala interacts directly with the BNST via regions of the bed nucleus, which directly project to the PVN to regulate the HPA axis; more generally, regions of the bed nucleus of the stria terminalis are known to influence HPA function (Gray et al. 1993; Herman et al. 2003; Choi et al. 2007). And some regions within the PVN itself are tied to both anticipatory and reactive forms of regulatory functions (Herman et al. 2003;

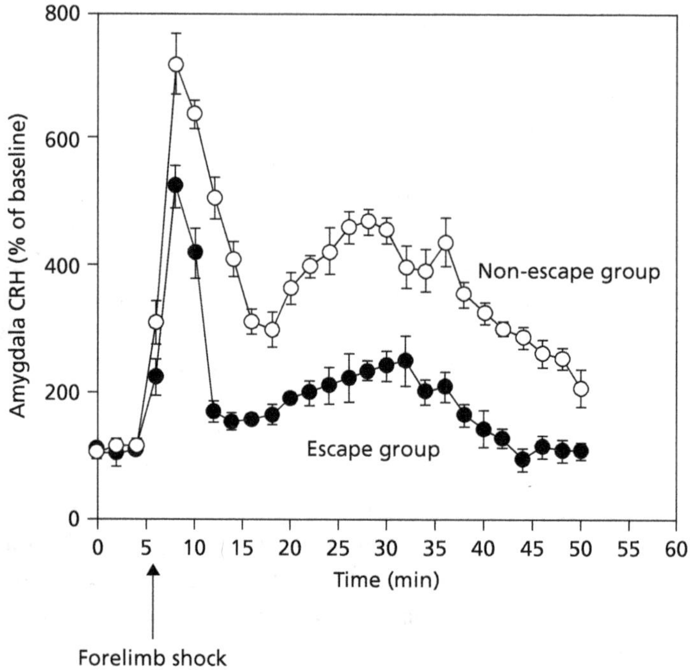


**Figure 5.4** The sight of a predator and activation of CRF. Effects of a forelimb shock on amygdala CRF in the escape group (black symbols) and the nonescape group (white symbols) animals. Data are exhibited as percent change (mean ± S.E.) from baseline. Baseline CRF were 63.8 ± 14.9 and 89.5 ± 19.4 pg/ml for the escape and nonescape groups, respectively.

Reprinted from *Physiology & Behavior*, 75 (4), Christian J Cook, Glucocorticoid feedback increases the sensitivity of the limbic system to stress, pp. 455–464, doi:10.1016/S0031-9384(02)00650-9 Copyright © 2002 Elsevier Science Inc., with permission from Elsevier.

Sawchenko et al. 2000). The BNST may be linked to anxious/aberrant behaviors through CRF expression (Jasnow et al. 2004, 2006; Erb and Stewart 1999; Erb et al. 1998; Davis et al. 1997; Walker and Davis 2008; Lebow and Chen 2016), in which GABA influences HPA functions (Cullinan et al. 2008; Herman 2013; Cabral et al. 2016).

Exposure to a predator odor elevated plasma levels of corticosterone and adrenocorticotropic hormone (ACTH), and stimulated c-fos mRNA expression in several brain areas, including the BNST, medial amygdala, and mPFC (Masini et al. 2005). Predatory odor has a profound effect on the endocrine system (Wallace and Rosen 2000; Blanchard et al. 1994, 1998; Takahashi 2014). Temporary inactivation of the left BNST by local infusion of lidocaine blocked the unconditioned freezing response to a predator odor; inactivation of the

CEA had no such effect (Fendt et al. 2003), suggesting that the BNST is tied to unconditioned responses, whereas the CEA is tied to conditioned responses (Asok et al. 2016b).

Other data suggest that chronic systemic corticosterone administration enhances freezing in response to predator odor (Kalynchuk et al. 2004; Rosen et al. 2004); further, 2-week-old adrenalectomized rat pups showed deficits in unconditioned fear, and corticosterone replacement restored the response and facilitated CRF gene expression (Takahashi and Rubin 1993).

Taken together, the results suggest that corticosterone contributes to facilitation of unconditioned fear responses to predator odor, possibly through the positive induction of CRF expression in the left BNST. However, it is not known whether the CRF-expressing neurons in the left BNST are responsible for these effects.

We do know, however, that contextual fear conditioning in rats has been linked to elevated CRF expression induced by treatment with corticosterone (Thompson et al. 2000). One week after the completion of the conditioning and the last corticosterone injection, the rats were tested for the retention of conditioned fear. The corticosterone-treated rats displayed more fear conditioning than the vehicle treated rats did. The amount of CRF mRNA in the central amygdala is significantly increased when corticosterone is repeatedly administered. This administration also seems to enhance long-term memory for contextual fear conditioning. In tests involving shock, retention of the memory of the shock was displayed by maintenance of freezing behavior a day after conditioning, owing to repeated corticosterone administration (Thompson et al. 2004).

If CRF is elevated, there is cautious avoidance, (i.e., a reluctance to enter an open field). Indeed, this is all the more remarkable, since one other consistent finding is that central CRF facilitates motor activity (Stenzel-Poore et al. 1994; Rosen et al. 1994; Moore and Rose 2002). CRF activation has been consistently and reliably linked to the induction of a number of behaviors including cautious avoidance (e.g., Kalin and Takahashi 1990; Stenzel-Poore et al. 1994; Koob et al. 1993; Skutella et al. 1994; Koob and LeMoal 2005). Central infusions of CRF induce or potentiate a number of fear-related behavioral responses (e.g., Takahashi et al. 1989; Linthorst et al. 1997; Kalin 1985; Kalin et al. 1989; Davis et al. 2010). Infusions of CRF-receptor antagonists both within and outside the amygdala reduce fear-related responses and anxious behaviors (e.g., Koob et al. 1993; Takahashi 2001, 2005; Myers et al. 2005; Weninger et al. 1999; Smith et al. 1998; Timpl et al. 1998; Binder and Nemoeroff 2010; Tovote et al. 2010; Muller et al. 2003; Magalhaes et al. 2010; Sztainberg et al. 2010; Heim et al. 2009).

Corticosterone facilitates the central nucleus amygdala/CRF and behavioral activation of fear and anxious behaviors. Indeed, an advance of the argument with regard to the differential regulation of CRF by glucocorticoids is that when the central glucocorticoid receptor is deleted specifically in the central region of the amygdala, one result is a reduction of CRF expression in the CEA, and the other result is decreased contextual fear-related behaviors. CRF infusions into the lateral ventricle reversed this effect, as depicted in Figure 4.5 (Kolber et al. 2008).

In CEAGRKO mice (central amygdala glucocorticoid receptor knockout mice), a reduced number of cFos positive cells were present in the central amygdala after undergoing conditioned fear paradigm and during basal conditions. In the dorsal BNST and BLA, normal basal levels and reduced conditioned fear levels of cFos were found. Following conditioned fear training, CRF mRNA levels in CEAGRKO mice show no upregulation in the CEA or PVN, whereas there was upregulation of CRF mRNA in the CEA of control mice (Kolber et al. 2008; Figures 5.5 and 5.6).

Moreover, CRF in the PVN and CEA is tied to a transcriptional factor of the glucocorticoid receptor (coactivator-1). Research has shown that switching the isoform of steroid receptor coactivator 1 (SRC-1) toward a more repressive variation results in a reduction in the CRF mRNA expression induced by glucocorticoids (Zalachoras et al. 2012, 2016). As shown in Figure 5.7, SRC-1 knockout mice were found to be linked to both the restraint of CRF by glucocorticoids and the elevation of CRF (Lachize et al. 2009). When SRC-1 is knocked out of mice, under basal conditions, there is a significant decrease in CRF mRNA expression in the central nucleus of the amygdala, and an increase in the paraventricular nucleus. The SRC-1 knockout mice CRF expression appears resistant to dexamethasone treatment; an increase in CRF mRNA expression after dexamethasone treatment is found in the central nucleus of the amygdala of wild-type mice (Lachize et al. 2009; Figure 5.7).

Site-specific cerebral injections of glucocorticoids increase CRF expression and cautious avoidance behaviors. For instance, implants of corticosterone into the CEA increase CRF in the CEA; this has now been demonstrated a number of times (Myers et al. 2005; Shepard et al. 2000; 2006). In addition to increasing CRF in this region, the implants increased a reluctance to explore open arms in a T-maze (Shepard et al. 2000; Myers et al. 2005). Systemic treatment of a CRF-R1 receptor antagonist known to influence cautious avoidance behaviors in several species (rat, Deak et al. 1999; macaque, Habib et al. 2000; marmoset, French et al. 2007) ameliorated the effects of the amygdala implant of corticosterone (Myers et al. 2005; Greenwood-Van Meerveld et al. 2006).

Moreover, corticosterone implanted into the region that surrounds the CEA provokes visceral motor distress as the hormone affects this region of the brain over several days and increases CRF in both CEA and lateral BNST (see

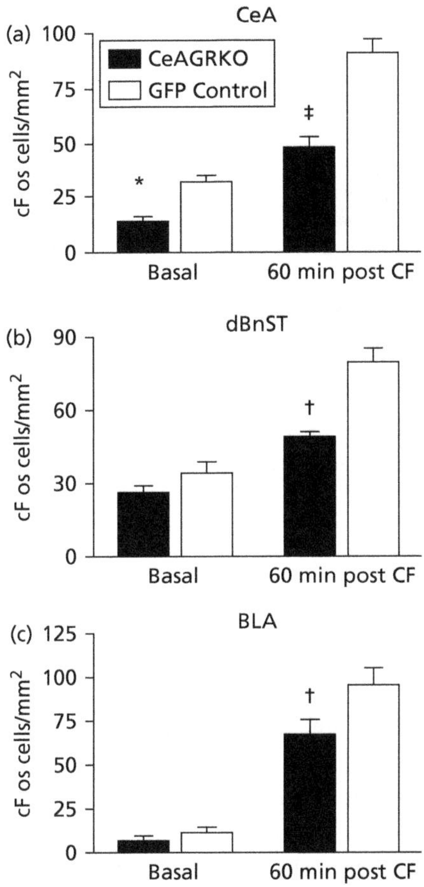


**Figure 5.5** cFos expression is altered in CEAGRKO mice in the CEA, bed nucleus of the stria terminalis (BNST), and BLA. (a) In the CEA, CEAGRKO mice (n 7) exhibit a reduced number of cFos-positive cells under basal conditions and 60 min following conditioned fear (CF) compared with GFP controls (n 8). (b) In the dorsal bed nucleus of the stria terminalis (dBNST), CEAGRKO mice exhibit normal basal cFos expression but reduced cFos expression following CF compared with GFP controls. (c) In the BLA, CEAGRKO mice exhibit normal basal cFos expression but reduced cFos expression following CF compared with GFP controls. (*, P 0.05 vs. basal GFP group; †, P 0.01 vs. GFP control group after conditioning; ‡, P 0.001 vs. GFP control group after conditioning.)

Reproduced from Benedict J. Kolber, Marie S. Roberts, Maureen P. Howell, David F. Wozniak, Mark S. Sands, and Louis J. Muglia, Central amygdala glucocorticoid receptor action promotes fear-associated CRH activation and conditioning, *Proceedings of the National Academy of Sciences of the United States of America*, 105 (33), pp. 12004–12009, Figure 3, DOI: 10.1073/pnas.0803216105, Copyright © 2008, National Academy of Sciences.

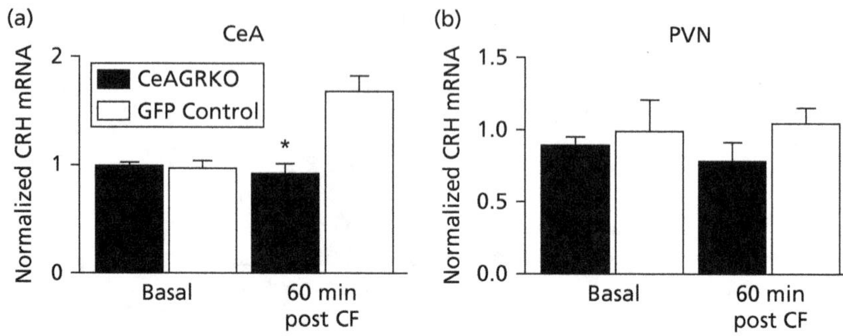


**Figure 5.6** CRF mRNA is decreased in CEAGRKO mice after conditioned fear (CF). (a) Although basal levels of CRF mRNA in CEA are equivalent between groups, CRF mRNA is significantly upregulated in GFP controls (n 6–8 per time point) but not in CEAGRKO mice (n 6–7) after CF. (b) There are no significant changes in CRF expression in the PVN in either group under basal conditions or after CF training. (*, P 0.001 vs. GFP control group after conditioned fear.)

Reproduced from Benedict J. Kolber, Marie S. Roberts, Maureen P. Howell, David F. Wozniak, Mark S. Sands, and Louis J. Muglia, Central amygdala glucocorticoid receptor action promotes fear-associated CRH activation and conditioning, *Proceedings of the National Academy of Sciences of the United States of America*, 105 (33), pp. 12004–12009, Figures 4c and 4d, DOI: 10.1073/pnas.0803216105, Copyright © 2008, National Academy of Sciences.

Tran et al. 2014). In experiments in which antalarmin (a CRF type 1 receptor antagonist) was also used, one result was a reduction of the distress from colonic inflammation (Greenwood-Van Meerveld et al. 2001, 2005; Myers et al. 2005; Tache et al. 2001, 2005; Tache and Perdue 2004).

Rat strains have different degrees of endocrine responses to environmental challenges (Stöhr et al. 2000) with high or low anxiety or measurable levels of cautious behavior that have been linked to CRF levels in the brain (Sternberg et al. 1992; Wood et al. 2010; Ergang et al. 2015). Strains of inbred rats, revealed by avoidance of open-space tests, displayed both greater responsiveness to corticosterone implants in the central amygdala and greater CRF gene–related expression; as well as greater visceral or pain-related responses (Johnson et al. 2012; Ji et al. 2007; Lee et al. 2008).

Thus many experiments have shown the important interactions between glucocorticoids and CRF, linking the BNST in the regulation of adversity (e.g., Gray et al. 1993; Hammock and Young 2004; Jasnow et al. 2004; Glowa and Gold 1991). The BNST is linked to CRF-related effects on anxiety-like behavioral responses (e.g., Davis et al. 1997; Walker et al. 2003; Walker and Davis 2008; Shepard et al. 2009), and corticosterone may facilitate and sustain the anxious behaviors (Schulkin et al. 2005). Recall, the useful

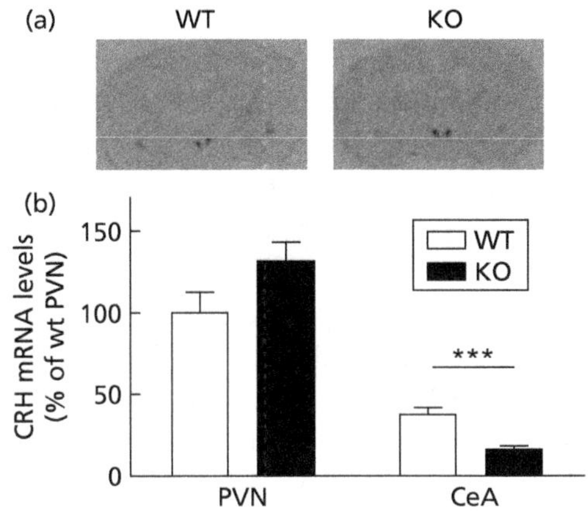


**Figure 5.7** SRC-1 affects basal CRF mRNA expression in a region specific manner. (a) Autoradiograms showing CRF mRNA signal in the brain at the level of PVN and CEA. (b) CRF mRNA is lower in the CEA in SRC-1 knockout mice under basal conditions. CRF expression is resistant to dexamethasone treatment in SRC-1 knockout animals (bottom). Upregulation in the CEA occurs only in wild-type mice. Data are expressed as percentage of vehicle-treated wild-type mice.



perspective provided by Davis and his colleagues has contained data for different neuroanatomical circuitries (Davis et al. 1997; Lee and Davis 1997) for the classic distinction between fear as a learned, stimulus-specific process, and anxiety as a nonspecific, generalized phenomenon. Looking at the mean startle amplitude before and after intraventricular infusion of CRF in animals given either sham or chemical lesions of either the BNST or amygdala provides a good idea of the effect of CRF on these regions of the brain. Post CRF, the mean startle amplitude actually decreased in those with a chemical lesion to the BNST. However, a chemical lesion in the amygdala resulted in a comparable amplitude of startle as found those animals given a sham lesion (Davis 1998). These regions, and in particular CRF, are involved in the acquisition and/or expression of conditioned fear in conjunction with, for instance GABA or NMDA receptor subtypes (Gafford and Ressler 2015).

Indeed, the data supported this idea. Inactivation of the BNST can interfere with the unconditioned startle response and with long-term CRF effects on behavior (Walker et al. 2008). For instance, light-enhanced startle as opposed to cue-enhanced startle is linked to BNST function, whereas the amygdala is linked to cue-related fear (Lee and Davis 1997; Walker et al. 2008). CRF has different effects in the amygdala and BNST (Partridge et al. 2016; Flandreau et al. 2012).

Optical stimulation is one way in which to separate the pathways of specific regions of the brain (Stuber et al. 2011). We know that enhanced CRF expression is impacted by lentiviral vectors within the BNST, and affected CRF-mediated effects on memory (Sink et al. 2013). In experiments using optogenetics, fear-related events were studied in a model that underwent BNST stimulation (Partridge et al. 2014, 2016; Asok et al. 2016a). Optogenetic activation via laser of CRF neurons in the BNST increases anxiety-like behavior in animals in the elevated plus maze.

For instance, a decrease in the total number of entries into the open arms of the maze, as well as total time spent in the open arms, suggests that the addition of a laser and its effects on the BNST played a role in this shift in behavior. Total distance traveled within the maze was unaffected by the addition of a laser in mice injected with ChR2 virus and control vector, suggesting nonspecific effects on locomotion were not a contributing factor (Partridge et al. 2014; Figure 5.8).

In another series of experiments using both optogenetic and pharmacological manipulations, activation of this CRF pathway from the bed nucleus to the

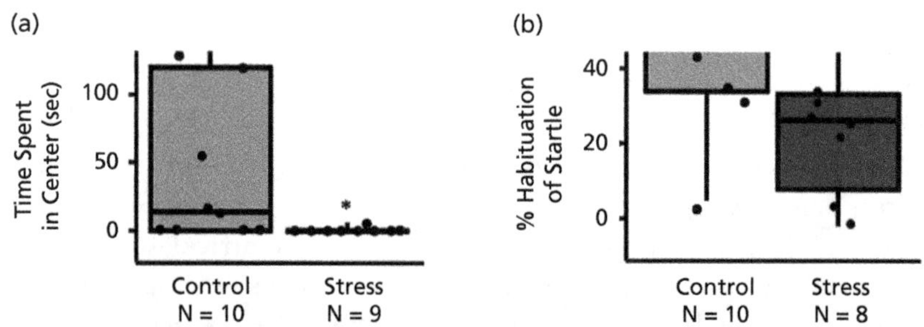


**Figure 5.8** Chronic Unpredictable Stress induces anxiety-like behavior in wild type mice and affects synaptic connectivity in CRF+ neurons in BnST and CeA. (A) Two days after cessation of chronic stress (consistent with the timing used in our electrophysiological experiments), stressed animals showed increased anxiety-like behavior in the open field. This is evident in a significant decrease in the time spent in the center of the arena. *$p < 0.05$, Mann-Whitney test. (B) Two weeks after cessation of chronic stress, anxiety-like behavior was assessed by habituation to acoustic startle (120 dB white noise). Animals exposed to chronic unpredictable stress displayed significantly less habituation of the acoustic startle response. *, $p < 0.05$, Mann-Whitney test.
Reprinted from *Neuropharmacology*, 107, John G. Partridge, Patrick A. Forcelli, Ruixi Luo, Jonah M. Cashdan, Jay Schulkin, Rita J. Valentino, and Stefano Vicini, Stress increases GABAergic neurotransmission in CRF neurons of the central amygdala and bed nucleus stria terminalis, pp. 239–250, doi:10.1016/j.neuropharm.2016.03.029, © 2016 Elsevier Ltd., with permission from Elsevier.

amygdala enhanced contextual conditioning and pharmacological manipulations via a CRF type receptor antagonist–inhibited contextual fear (Asok et al. 2016b).

In addition, the BNST is knotted to the general anxiety associated with drug abuse (Erb and Stewart 1999; Koob and LeMoal 2001) and to symptoms associated with pathological generalized anxiety disorder (Davis et al. 2010). Infusions of CRF directly into the BNST facilitate fear/anxiety-related behavioral responses, whereas antagonists of CRF do the opposite (Lee and Davis 1997; Walker et al. 2008; Holsboer 1999). Implants of corticosterone into the CEA increase CRF not only in this region (Myers et al. 2005; Shepard et al. 2000) but also in the lateral BNST (Shepard et al. 2006), and promote cautious avoidance (Shepard et al. 2000; Myers et al. 2005).

Systemic injections over several days can increase CRF gene expression in the dorsal region of the lateral BNST (Makino et al. 1994a, b; Henckens et al. 2016), though the effect is less consistent than the increase in CRF in CEA (Thompson et al. 2004). However, corticosterone implants aimed at the dorsal lateral BNST, a primary source of CRF (Makino et al. 1994b; Watts 2005; Watts and Sanchez-Watts 1995) and the region linked to the increase in CRF to systemic injection can also increase cautious avoidance behavior (e.g., decrease in exploration in an open-field maze in rats) (Shepard et al. 2009), but not unconditioned fear responses to TMT (Rosen et al. 2008).

**Glucocorticoid facilitation of CRF in frontal/infralimbic cortex:** The infralimbic cortex is linked to extinction fear learning (Morgan and LeDoux 1995; Morgan et al. 2003; Phelps et al. 2004; Quirk et al. 2000; Delgado et al. 2008). The region also contains adrenal steroid–receptor sites in addition to CRF-receptor sites (Imaki et al. 1991; Jaferi and Bhatnagar 2007) and has direct neural access through connectivity to and from the amygdala (Price 1999; Aggleton 1992/2000).

Infusions of CRF into this neocortical region are known to influence fear events (Sullivan and Gratton 1999). There is also evidence that the ventromedial prefrontal cortex accentuates autonomic and glucocorticoid responses, whereas the dorsomedial prefrontal area restrains these regulatory physiological responses (Sullivan and Gratton 2002).

Moreover, regions of the dorsal and ventral medial prefrontal cortex have differential effects on the regulation of PVN CRF (Sullivan and Gratton 1999; Jaferi and Bhatnagar 2007). For example, lesions of the dorsal medial prefrontal region enhanced CRF expression in the PVN following duress (Radley et al. 2006a, b; Radley 2012). And this region in primates, in which there are elevated cortisol and CRF, is linked to temperamental differences in response to social or unfamiliar events (Schmidt et al. 1997; Schwartz et al. 2003; Kalin et al. 1998,

2011; Habib et al. 2000). CRF is tied to temperamental differences (Kalin et al. 1998; Davis et al. 1997).

Glucocorticoids can increase the release of CRF at the level of the frontal cortex in addition to the amygdala. Rats systematically treated with corticosterone for several days (who had indwelling catheters to measure CRF via microdialysis) released more CRF in response to an aversive stimulus than vehicle or non-corticosterone-treated rats (Merali et al. 2008). In addition, cortical implants of corticosterone aimed at the infralimbic regions of the prefrontal cortex impacted exploration in an open-field test perhaps via CRF expression (Croteau et al. 2017).

**Extinction of fear and possible effects of corticosterone and CRF in the medial prefrontal cortex (mPFC):** In addition to facilitation of mechanisms for enhancing and sustaining fear responses, other processes also regulate fear-related behavior (Stockhorst and Antov 2016). Whereas facilitated or sustained fear can be adaptive when threat and danger linger, fear must be reduced when it is no longer necessary. Sustained and facilitated fear behavior can become chronic, hyperexcitable, and maladaptive, producing several physiological and psychological disorders. Processes for shutting off or reducing fear include habituation and extinction mechanisms.

Chronic exposure to corticosterone has been shown to potentiate freezing responses to a conditioned fear stimulus when the stimulus no longer predicts danger (Corodimas et al. 1994). In this study, rats were chronically implanted with a corticosterone pellet 2 days after fear conditioning (tone–foot-shock pairing), and then were exposed to the tone alone, starting on the fifth day of hormone exposure. Treated rats did not differ from placebo-treated controls in levels of freezing during the first tone-alone presentation, but they did demonstrate higher levels of freezing in response to the conditioned stimulus (CS) during subsequent CS-alone exposures 4, 24, and 48 hours after the first tone-alone presentation. These results indicate that corticosterone did not affect the initial expression of fear, but that corticosterone-treated animals demonstrated an enhanced, sustained fear response compared with controls.

The prefrontal cortex is involved in extinction of conditioned fear, perhaps by a pathway from the infralimbic prefrontal cortex to the intercalated nucleus in the amygdala (Quirk and Gehlert 2003). This prefrontal pathway can interrupt and alter information passing from the basolateral amygdala to the CEA and inhibit fear behavior. The prefrontal cortex has relatively dense expression of glucocorticoid receptors in most regions, including the infralimbic, cingulate, and orbital cortical area; CRF neurons are also located in most regions of the prefrontal cortex (Radley et al. 2006b; Swanson et al. 1983).

Restraint-stress experiments resulted in an increase in CRF mRNA expression in the PVHmpd of intact animals, as well as in mPFCd lesioned rats. This enhancement was not shown in mPFCv lesioned animals. Stress exposure repeatedly shows increased levels of CRF mRNA, and is further increased by PL lesions. IL-lesioned animals display levels that fall between unlesioned stressed and nonstressed controls (Radley et al. 2006b).

These regions are part of the circuitry involved in modulating cardiovascular (Powell et al. 1994) and ACTH–corticosterone-mediated (Diorio et al. 1993) responses to aversive stimuli. It has been shown that the inhibitory influence of the mPFC on the PVN stress response involves a relay through the BNST (Spencer et al. 2005), similar to the indirect connectivity between the amygdala and PVN. Moreover, the mPFC projects to nuclei of the amygdala and anatomically related brainstem areas (Price 1999).

The medial prefrontal cortex may act through separate and different pathways to regulate the paraventricular nucleus of the hypothalamus in response to stress. Dorsal and ventral components of the medial prefrontal cortex possibly contribute differential regulation of neuroendocrine and autonomic responses to acute emotional stress. Ultimately, both routes lead to the PVN and trigger a response that goes on to affect the pituitary or the brainstem (Radley et al. 2006b, Radley and Sawchenko 2015).

The mPFC has been implicated in the extinction of conditioned fear; mPFC-lesioned rats exhibit a greater tendency than control rats do to respond to a conditioned fear stimulus during extinction trials, but not during acquisition training (Morgan et al. 1993; Phelps et al. 2004). Medial PFC lesions disrupt extinction performance, regardless of whether lesions were made before or following acquisition training (Morrow et al. 1999). Lesions of ventral mPFC have also been shown to increase reacquisition of a conditioned fear response following prior extensive extinction trials (Morgan et al. 2003), further suggesting a reduced ability to benefit from extinction training. In concert with findings that mPFC is important for retention of extinction learning (Quirk et al. 2000), the data suggest that alterations within the mPFC might predispose one to develop fear responses that are difficult to extinguish, and thus sustain fear beyond the time when it is adaptive.

Implants of corticosterone into the mPFC have been shown to decrease plasma levels of corticosterone with restraint stress, and lesions of the mPFC result in significantly increased plasma levels of ACTH and corticosterone after restraint stress, suggesting reduced inhibitory control of the HPA axis (Diorio et al. 1993; Akana et al. 2001). One study suggests that a single injection of corticosterone immediately before extinction training can facilitate extinction (Quirk et al. 2004). Interestingly, chronic glucocorticoid treatment has been shown to alter apical dendrites of mPFC neurons, and perhaps glucocorticoids are modulating the expression of conditioned fear during extinction, as was shown recently in both rats and humans (Quirk et al. 2003; Phelps et al. 2004). Alternatively, or conjunctively, chronic duress that elevates corticosterone levels increases dendritic growth and sprouting in the amygdala, indicating that chronic duress and high glucocorticoid levels also induce neuronal processes that could facilitate and sustain fear directly in fear circuits of the amygdala (Vyas et al. 2002).

There is some evidence that corticosterone regulates expression of CRF in the mPFC (Merali et al. 2008; Gray et al. 1993; Shepard et al. 2010). There is evidence that inhibiting CRF receptors blocks the reinstatement of extinguished fear (Waddell et al. 2004). Central injections of the CRF antagonist a-helical CRF have been shown to interfere with fear conditioning, and with reinstatement of extinguished fear (Waddell et al. 2004; Takahashi 2001). Whether these effects are because of inhibition of CRF expression in the mPFC or amygdala is not known.

But given the involvement of the mPFC in extinction and sustained fear, its interactions with the amygdala (perhaps through inhibitory control), and the expression of corticosterone receptors in the mPFC, it is an ideal region for investigating the role of corticosterone-mediated regulation of CRF in sustained fear. In other words, the infralimbic mPFC could contain CRF-expressing neurons that are both inhibited by glucocorticoids and induced by glucocorticoids; in addition to fear, this could underlie a wide variety of behavioral responses, both aversive and appetitive (Merali et al. 2004, 2008; Pecoraro et al. 2005b).

**Devolution of function: from chronic fear to anxiety:** Regions of the brain, such as the amygdala, have long been linked to auras, including olfactory auras of impending doom. Excessive fear is another expression of a hyperactive amygdala. Diverse studies have linked regions of the infralimbic cortex to fear learning, fear monitoring, and extinction. One putative role of this region is to restrain amygdala expression in the organization of fear: fear when appropriate, fear when necessary, and the avoidance of events that are fearful.

Of course, regions of the amygdala are not simply restrained by neocortical regions such as the infralimbic cortex; the amygdala generates and regulates behavioral expressions of fear. In the midst of breakdown of normal function, coping with fear through both behavioral strategies and neuroendocrine events that combat fear and provide anticipatory regulation of adaptation to change leads to the wear and tear of systems. Fear is an expensive bodily event. The price of excessive fear as it turns to anxiety is the devolution of normal function.

Consistent with this, CRF is elevated in the cerebrospinal fluid, together with elevated systemic levels of cortisol, in people who are chronically fearful/anxious and depressed (Nemeroff et al. 1984; Gold et al. 1988; Kling et al. 1993; Nemeroff et al. 1992; Parker et al. 2003; Binneman et al. 2008). Indeed, in the macaque, the CRF-R1 genotype has been linked to vulnerabilities of anxiety and depression

(Rogers et al. 2013). Finally, depression with elevated cortisol is also one in which there is a vulnerability to bone loss (Michelson et al. 1996), or neuronal deterioration (Sheline et al. 1996; Bao et al. 2008), which then impacts CRF. Indeed most conditions alter CRF expression across diverse regions of the brain (Raadsheer et al. 1993) including pain (Shenoda et al. 2016; McLean et al. 2006; Hubbard et al. 2011), rheumatoid arthritis (Baerwald et al. 1999, 2000), multiple sclerosis (Huitinga et al. 2000; Erkut et al. 1995), postpartum brains of depressed individuals (Bao and Swaab 2010), seizures (Baram and Schultz 1991; Baram et al. 1999), suicide (Zhao et al. 2015), and Alzheimer's disease (Le et al. 2016; Figure 5.9).

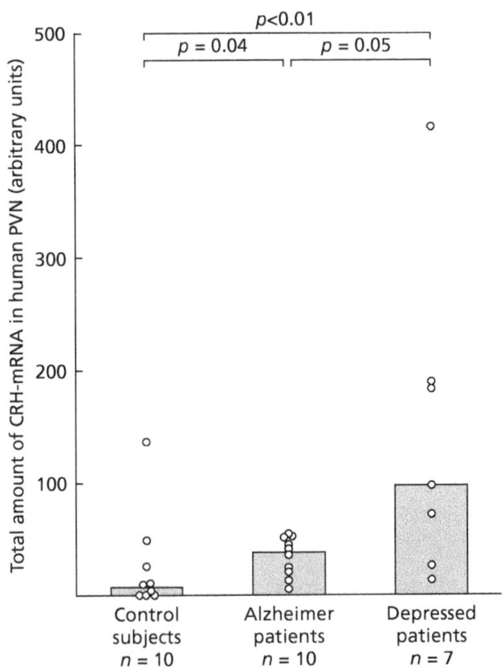


**Figure 5.9** Total hybridization signal for human CRF mRNA (arbitrary units) in the PVN. Bars indicate median values per patient group. The PVN of the Alzheimer's patients (n = 10) contained significantly more (MW; U = 23.0, W = 0.78, Z = −2.0, p = 0.04) CRF mRNA than that of comparison subjects (n = 10). The amount of radioactivity in depressed patients (n = 7) was significantly higher than in comparison cases (MW; U = 7.0, W = 91.0, Z = −2.7, p = 0.006) and Alzheimer's disease patients (MW; U = 23.0, W = 0.78, Z = −2.0, p = 0.05).

Reproduced from F.C. Raadsheer, J.J. Van Heerikhuize, P.J. Lucassen, W.J.G. Hoogendijk, F.J.H. Tilders, and D.F. Swaab, Corticotropin-releasing hormone mRNA levels in the paraventricular nucleus of patients with Alzheimer's disease and depression, *American Journal of Psychiatry*, 152 (9), pp. 1372–1376, Figure 2, doi.org/10.1176/ajp.152.9.1372 © 1995, American Psychiatric Association.

Patients suffering from Alzheimer's disease display significantly higher levels of CRF mRNA in the PVN (shown through radioactive hybridization experiments) than control patients. Patients suffering from depression display significantly higher levels of CRF mRNA in the PVN than both control patients and those suffering from Alzheimer's disease (Bao and Swaab 2010).

Interestingly, in a study considering the effect of physical exercise on a possible model of psychiatric disorders, like depression, CRF levels in the PVN of male rats increased during 30 minutes of high-speed running on a treadmill. Results from this study suggested that low-speed running (i.e., physical exercise with mild and low intensity) increases serotonin in the dorsal raphe nucleus, and consequently, rats exhibited decreased anxiety and depressed behavior (Otsuka et al. 2016).

## Conclusion

The role of glucocorticoids and CRF, from a neurobiological view, is to enhance attention to biologically significant events, including those that are novel or threatening. An animal being chased through the woods by a predator needs to sustain the central state of fear as long as is necessary to avoid being a meal. Induction of CRF is part of the orienting response to something potentially dangerous, whereas a sustained CRF response depends on glucocorticoids. In other words, CRF is involved in both short-term and sustained fear responses, where short-term responses are modulated by glucocorticoid-independent CRF but sustained fear responses are regulated by glucocorticoids. Glucocorticoids acting in the amygdala, and perhaps the BNST, increase the availability of CRF and prefrontal cortex gene products to maintain fear-related behaviors.

The HPA axis interacts within itself, as well as with key forebrain regions. CRF release from the PVN of the hypothalamus stimulates the release of adrenocorticotrophin hormone (ACTH) from the pituitary gland. ACTH then stimulates the adrenal glands, which release cortisol. Cortisol interacts with the PVN through a negative-feedback mechanism. Cortisol also acts on the medial prefrontal cortex and amygdala during bouts of fear, triggering a feedback response to the PVN via the lateral bed nucleus of the stria terminalis.

One adaptive response within a fear-provoking event is to be able to turn off the fear once the danger subsides. Because chronic high levels of glucocorticoids enhance retention of fear-conditioned memory such that fear is elevated after an extended period, chronic glucocorticoids might also influence extinction of fear through regulation of CRF (Thompson et al. 2004). The infralimbic mPFC is important during the extinction phase of fear, perhaps by inhibiting the activation of CRF expression in the amygdala and left BNST. Chronic fear responses that out-last their usefulness could develop into pathological states associated with anxiety and depressive disorders that have their neural basis within the fear circuitry (Eede et al. 2007; LeDoux 1996; Drevets et al. 2002; Price 1999; Rosen and Schulkin 1998). Indeed childhood trauma, depression, and suicide are both linked to CRF and CRF receptors (Roy et al. 2012; Ben-Efraim et al. 2011; Bao and Swaab 2007; Waters et al. 2015) and to placental fetal CRF (Moog et al. 2016). Indeed extreme low birth weight via in part placental CRF may affect the brain at 28 weeks and that may increase the risk for brain capability and normal function (Leviton et al. 2016).

Aberration of cortisol and CRF function is a well-known finding in anxious and/or depressed patients (Gold et al. 2002; Nemeroff and Owens 2004). Less is known about the role of gluococorticoids in the amygdala and BNST in anxiety and depression; however, enhanced activity (i.e., increased glucose metabolism) in the left amygdala is positively correlated with plasma cortisol levels in patients with unipolar and bipolar disorders (Drevets et al. 2002). Interesting genetic overexpression may decrease PTSD-related symptoms tied to fear (Elharrar et al. 2013). Dysfunction of the amygdala and prefrontal cortex is associated with anxiety and depressive disorders, and perhaps dysregulation of glucocorticoid-mediated regulation of CRF within these brain regions is linked to such disorders (Holsboer and Ising 2008; Nemeroff et al. 1984). CRF and CRF-R1 genetic models are consistently linked to anxiety-related behaviors, some of which reveal important interaction of NMDA-mediated glutamatergic regulation of CRF (Samaco et al. 2012; Gilman et al. 2015).

Achieving a better understanding of glucocorticoid-mediated regulation of CRF in the amygdala, BNST, and prefrontal cortex is clearly important for understanding the adaptive functions of fear and its aberrations that lead to pathology. In sum, fear-related events, perhaps via stimulation of CRF gene expression in the brain, may increase vulnerability to later experiences of unusual stress. These experiences increase production of CRF in the brain, thereby influencing the vulnerability to anxiety-like behavioral responses. Animal studies have readily demonstrated that early adverse experiences have long-term consequences for production of CRF in the brain and the propensity to express fear-related behaviors (Radulovic et al. 1999; Coplan et al. 1996; Kalin et al. 1989).

Chapter 6

# Social Contact, Gonadal Steroids, and CRF

## Introduction

Hormones of reproduction, such as estrogen, prolactin, oxytocin, and testosterone, are involved in diverse forms of behaviors beyond reproductive behaviors alone. But reproduction is an essential feature in our evolution. For instance, if one is an egg-laying animal, prolactin is secreted in anticipation of incubation, for which biparental contribution is a feature in many birds. Indeed, the incubation process of eggs reflects the panoply of information molecules; prolactin, for instance, stands out.

Importantly, in penguins, mice, and in primates like marmosets, prolactin is tied to diverse forms of biparental behaviors; it is elevated in both males and females since both contribute to child rearing. Moreover, infusions of prolactin into sexually dimorphic regions of the brain can elicit parental behaviors. Prolactin levels are indeed elevated in the brains of both males and females during parental behaviors, as is oxytocin. Oxytocin is elevated in expectant fathers who participate in parental care (Carter et al. 1999). Indeed, oxytocin is differentially elevated in voles that are monogamous and involved in biparental care from those that are not (Young et al. 1997).

Lactation and parental care affect many information molecules, most notably oxytocin, prolactin, estrogen, and progesterone. They also affect CRF. Interestingly, estrogen-related compounds also increase CRF in the placenta (Jasnow et al. 1999; Byrnes et al. 2012; Magiakou et al. 1996).

Changes in gene expression without altering the underlying DNA is a workable definition of epigenetics and is, perhaps, at the heart of phenotypic alterations in development, contingent upon context and circumstance. Behavioral adaptations for which CRF contributes importantly are long noted to be fundamental to evolutionary change and indeed underlie any definition of evolutionary change (Crews 2005).

Importantly, the rapid-fire expression of CRF would serve essential for differing social/ecological demands; parenting is one, responding to conspecifics is another. What evolved is a CRF signature ready for action, responding to changing demands of importance, part of the neural armor in foraging for coherence.

Changes in social contact affect central CRF, in that separation from the female partner increases CRF in the bed nucleus of the stria terminalis (Bosch et al. 2009) and may enhance approach and other pair bonding; both CRF receptors seem to underlie this adaptation. CRF serves to facilitate social comfort and social attachment (Moriceau et al. 2009; Hostetler and Ryabinin 2013; Bosch et al. 2009). CRF is involved in two ends of the same spectrum: activity for attention and activity for avoidance and fear.

In our species, and indeed in all mammals, first steps are social; contact with mother, and contact with others. These fundamental attachments (Bowlby 1977) trigger diverse forms of rich information processing in many sensory systems that contribute to this formation; olfaction and gustatory systems are two particularly important sources of information early on (Kondoh et al. 2016), but vision and the auditory system also play a part. The mother's linguistic expression ("motherese") is a significant, affectively rich source of information for the neonate. The regulation is not homeostatic between mother and baby during this time, though it has been characterized that way; but it is regulatory. Metabolic and social contact are crucial events (Hofer 1973); the fluids that are retained, the comfort that is afforded, and the anticipatory behaviors for longer term viability are apparent across a wide array of mammals and animals in general (Power 2004).

Below are pictured a few of the forms of parental behavior and care strategies found within the animal kingdom: female uniparental, male uniparental, and biparental (Royule et al. 2014; Lehrman 1958, 1965; Rosenblatt 2003; Dulac et al. 2014; Figure 6.1).

Steroids can either facilitate or repress the expression of diverse neuropeptide or neurotransmitters in the brain (Herbert 1993). Estrogen, progesterone, and testosterone influence CRF, as we will see in the first part of this chapter. The evolution of steroid hormones is as ancient as is the original chordate ancestor. Synthesis of estrogens via progestogens and androgens can be traced back that far. Corticosteroid synthesis can only be traced back as far as the origin of vertebrates. A phylogeny of the steroid-receptor gene family can be created by looking to ancestral steroid receptors (Eick et al. 2012).

Steroids secreted during pre- and postnatal development have many different effects on the brain (Goy and McEwen 1980; Arnold and Breedlove 1985). We have known for some time about the defeminization of the brain by the secretion of testosterone during postnatal development (Goy and McEwen 1980; Phoenix et al. 1959; McCarthy 2008) through aromatization to estrogens is fundamental, and that in fact can occur in utero. The gonadal steroid hormones

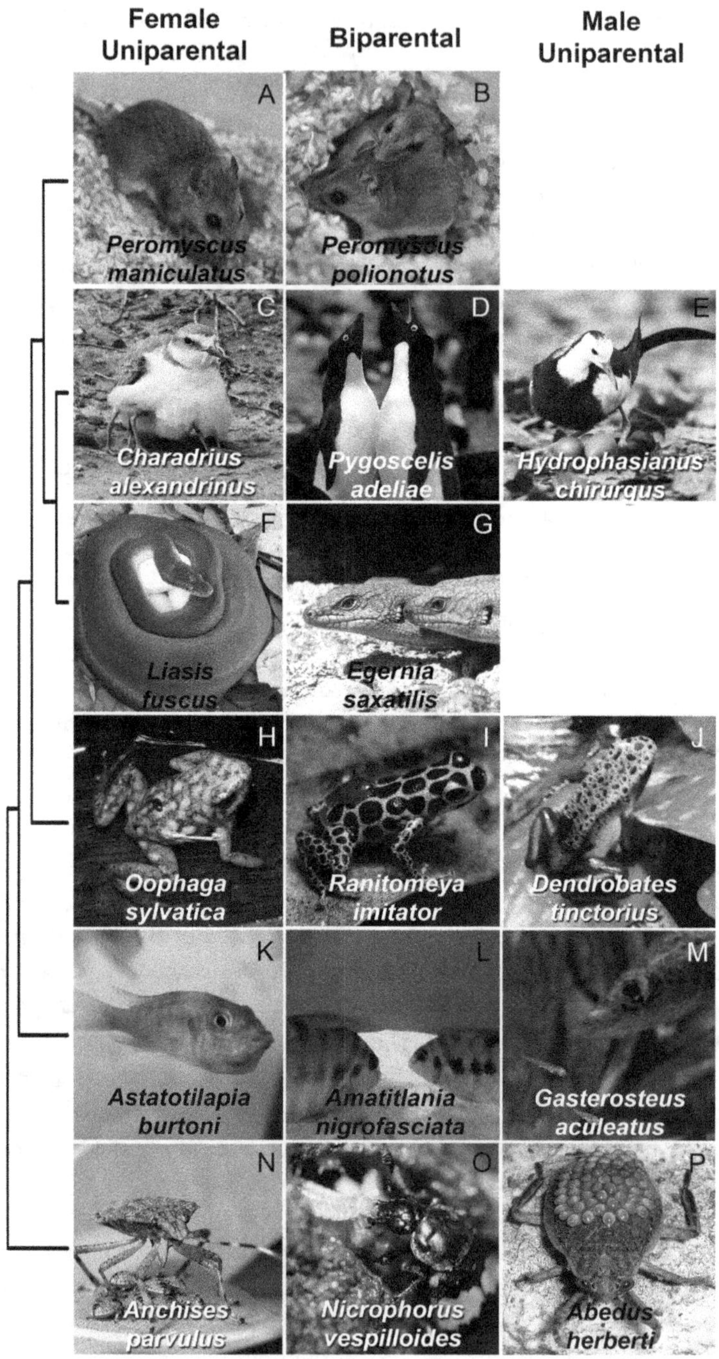
Female Uniparental
Biparental
Male Uniparental
A
Peromyscus maniculatus
B
Peromyscus polionotus
C
Charadrius alexandrinus
D
Pygoscelis adeliae
E
Hydrophasianus chirurgus
F
Liasis fuscus
G
Egernia saxatilis
H
Oophaga sylvatica
I
Ranitomeya imitator
J
Dendrobates tinctorius
K
Astatotilapia burtoni
L
Amatitlania nigrofasciata
M
Gasterosteus aculeatus
N
Anchises parvulus
O
Nicrophorus vespilloides
P
Abedus herberti

affect postnatal development and expression (Balthazart and Ball 1998), and these postnatal effects continue to influence behavioral events; the neural conditions that render them possible are in part due to the induction and regulation of information molecules such as CRF (Arnold and Breedlove 1985).

Importantly, the steroid induction of neuropeptide regulation is a common theme and it underlies social communicative behaviors; the responses to acoustic signals or particularly olfactory and gustatory signals are prepotent information laden with a rich communicative context of social modulation (Hofer 1973); for instance, the quiescence of behaviors after the appetitive calls are mediated by diverse information molecules. Social attachment of course has real effects on internal hormonal systems (Levine 1975; Meaney 2001); for instance, growth hormone declines and cortisol increases with decreased social attachment (Kuhn et al. 1990). Social context of course influences a number of neuropeptides (e.g., CRF, vasopressin, and oxytocin; Ruscio et al. 2007). Interestingly, the CRF knockout mice produce the normal numbers of pups (Muglia 2000), but the CRF type 1 receptor seems critical and not the particular ligand, and may be the essential factor (see Majzoub and Karalis 1999).

In this chapter, I first begin with a brief discussion of CRF in approach/avoidance behaviors across pre and postnatal events—a theme throughout the chapter—followed by a description of steroids (in particular gonadal steroids) and their effect on CRF, as well as sex differences in the expression of CRF in the brain and differential regulation of CRF in various brain regions. Finally, the chapter concludes with an overview of biparental behaviors and epigenetic events and social contact.

---

**Figure 6.1** (See color plate section) There are examples of different parental care strategies across vertebrates (depicted here). Male uniparental care is lacking only in the mammalian and reptilian lineages, although there are male-biased parental care systems in some canids and primates. (a) Common deer mouse (Peromyscus maniculatus). (b) Oldfield mouse (Peromyscus polionotus). (c) Kori bustard (Ardeotis kori). (d) Adélie penguin (Pygoscelis adeliae). (e) Pheasant-tailed jacana (Hydrophasianus chirurgus). (f) Water python (Liasis fuscus). (g) Black rock skink (Egernia saxatilis). (h) Diablito frog (Oophaga sylvatica). (i) Mimic poison frog (Ranitomeya imitator). (j) Dyeing poison frog (Dendrobates tinctorius). (k) Burton's mouthbrooder (Astatotilapia burtoni). (l) Convict cichlid (Amatitlania nigrofasciata). (m) Three-spined stickleback (Gasterosteus aculeatus). (n) Golden brown stink bug (Anchises parvulus). (o) Burying beetle (Nicrophorus vespilloides). (p) Giant water bug (Abedus herberti).

**Variants on social attachment:** CRF and urocortin receptors are differentially expressed. In the vole brain with regard to whether they are monogamous or promiscuous (Lim et al. 2005, 2006). Interestingly, CRF infusion in the nucleus accumbens activates approach behaviors in monogamous prairie voles. In the nucleus accumbens (NAcc), both CRF-R1 and CRF-R2 receptors facilitate partner preference formation in prairie voles. CRF injections led to increased contact time with partner vole, rather than with strangers, and made voles were more likely to develop a partner preference. CRF antagonists negate these effects (Gammie et al. 2005; Lim et al. 2007; Bosch et al. 2009).

The facilitation of social contact may be mediated in part by estrogen-receptor regulation and oxytocin expression in the evaluation of social stimuli. Genes encoding for several estrogen-receptor subtypes and oxytocin expression and receptors are essential for social affiliative behaviors in mice (Choleris et al. 2003, 2006), and for that matter, CRF expression (Bosch et al. 2009). CRF receptors in the regions of the nucleus accumbens affect partner preferences in voles (Lim et al. 1997, 2006; Olazabal and Young 2006). A putative role for CRF is attention to salient events (Peciña et al. 2006). After all, the world is littered with unfamiliar, potentially dangerous events-contexts in which CRF regulation is vital (Habib et al. 2000; Merali et al. 2003). One experiment showed that rats with high levels of CRF were less able to discriminate signaled from nonsignaled events during a sustained attention task, though they had been previously trained to do so successfully (Cole et al. 2016). The administered CRF infusions lead to decreased capability, owing to aberrant levels, whereas normal adaptive levels allow for appropriate function and attention to signals.

Regions of the brain such as the medial amygdala, medial bed nucleus of the stria terminalis, and regions of the basal ganglia are importantly knotted to attachment behaviors and to oxytocin or vasopressin expression (Young et al. 1997; Liu et al. 1996). Many contexts in which social stimuli are evaluated activate CRF and oxytocin expression (Kramer et al. 2006; Choleris et al. 2003, 2006, 2008; Young et al. 1997). Indeed infusions of oxytocin into the shell of the nucleus accumbens facilitates social contact and partner preferences in female prairie voles via dopaminergic mechanisms (Liu and Wang 2003), and it minimizes the effects of partner loss through passive stress coping in a monogamous species (e.g., male prairie vole, Bosch et al. 2009). In addition, the greater the approach-related behavior, the higher the expression of oxytocin and CRF receptor sites within the region of the nucleus accumbens (Olazabal and Young 2006; Liu et al. 1996; Millan et al. 1986; Sánchez et al. 1999, 2000; Van Pett et al. 2000), supporting the claim that CRF is involved in approach and avoidance behaviors (Merali et al. 2003, 2008).

Three sexually dimorphic nuclei of the forebrain are important in prosocial parental behaviors; they include the medial nucleus of the amygdala, medial bed nucleus of the stria terminalis, and the medial region of the hypothalamus (Newman 2002; Chloresis et al. 2006; Keverne and Curley 2004; Crews et al. 1993). All three direct different aspects of hormone-related parental behaviors and control behaviors for lower species. They also participate in us.

While sexually dimorphic hormonal systems underlie core social contact, the evolution of our species is an "emancipation" of this strong link: the strong olfactory dependence from evolution corresponds to our corticalization of function and expansion of visual systems (Broad et al. 2006; Van Essen et al. 1992). Interestingly, the size of the sexually dimorphic region of the preoptic region is significantly smaller in relative size to our brain (Keverne and Curley 2004; Broad et al. 2006). Our species' dependence on these regions and the hormonal fluctuations reduces with encephalization of function (Keverne 1997).

The medial nucleus of the amygdala, as indicated, is a sexually dimorphic region of the amygdala (DeVries and Simerly 2002), and it is tied to olfactory stimulation (Hofer 1973). Therefore (and not surprisingly), oxytocin infusions into the brain in several species facilitates approach to pleasant odors (Nelson and Panskepp 1996), and to social stimuli that facilitate approach-related behavioral responses (Febo et al. 2005). In one experiment, for instance, oxytocin was found to underlie the odor preferences of neonatal rat pups (Nelson and Panksepp 1998). And oxytocin is known to influence fear-related behaviors as well (Missig et al. 2010). Social memory is tied to oxytocin expression in the medial amygdala (Ferguson et al. 2001).

Exposure to various predatory olfactory cues such as cats, weasels, and ferrets results in defensive responses in laboratory rodents (Takahashi et al. 2008; Rosen et al. 1994). CRF antagonists attenuate these avoidance behaviors (Takahashi et al. 1990); CRF increases in regions of the brain including the amygdala (Asok et al. 2013). Indeed, overexpression of CRF over small periods can aid memory in social recognition (Kasahara et al. 2011).

**Estrogen and CRF:** Estrogen interacts with many neuropeptides and neurotransmitter systems (e.g., enkephalin and dopamine). Estrogen can increase oxytocin gene expression under diverse contexts in extrahypothalamic sites, and is well known to facilitate the induction of oxytocinergic neurons in regions of the brain that underlie sexual motivation (Pfaff 1980, 1999; McCarthy 2008; Kendrick 2004). Classical examples abound from the hormonal literature detailing its effects on the brain in the expression of behavior: high levels of estrogen can lower the threshold for diverse stimuli to be noted related to sexual contact (Pfaff 1980, 1999). Estrogen influences nulliparous and primiparous females, as well as CRF expression (Byrnes et al. 2012). Broad-based steroids affect arousal systems important for a wide range of behavioral and cognitive adaptations (Pfaff 1980, 1999). In a sustained attention task, female rats fared worse than males did after CRF infusion, and female rats in the estrous stage after CRF infusion had the most diminished performance compared with males and females in other cycle stages (Cole et al. 2016).

Further, estrogen-induced increases in oxytocin in regions of the brain that underlie behavioral functions that range from sexual function, some of which can be rapid (Shumacher et al. 1990), suggesting both membrane (Orchinik et al. 1992, 1994) and genomic effects (McEwen 1995), to aggressive behavior in relatively safe environment (Morgan et al. 2004). Estrogen can promote active exploration (Morgan et al. 2004). Estrogen has diverse effects on a number of other neuropeptides. It facilitates prolactin expression and promotes parental behavior (Bridges and Freemark 1995; Bridges et al. 2001; McCarthy et al. 1994). And estrogen is known to affect CRF gene expression in the brain along with the direct effects of pregnancy (see Jasnow et al. 2006; Byrnes et al. 2012; Magiakou et al. 1996).

Maternal behaviors affect CRF type 1 and type 2 receptors (Klampfl et al. 2013, 2014). In the paraventricular nucleus, for instance, repeated administration of propyl-pyrazole triol (PPT), an estrogen-receptor α agonist, leads to significant differences in CRF mRNA levels in nulliparous (no births) and primiparous (single birth) females. In the lateral amygdala, PPT administration again displayed significant differences in mRNA levels. Administration of the vehicle injections showed no significant differences (Byrnes et al. 2013).

CRF facilitates motor activity (Rosen et al. 1994; Liang et al. 1992b); thus an effect of estrogen, of increasing arousal and motor activity, may be mediated via CRF. Under normal conditions, CRF promotes adaptive motor activity, or attention; under conditions of pathology, it generates seizures (Baram and Schultz 1991; Lowry and Moore 1991; Sansoni et al. 2013; Forcelli et al. 2007); seizures are the outer end of motor activity. Infusion of CRF into the brain results in behaviors that range from fear, as noted previously, particularly in the last chapter, to aggression, and finally seizures, if the duration and doses of CRF is beyond the adaptive context (Rosen et al. 1994; Weiss et al. 1986).

CRF facilitates diverse forms of activity. Estrogen increases running behaviors or activity perhaps in part via CRF. Indeed, we know that estrogen can also increase CRF expression in rats in unfamiliar environments and fear-related conditioning (Jasnow et al. 2006; Patchev and Almeida 1996), and it underlies the regulation of dopamine, essential for the organization of action. In fact, estrogen is known to reduce fear and increase activity (e.g., Richter 1943; Morgan et al. 2004). However, it does so contextually; in familiar environments, it increases activity and reduces fear, in unfamiliar, it does the opposite (Morgan and Pfaff 2001). In non-fear-related events, no association exists between elevated CRF in either the CEA or the bed nucleus of the stria terminalis (BNST) after social defeat (Jasnow et al. 2004). More generally, estrogen acts to promotes activity (Richter 1943; Morgan et al. 2004).

Mice that received a subcutaneous injection of estrogen displayed increased CRF mRNA expression in the central amygdala, as measured by in situ hybridization and qPCR. These data suggest that estrogens can modulate fear conditioning in female mice, as well as CRF gene expression within the central amygdala, a region that is important for fear conditioning (Figure 6.2).

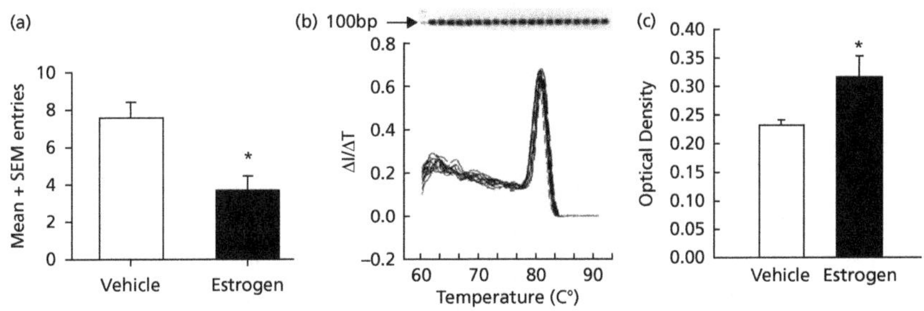


**Figure 6.2** (a) Mean entries into an open space. (b) CRF gene expression was determined by quantitative PCR. The resulting reaction was run on a 2% agarose gel to determine the size a specificity of the amplicon. Representative electrophoretic lanes of the PCR reaction produced a single amplicon that was approximately 100 bp, as predicted. In addition, a dissociation analysis of the CRF PCR reaction demonstrated that there was a single amplicon with a melting temperature equivalent to that predicted by the PRIMER EXPRESS software. This indicates that the fluorescent signal was used to quantify CRF amplification and was due to a single, specific product. (DI/DT = change in fluorescence intensity over the change in time). (c) Ovariectomized (OVX) EB-treated mice displayed significantly more CRF mRNA expression within the CEA compared with OVX vehicle-treated mice. Significant differences are denoted by an asterisk (*).

(a) Adapted from *Hormones and Behavior*, 40 (4), M.A. Morgan and D.W. Pfaff, Effects of Estrogen on Activity and Fear-Related Behaviors in Mice, pp. 472–82, doi:10.1006/hbeh.2001.1716, Copyright © 2001 Elsevier Science (USA), with permission from Elsevier.

(b) Adapted from *Hormones and Behavior*, 49 (2), Aaron M. Jasnow, Jay Schulkin, and Donald W. Pfaff, Estrogen facilitates fear conditioning and increases corticotropin-releasing hormone mRNA expression in the central amygdala in female mice, pp. 197–205, doi:10.1016/j.yhbeh.2005.06.005, Copyright © 2005 Elsevier Inc., with permission from Elsevier and from *Hormones and Behavior*, 40 (4), M.A. Morgan and D.W. Pfaff, Effects of Estrogen on Activity and Fear-Related Behaviors in Mice, pp. 472–482, doi:10.1006/hbeh.2001.1716, Copyright © 2001 Elsevier Science (USA), with permission from Elsevier.

(c) Adapted from *Hormones and Behavior*, 49 (2), Aaron M. Jasnow, Jay Schulkin, and Donald W. Pfaff, Estrogen facilitates fear conditioning and increases corticotropin-releasing hormone mRNA expression in the central amygdala in female mice, pp. 197–205, doi:10.1016/j.yhbeh.2005.06.005, Copyright © 2005 Elsevier Inc., with permission from Elsevier.

Specifically, CRF gene expression was elevated 24 hours after estrogen treatment, but not after 2 hours of chronic treatment.

During the estrous cycle, females are exposed to increasing levels of circulating estrogens, reaching peak levels during proestrus. The time course of the CRF gene expression response observed in the present study could fit into the context of the natural estrous cycle. With estrogens reaching peak levels on the morning of proestrus, the absence of an early effect might allow the female to engage in courtship behaviors. Following this, the female would become much more conservative and fearful after mating. Currently, it is not clear why CRF gene expression was not increased following 10 days of estrogen treatment (Jasnow et al. 2006). In sheep, estrogen treatment increases CRF mRNA expression within the BNST, but not within the PVN (Broad et al. 1995).

Direct transcriptional regulation of CRF has been demonstrated by estrogen in vitro (Vamvakopoulos and Chrousos 1993), and an indirect pathway has been suggested (Miller et al. 2004). CRF mRNA is increased in the ventral paraventricular nucleus (PVN) of female rats on the day of proestrus when estrogen is at its peak (Bohler et al. 1990). These data suggest that the CRF gene is under direct transcriptional regulation by estrogens, thereby providing a potential mechanism within the brain through which estrogens modulate emotional learning. Indeed exogenous estrogen can facilitate fear learning and increase CRF gene expression within the central nucleus of the amygdala. Both a-estrogen-receptor and h-estrogen-receptor genes are widely distributed throughout the amygdala in both rodents and humans (Laflamme et al. 1998; Merchenthaler et al. 2004), suggesting that this region is a likely site at which estrogens exert their actions on fear.

## Estrogen and Progesterone

Estrogen and progesterone across pregnancy and lactation were noted previously (Smith et al. 2012), and diverse changes in CRF expression in the brain have been noted in pregnant and lactating rodents. One behavioral feature is the reduction in startle responses in lactating females (Toufexis et al. 2006). Indeed, progesterone levels reduce CRF-related startle. A significant decrease in the startle response in female rats that have been infused with CRF and who are lactating was shown compared to female control rats that were not lactating but had also been infused with CRF. Ovariectomized (OVX) females with progesterone replacement showed a decrease in their CRF-enhanced startle compared with OVX females, OVX females with estrogen replacement, and OVX females, which were estrogen primed. OVX rats with an implanted blank or progesterone-filled capsule showed a significant difference only around minute 90. And rats with tetrahydro-progesterone injections showed a decrease in their CRF-enhanced startle (Toufexis 2007; Figure 6.3).

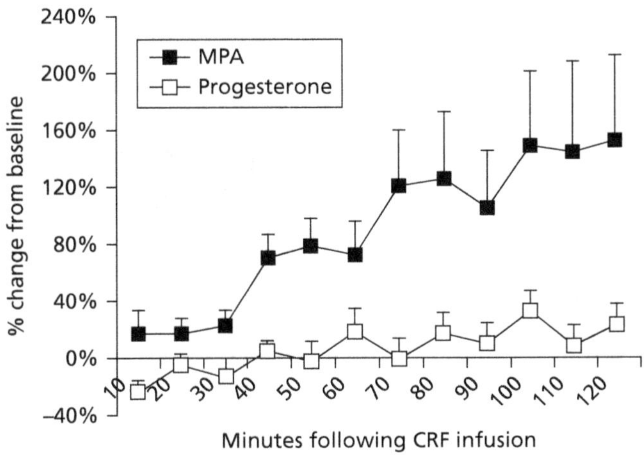


**Figure 6.3** Percentage change from baseline startle (SEM) after ICV CRF infusion in OVX rats implanted with MPA (dark squares; n6) or P (white squares; n7). Data were collapsed into 10 minute intervals of 20 startle stimuli.

Reproduced from Donna J. Toufexis, Carrie Davis, Alexis Hammond, and Michael Davis, Progesterone Attenuates Corticotropin-Releasing Factor-Enhanced But Not Fear-Potentiated Startle via the Activity of Its Neuroactive Metabolite, Allopregnanolone, *The Journal of Neuroscience*, 24 (45), pp. 10280–10287, Figure 4, doi: 10.1523/JNEUROSCI.1386-04.2004 Copyright © 2004 Society for Neuroscience.

Lactating female rats displayed lower levels of CRF-enhanced startle responses than did cycling female controls. In OVX females, progesterone and estrogen prevented a CRF-enhanced startle response. Constant progesterone without estrogen simply disrupted the startle. Intact females showed a significant level of light-enhanced startle compared with their male counterparts. But castrated males displayed nearly significant differences in response to light-enhanced startle (Walker et al. 2003).

Elevated levels of progesterone or the neuroactive metabolite allopregnanolone also tend to decrease startle responses and perhaps diverse forms of fear (Toufexis 2007). Ovariectomized (OVX) rats underwent CRF infusion under varying conditions. OVX rats with no hormone replacement, OVX rats with estrogen replacement, and estrogen-primed OVX rats exhibited variable deviations from baseline startle. Those OVX rats that underwent allopregnalone injections experienced a smaller change from baseline than those OVX rats that had no injections (Toufexis et al. 2006). Ovariectomized rats infused with intracerebroventricular CRF were then implanted with either MPA (medroxyprogesterone) or progesterone. Data suggest that MPA increases the CRF-enhanced startle response in these rats as compared to those rats that were infused with progesterone (Toufexis et al. 2004). Estrogen, when coupled with progesterone, also resulted in CRF-enhanced startle responses.

Steroids are not typically produced in the brain but there may be small amounts of progesterone being produced. Progesterone is not simply inhibitory on end organ tissue, or progesterone activates a number of behaviors (e.g., sexual receptivity). Nonetheless, progesterone is tied to GABA and is known for its quiescent effects; in human parturition, high progesterone delivery tends to slow parturition while also decreasing placental CRF and most directly inhibiting the production of CRF.

## Testosterone and CRF

Both estrogen and testosterone affect the HPA axis (Weiser et al. 2008; Bale and Epperson 2015). Anabolic androgenic steroids (AAS) cause a wide range of psychological effects, facilitating a range of emotions from euphoria to despair and anxiety, as well as aggression levels, libido, fear, and appetite. In experiments with mice, anabolic steroids potentiate fear-related behavior; mice show a higher startle response in anticipation of shock as a function of anabolic steroid treatment. Anabolic steroids appear to increase CRF gene expression in regions of the brain such as the central nucleus of the amygdala (Oberlander and Henderson 2012a, b). Mice that received AAS injections displayed significantly greater startle responses than did those injected with oil. In mice who underwent various infusion and surgical procedures, mice who were injected with AAS exhibited greater startle amplitude after no surgery and no infusions, sham surgery and no infusions, and vehicle ventricular infusion of saline than mice injected with oil (Oberlander and Henderson 2012a,b). Mice treated with anabolic-androgenic steroids displayed significantly greater levels of CRF mRNA in the central amygdala compared with mice treated with oil (Costine et al. 2010). In other words, anabolic steroids can increase fear-related behaviors such as startle responses (Figure 6.4).

The BNST, part of the head ganglia of the HPA axis, is active in regulation of the internal milieu (Herman et al. 2003). Many formation molecules are produced in this region, including CRF and endorphins, which play roles in various behaviors. Steroids, including the anabolics, induce the production of neuropeptides such as CRF, which can result in anxiety; they affect CRF neuronal excitability directly (Oberlander and Henderson 2012a; Partridge et al. 2014, 2016). For instance, on-cell recordings from the lateral part of the central amygdala from AAS-treated mice were much more frequent than in oil-injected mice. AAS-treated mice also displayed significantly higher action potential and a lower cumulative probability distribution of action potential frequencies (Oberlander and Henderson 2012a, b). Naïve mice provided with anabolic-androgenic steroids and acute application of CRF on neurons produced the

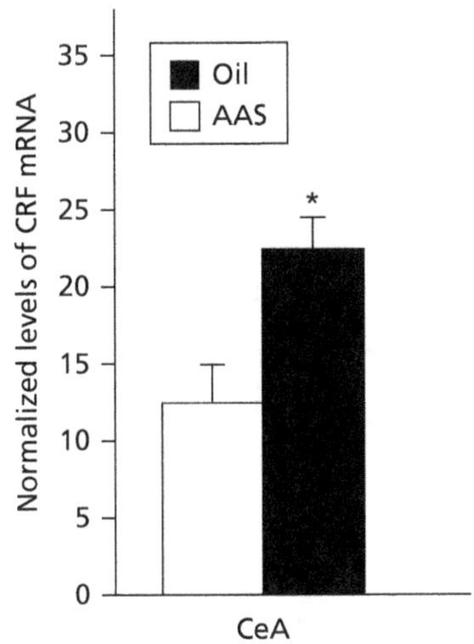


**Figure 6.4** Anabolic-androgenic steroid (AAS)-dependent changes in corticotrophin-releasing factor (CRF) mRNAs. Asterisk indicates CRF mRNA in the CEA was significantly greater in AAS-treated mice than in oil-treated mice. CEA, central amygdala; AAS, anabolic-androgenic steroid; CRF, corticotrophin-releasing factor.

Reprinted from *Psychoneuroendocrinology*, 35 (10), Beth A. Costine, Joseph G. Oberlander, Matthew C. Davis, Carlos A.A. Penatti, Donna M. Porter, Robert N. Leaton, and Leslie P. Henderson, Chronic anabolic androgenic steroid exposure alters corticotropin releasing factor expression and anxiety-like behaviors in the female mouse, pp. 1473–1485, doi:10.1016/j.psyneuen.2010.04.015, Copyright © 2010 Elsevier Ltd, with permission from Elsevier.

most active current when compared with mice only given saline, saline and CRF, or AAS alone. CRF exposure in addition to saline or AAS resulted in an increased frequency (Oberlander and Henderson 2012a, b; Figure 6.5).

The various forms of aggression share some common elements. One is steroid facilitation of neuropeptide gene expression, for which testosterone, or estrogen-induced vasopressin in the medial amygdala, is one model. In this, the aggression is for territorial defense (Neumann and Landgraf 2012). Testosterone and their metabolites, which have long been known to enhance aggression and impulsivity, are linked to steroid dependence, a form of addictive behavior (Herbert 1993; Herbert and Ryu 2015).

It is common knowledge that testosterone can enhance aggression (Wood and Stanton 2012), but independent of testosterone-induced changes in morphological features (muscle mass and facial features), enhanced physical aggression in general is expressed early on in differences between boys and girls in rough-and-tumble play. But, of course, one fundamental function of testosterone is reproductive capabilities (Herbert 1993, Herbert and Ryu 2015).

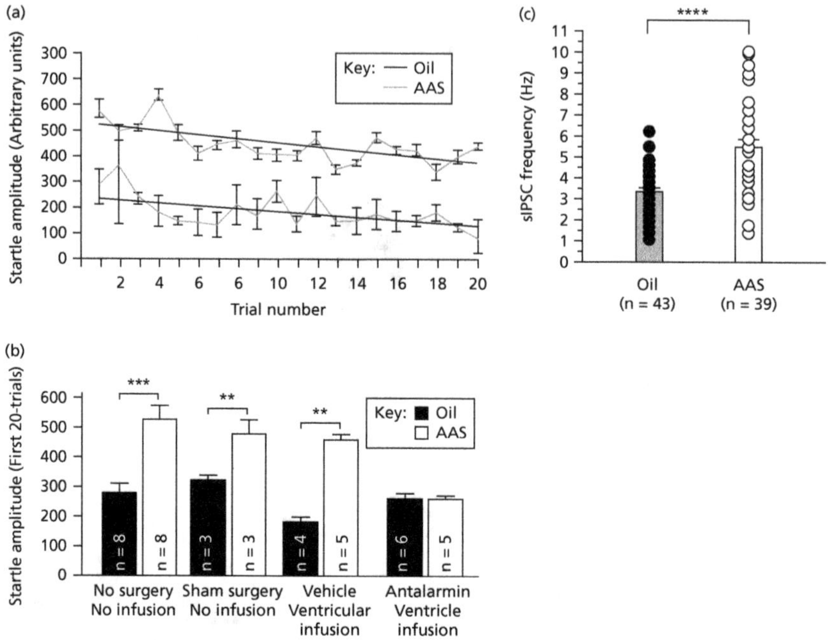


**Figure 6.5** AAS enhances anxiety and GABAergic transmission in the bed nucleus of the stria terminalis (BNST) by a CRF-dependent mechanism. (a) Data demonstrating that the amplitudes of the acoustic startle response were significantly greater on all trials in animals receiving a chronic mixture of three commonly used AAS (testosterone cypionate, nandrolone decanoate, and methandrostenolone; 7.5 mg/kg/day) than in control animals receiving a comparable volume of sesame oil (vehicle) (n = 8 in each group). (b) Mean startle amplitudes over the first 20 trails for oil- and AAS-treated animals that received no infusions or surgery, sham surgery, ventricular infusion of saline, or ventricular infusion of the CRF-R1 antagonist, antalarmin. n values represent the number of animals per group/condition (**P <0.01; ***P <0.001). (c) sIPSC frequencies from individual dorsolateral bed nucleus of the stria terminalis (BNST) neurons from oil-injected (black circles) and AAS-treated (white circles) mice. Histograms display the same data as means with error bars (SEM); n values represent the number of cells (****P < 0.0001).

Reprinted by permission from Macmillan Publishers Ltd: *Neuropsychopharmacology*, 37 (6), Joseph G Oberlander and Leslie P Henderson, Corticotropin-Releasing Factor Modulation of Forebrain GABAergic Transmission has a Pivotal Role in the Expression of Anabolic Steroid-Induced Anxiety in the Female Mouse, pp. 1483–1499; doi:10.1038/npp.2011.334, Copyright © 2012, Nature Publishing Group.

**Sex differences in affective disorders:** Sex differences in CRF-related events are a feature of the literature. Diverse forms of sex differences in CRF have been noted under different conditions, and specific effects on CRF in sexually dimorphic regions have been noted (Babb et al. 2013). And perhaps these sorts of findings reflect differences in several key features of reported human experience: one of which is depression. I say reported since men may just not report their experiences of this sort as women do. In other words, women are more likely to report affective disorders than males are (Iwasaki-Sekino et al. 2009; Bao and Swaab 2007). The lifetime prevalence of many psychiatric disorders can be broken down by sex, and a female-to- male ratio can be easily determined for disorders like generalized anxiety, bipolar disorder, and conduct disorder.

CRF receptors sites described earlier have been depicted in multiple brain regions (Sawchenko 1987). There is about 70% overlap of the type 1 and type 2 receptors (Bale and Vale 2004). A number of neuropeptides generate arousal and they seem in part to do so via the type 1 CRF receptor (Paneda et al. 2009). The type 1 receptor has been linked to a number of behaviors in addition to arousal, such as attention, which should not be surprising; the type 2 receptor is thought to counteract some of the actions of the type 1 receptor. Arousal generator hypocretin (orexins) cells (Carter et al. 2009) express both types of CRF receptors (Winsky-Sommerer et al. 2004; de Lecea et al. 2006), and may play a role in aversive learning and the etiology of fear and anxiety (Sears et al. 2013).

CRF is extensively linked to locus coeruleus noradrenergic activation and perhaps to fearful or anxious states (Koob 1999; McCall et al. 2015). CRF input from the amygdala facilitates tonic LC noradrenergic activity and anxious or avoidant-related behaviors; inhibition of this pathway reverses these effects and the effects on anxious behaviors; CRF type 1 antagonism reduced or abolished these events (McCall et al. 2015; Figure 6.6).

Interestingly, sex differences in CRF are apparent at a number of instances: first at the level of morphology within the LC (Valentino et al. 1993) and at the cellular and molecular levels of CRF signaling (Bangasser and Valentino 2012; Wiersielis et al. 2016). One plausible view is in the differences in arousal systems (Valentino et al. 2013; Bangasser et al. 2011) and that the tie to the interaction of norepinephrine and CRF from the brainstem to the forebrain may underlie some of these sexually dimorphic differences (Curtis et al 2006; Bangasser et al. 2013). Depicted below are projections from the locus that are abundant from the brainstem to the forebrain site, in addition to other brainstem functions (Valentino and Van Bockstaele 2008; Figure 6.7).

Various brain regions interact through CRF afferents. CRF neurons extend from the Barrington nucleus and the nucleus of paragigantocellularis, and end in the nuclear core of the locus coeruleus; these nuclei may play a role in coordinating autonomic activity with cognitive functions. The central nucleus of the amygdala, BNST, and paraventricular nucleus of the hypothalamus end outside of the nuclear core (Valentino and Bockstaele 2008). CRF immunoreactivity was more concentrated in both the core and the dorsolateral perilocus coeruleus.

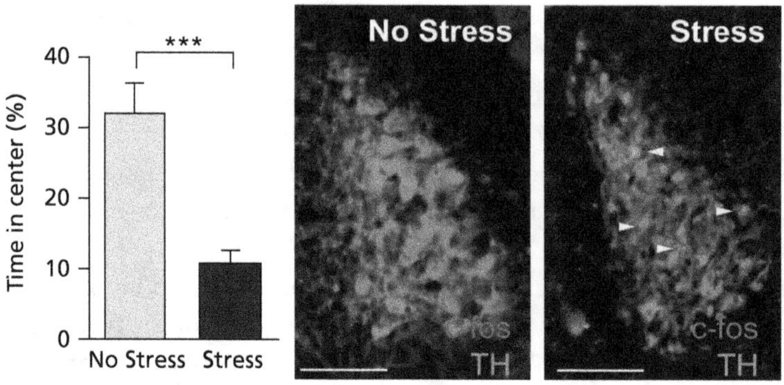


**Figure 6.6** (See color plate section) Stressed animals spend significantly less time exploring the center of the open-field test than nonstressed control (left). Representative immunohistochemistry (IHC) and quantification of CRH expression (right). TH, tyrosine hydroxylase.



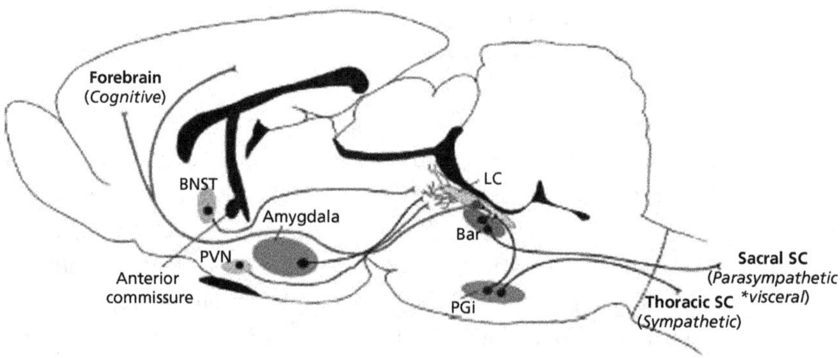


**Figure 6.7** Schematic depicting the localization of CRF afferents to the locus coeruleus and sites of termination in the locus coeruleus region. CRF neurons from the Barrington nucleus (Bar) and the nucleus paragigantocellularis (PGi) terminate in the nuclear core of the locus coeruleus. These nuclei also project to preganglionic parasympathetic and sympathetic neurons, respectively, and thereby may coordinate autonomic activity with cognitive functions. The central nucleus of the amygdala, bed nucleus of the stria terminalis (BNST) and paraventricular nucleus of the hypothalamus (PVN) terminate outside of the nuclear core in the pericoerulear region where locus coeruleus dendrites extend.

Photomicrographs show coronal sections where CRF-immunoreactive fibers are fluorescently visible (Bangasser et al. 2013).

And this is very much in contrast to CRF expression in forebrain sites such as the sexually dimorphic limbic sites (amygdala and BNST); these regions are larger in males than females, with greater levels of diverse neuropeptides (e.g., angiotensin, CRF) gene expression. Importantly, CRF expression may be greater in females than in males under diverse conditions of adversity; prenatal events may contribute toward this response (Bale 2011, 2014).

CRF fits this pattern in the BNST, and its projections to the ventral tegmental region, or dopaminergic region may represent a way in which CRF and dopamine interact with regard to excessive incentive salience (Berridge 2004; Koob and LeMoal 2008).

For now, note this sexually dimorphic pattern, but now the focus is on the locus coeruleus, (LC) one region for which the structure is larger in females than in males (Valentino et al. 1993). Interestingly, unlike forebrain regions, the sexually dimorphic region in the brainstem in the LC is greater in females than that in males (Valentino et al. 1993). Locus coeruleus neurons may be longer in female rats, suggesting that the dendrites will have a higher probability of making contact with limbic afferents that convey emotion-related information. Male rats may not have as long and complex dendrites, suggesting a lesser arousal in response to emotion-related stimuli (Valentino et al. 2013; Schussler et al. 2016).

Perhaps this sexual dimorphic response has its roots in part because of the demands of reproductive and parental behaviors. Rapid-fire CRF regulation and expression within various social/ecological contexts of demands is a feature of the evolution and diversification of use in neural capability. For instance, researchers showed that activation of CRF type 1 receptors within the BNST of lactating mothers could lead to maternal neglect of pups (Klampfl et al. 2016). Incidentally, activation of CRF type 2 receptors resulted in significantly increased nursing, 5 hours after treatment. Perhaps minimal levels of CRF are required for normal maternal behavior; therefore, it is not the presence of CRF that results in devolution of function in this case, but rather the effect of receptor manipulation through overactivation.

## Biparental Behaviors and CRF

Studies involving macaques have shown that rearing conditions (and attachment) have long-term implications in terms of neuropeptide steroids and neurotransmitters expression (e.g., CRF, monoamines, and adrenal steroids), and of course, behavior (Harlow and Soumi 1971; Higley et al. 1992, 1993; Erickson et al. 2005; Winslow et al. 1993).

For instance maternal deprivation in macaques results in elevated levels of cortisol (Coe et al. 1985); and elevated levels of CRF in the cerebrospinal fluid

(Coplan et al. 1996). Rearing conditions matter. Macaques in normal maternal-reared conditions developed higher elevated levels of oxytocin compared with the nursery-reared macaques (Winslow et al. 1993, 2003; Rosenblum et al. 2002) and lower levels of CRF; nursery-reared animals also have higher elevated levels of serotonin (Habib et al. 2000; Erickson et al. 2005).

We are a species in which both parents are involved in varying degrees in parental care. In particular, marmosets, voles, and degus engage in biparental behavior, whereas the vole and degus stand out as the primary mammalian models that have been used to study the roles of central CRF.

In a number of species, reproductive success requires parents to invest a substantial amount of energy in feeding and caring for their offspring. Since, in the human species, parental care, specifically the paternal investment, is mediated by cultural and economic influences, the paternal contribution has varied across generations owing to the exposure to challenging factors including death, divorce, adoption, and the introduction of women into the workforce (Marsiglio et al. 2000).

Paternal engagement through direct interaction with the child includes play and other caregiving roles. Paternal and maternal care plays an essential role in the functional development of CRF expression and oxytocinergic and vasopressinergic neuronal circuits in the extended amygdala (including the BNST) and in hypothalamic regions.

Oxytocin, for example, is paramount to the process of muscular contraction during birth, and for milk letdown in postpartum mothers. Over the past decades, interest in oxytocin has begun to shift from a primarily female focus, to that of male affiliative behavior. Male mice lacking oxytocin receptors make fewer ultrasonic vocalizations during social isolation than wild-type controls do; as adults, oxytocin-receptor knockout males showed elevated aggression levels, similar to males deficient in the oxytocin ligand (see Takayanagi et al. 2005).

Next, following parturition and continuing into adulthood, prairie voles exhibited higher vasopressin receptor binding in the BNST and the amygdala than did montane voles (Wang and Young 1997). When the pup retrieval rate by fathers is experimentally manipulated, pups that are retrieved more often have a higher density of immunostaining in the dorsal fiber tracts of the BNST, and less in the ventral cell body-containing region (Frazier et al. 2006). This prior experience is correlated to a decrease of aggression in the resident–intruder paradigm, demonstrating that paternal behavior affects the neural organization and behavior of offspring.

Such early life experiences (e.g., handling) have differential effects on CRF expression in the CEA and the PVN. Glucocorticoid-receptor mRNA expression has been shown to be modified in rats who have undergone handling rather than being left undisturbed in the central amygdala, but levels were similar in both the PVN and the hippocampal CA1 region. In the CEA (and BNST) by postnatal day 9, CRF gene expression is enhanced, and in the PVN it is diminished by the handling experiences (Fenoglio et al. 2004, 2006).

Studies on humans and animal models revealed that paternal care significantly affected the development of cortico-limbic synaptic circuits (Helmeke et al. 2009; Braun et al. 2011; Seidel et al. 2011; Narita et al. 2010). In an animal model, the trumpet-tailed rat *Octodon degus*, a biparental South American rodent, has been used systematically and quantitatively to study the effect of paternal care on the neuronal and behavioral development of the offspring (Fuchs et al. 2010; Pinkernelle et al. 2009; Helmeke et al. 2009; Ebensperger et al. 2010).

Degus display some similarities to human and non-human primate behavioral, endocrine, and physiological characteristics. Developmentally, the relative advanced maturity of their sensory systems at birth matches the situation in human and non-human primate neonates. Similar to human babies (DeCasper and Fifer 1980), newborn degus learn to recognize and respond to their mothers' vocalizations within the first days of life (Braun and Poeggel 2001; Braun and Scheich 1997; Poeggel and Braun 1996). And similar to humans, this vocal communication is an important component for the establishment and maintenance of the emotional attachment to the parents.

Among other things, there is evidence that paternal care alters the developmental expression pattern of CRF-expressing interneurons in an age- and a region-specific manner. At weaning age (P21), paternal deprivation resulted in an elevated density of CRF-containing neurons in the orbitofrontal cortex (OFC) and in the basolateral amygdala (BLA) of male degus, whereas a reduced density of CRF-expressing neurons was measured in the dentate gyrus and stratum pyramidale of the hippocampal CA1 region. With the exception of the CA1 region, the deprivation-induced changes were no longer evident in adulthood (P90), which suggests a transient change, which in later life might be normalized by other socioemotional experience. The central amygdala, characterized by dense clusters of CRF-immunopositive neuropil, and the precentral medial, anterior cingulate, infralimbic, and prelimbic cortices were not affected by paternal deprivation.

Interestingly, the number of GABAergic neurons also changes in response to daily exposure to separation stress, in a neuron and region-specific manner (Braun et al. 2011). Moreover, selectively in the basolateral amygdala, the number of those GABAergic interneurons that co-express CRF is dramatically elevated (Seidel et al. 2011). On the pharmacological level, the stressed/deprived animals showed an enhanced sensitivity toward amphetamine and methylphenidate, indicating a monoaminergic "hypersensitivity." Most of these changes are detectable around the age of puberty, and future studies will analyze if, and through which manipulations (enriched environment), such an altered monoaminergic balance can be normalized.

In the hippocampal CA1 and dentate gyrus (DG) regions, immunostaining for CRF revealed spindle-shaped cells. Deprivation-induced changes of CRF-immunoreactive cells in the hippocampal DG are visible between postnatal day 21 and postnatal day 90. CRF-positive neurons are scattered throughout the OFC. Deprivation-induced changes of CRF-immunoreactive cells are highly significant in both the lateral orbitofrontal and ventromedial cortex at postnatal day 21 and between this time point and postnatal day 90 (Seidel et al. 2011; Figure 6.8).

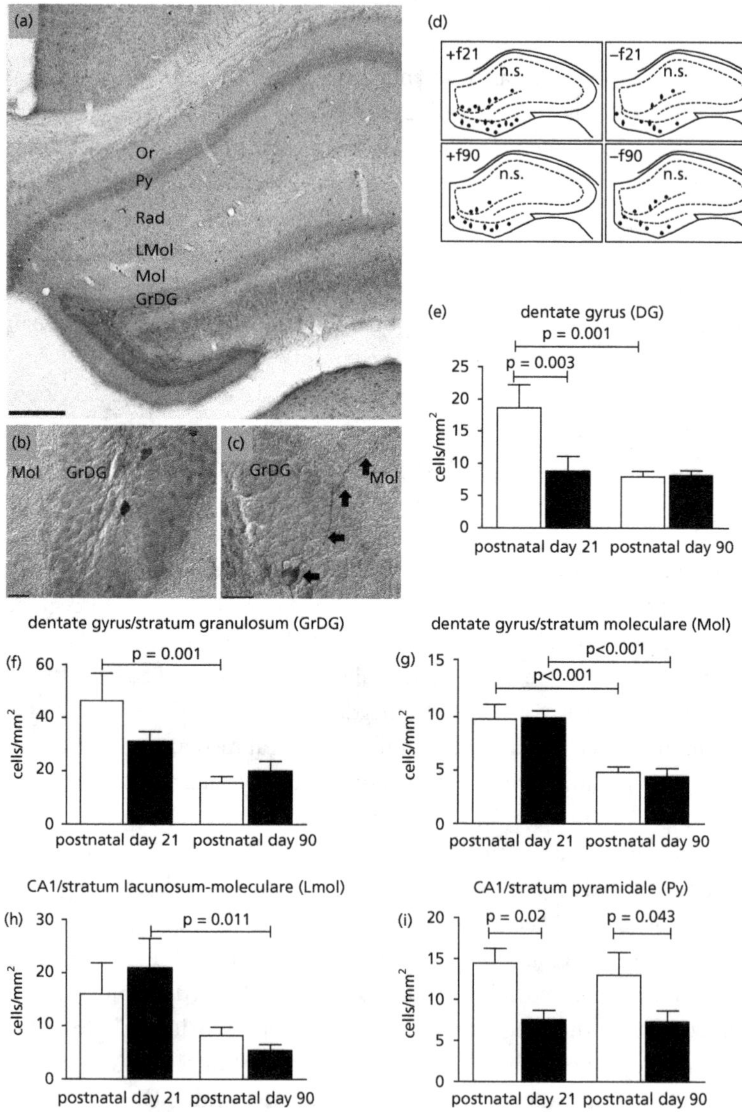

In the *degus*, it was observed that paternal care interfered with the developmental expression patterns of CRF-expressing neurons in prefrontal and limbic brain regions of the offspring (Braun et al. 2011; Seidel et al. 2011). For instance, the father-deprived juvenile and adult animals displayed a significantly lower density of CRF-expressing neurons in the medial BNST compared with biparentally raised animals. In contrast, no significant difference was observed in the lateral BNST. Interestingly, no significant differences were found between the rearing and age groups for the density of CRF-expressing neurons in the PVN.

It is suggested that father-deprived animals display dysfunctions of "anxiety"-related subcircuits. This view is supported by previous findings, which demonstrate that, in addition to the effects on CRF-expressing neurons in the medial BNST, paternal deprivation induces region- and age-specific changes of CRF-expressing neurons in the prefrontal and limbic regions. Other experiments using the open field test indicate that the father-deprived offspring display elevated anxiety, indicated by a decrease in activity in the center of the open field arena (data not shown). In juvenile animals, the orbitofrontal cortex and the basolateral amygdala displayed an elevated density of CRF-positive neurons. These changes disappear until adulthood, indicating that social experience during later life may compensate for the juvenile deprivation.

On the other hand, even transient changes occurring within short developmental time windows can still result in long-lasting effects, including the "hypofunction" of the medial BNST. In contrast, the CA1 region of the

---

**Figure 6.8** Analysis of corticotrophin-releasing factor (CRF)-expressing neurons and fibers in the hippocampal formation of 3 week and adult Octodon degus. (a–c) Immunostaining for CRF in the hippocampal CA1 region and in the dentate gyrus (DG) revealed spindle-shaped cells. Or, stratum oriens; Rad, stratum radiatum. Scale bars: (A) = 250 lm; (B, C) = 20 lm. (f) The graph shows the deprivation-induced changes of CRF-immunoreactive cells (mean SEM) in the hippocampal DG. (d) Representative illustration of the spatial distribution of CRF-positive neurones in the DG. n.s.: non significant difference. (e–i) The graphs summarize deprivation-induced changes of CRF-immunoreactive cells (mean SEM) in the hippocampal formation (e, DG/total; f, DG/stratum granulosum; g, DG/stratum moleculare; h, CA1/stratum lacunosum-moleculare; i, CA1/stratum pyramidale) P-values indicate the results obtained from the Sidak–Holmes post hoc test. Black histogram bars = without father; white histogram bars = biparental.

hippocampal formation and the medial BNST show lasting reduction in the density of CRF-expressing neurons in father-deprived animals, whereas the subregions of the medial prefrontal cortex remain unaltered (Seidel et al. 2011). Statistical differences in the amount of CRF-expressing neurons are significant only in the medial region of the BNST, in response to paternal deprivation. Photomicrographs display the reduction in CRF expression in the medial BNST of these animals. Statistical differences in the amount of calbindin-D28k (CaBP-D28K) expressing neurons is significant only in the medial region of the BNST at postnatal day 21, in response to paternal deprivation, and in the lateral region of the BNST at postnatal day 90. Photomicrographs display the reduction in CaBP-D28K expression in these regions of the BNST in those who were reared without a father and those who were biparentally reared (Gos et al. 2014; Figure 6.9).

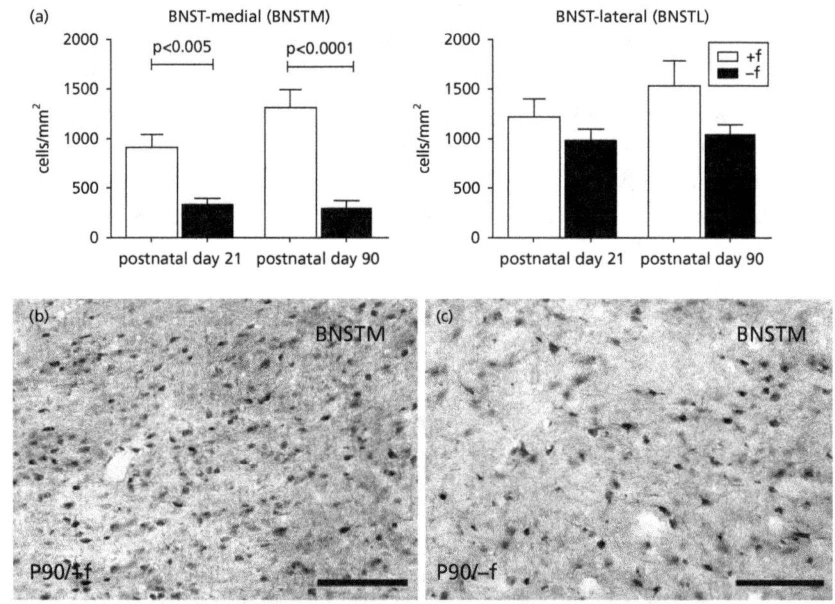


**Figure 6.9** Changes in CRF-expressing neuron density in the BNST in response to paternal deprivation. (a) Differential changes in the medial (left) and lateral (right) BNST. (b and c) Photomicrograph showing the reduction of CRF-positive neurons in the medial BNST (BNSTM) of adult father-deprived (+ f) compared to biparentally reared (–f) offspring. Bar in b, c = 100 um.

Reprinted from *Brain Structure and Function*, 219 (6), Tomasz Gos, Jay Schulkin, Anna Gos, Joerg Bock, Gerd Poeggel, and Katharina Braun, Paternal deprivation affects the functional maturation of corticotropin-releasing hormone (CRH)- and calbindin-D28k-expressing neurons in the bed nucleus of the stria terminalis (BNST) of the biparental Octodon degus, pp. 1983–1990, DOI: 10.1007/s00429-013-0617-4 © The Author(s) 2013.

## Epigenetic Events and Social Contact

Importantly, similar to glucocorticoids and CRH, testosterone and estrogen are tied to epigenetic events: enhancement and decrement of diverse neuropeptides and neurotransmitters in the brain and in the body in general. Epigenetics is an old concept that precedes our modern understanding of genetic mechanisms. The term and original concept was introduced by C. H. Waddington (1942), combining epigenesis from embryology with genetics, to describe how genotype interacting with environment gives rise to phenotype. A more recent conception of epigenetics is heritable changes in gene expression without underlying change in DNA sequence, and it refers primarily to mechanisms such as DNA methylation and histone modifications that alter the functional qualities of the DNA sequence (e.g., ability of transcription factors to bind) but leave the DNA sequence unchanged. This may be a fundamental mechanism for changing set points, as it changes gene expression in tissue.

Regulation is the key to survival. Animals are constantly adjusting their physiology and metabolism in order to remain within the bounds of viability. The new and exciting understanding of genetics and epigenetics is that, at the level of DNA, regulation is also the key to viability. Our new understanding of genomics brings it closer to that of regulatory physiology. There is metabolism at all levels in an organism: the organism level, the organ level, the cellular level, and the genome level. Many of the changes to fetal metabolism in response to the in utero environment of maternal obesity are likely due to epigenetic changes (i.e., changes in DNA expression via DNA methylation or demethylation, histone modifications, or other changes to the structure of the chromosomal DNA) without any actual change of DNA sequence. These epigenetic changes can activate or silence genes by such mechanisms as recruiting methyl-CpG binding proteins, which then block transcription factors from binding to the promoter sites or change chromatin structure enhancing heterochromatin formation.

Maternal and paternal behavior activates a broad array of CRF-related events across diverse regions of the brain; one intriguing finding is the activation of predominantly type 2 receptor activation in the BNST (posterior region; Klampfl et al. 2014).

One example is the link between maternal licking and grooming (high or low), which is consistently transmitted to female offspring; moreover, decreases in social comforting contact has long-term implications for most mammals studied (e.g., Levine 1975; Meaney et al. 1989 1996; Liu et al. 1997). There is wide variation in this phenomenon that has long-term implications on cephalic systems; those rat pups comforted by social contact have greater regulatory capacity as adults on diverse systems in the brain, including neuropeptide and neurotransmitter systems (dopamine, serotonin, and CRF).

Diverse prenatal and postnatal events, in addition to genetic makeup, determine the development of the brain. Many events in utero, from maternal nutrition, ecological/social conditions, and medical conditions affect brain development and neurotransmitter and neuropeptide expression (e.g., Weinstock 2005; Takahashi et al. 1998; Uban et al. 2013). But even pregestational events can have long-term effects on behavior. For instance, female rats subjected to various adverse events before pregnancy resulted in a reduction of social behavior in their offspring, particularly in females (Schachar-Dadon et al. 2009). In other words, pregestational adversity (e.g., fear-related events) resulted in decreased social contact and social exploration in postgenerational offspring. These effects seem to occur via the regulation of CRF-R1 receptor subtypes (Zaidan et al. 2013; Zaidan and Gaisler-Salomon 2015).

Longer term effects recruit CRF: experiments have shown showed that in pregnant animals exposure to adverse events affected the activation of the HPA, glucocorticoid receptors in diverse regions of the brain including the hippocampus and extrahypothalamic CRF, or CRF in their offspring (Welberg et al. 2000, 2001; Welberg and Seckl 2001; Francis et al. 1999).

Moreover, diverse parental-related behaviors, grooming for instance, can buffer against the overactivation of an HPA axis when later exposed to adversity as an adult, which can be transgenerational (Diorio et al. 1993; Francis et al. 1999, 2000; Meaney 2001). But it is not just the gonadal steroids that have long-term effects on morphology and neuropeptide expression; the adrenal steroids have long-term effects as well (McEwen 1995, 1998).

The activation of cortisol or corticosterone in rats affects the expression of many information molecules such as CRF or neurotransmitter sites in addition to glucocorticoid receptor sites. The events also affect behavior; increased fear-related social behaviors have often been noted to result from adverse experiences in utero, or following postnatal development (e.g., Plotsky et al. 1993; 2005; Takahshi et al. 1998; Schachar-Dadon et al. 2009).

Parental regulation of corticosterone secretion, for instance, is modulated by social contact (Levine et al. 1967), a feature in mammals that appears to be long term and constant in social species. There is some evidence for differential developmental calendars for the regulation of the HPA axis in rats and more broadly, glucocorticoid regulation of CRF (e.g., Levine 2000; Plotsky and Meaney 1993; Brunson et al. 2001); parental responsiveness has profound effects on HPA and extrahypothalamic CRF expression (Seidel et al. 2011; Gos et al. 2014).

Parental responsiveness has been imitated in laboratory experiments. Animals were placed in four rearing-type groups: animal facility reared, nonhandled (HMS0), handling with maternal separation for 15 min (HMS15), and handling with maternal separation for 180 min (HMS180). Cerebrospinal fluid samples were collected and those rats who underwent handing, and maternal separation for 180 minutes displayed the highest levels of CRF-like immunoreactivity (Plotsky et al. 2005).

These events affect subsequent behavior. Increased fear-related social behaviors have often been noted to result from diverse adverse-experiences preconception, in utero, or following postnatal development (e.g., Plotsky et al. 1993, 2005; Takahashi et al. 1998; Shachar-Dadon et al. 2009; Sandman and Glynn 2009; Avishai-Eliner et al. 2002). These are many forms of fetal programming; events in utero set the conditions for long-term vulnerabilities (Barker 1997; Welberg and Seckl 2001; Meaney 2001; Noll et al. 2007) that affect a range of behaviors such as obesity, fearful shyness, preterm low birth-weight babies, diabetes, and cardiovascular pathology (Barker 1997, 2004; Pesonen et al. 2006; Wadhwa et al. 1996, 1998, 2001), as well as hippocampal expression (Takahashi et al. 1998). Of course, hippocampal regulation by cortisol secretion and direct application during development can also affect fear-related behaviors (Takahashi and Kim 1994).

For instance, experimentally inhibiting an enzyme by a substance found in licorice fed to pregnant rats (11 b-hydroxysteroid dehydrogenase) transforms corticosteroid-receptor sites (Welberg et al. 2000, 2001; Francis et al. 1999; Vazquez et al. 2006); one result is elevated levels of CRF gene expression in the central nucleus of the amygdala, which, among other things, can affect the level of CRF neurons in the offspring of pregnant rats, perhaps increasing vulnerabilities later in adulthood (Seckl and Walker 2004; Welberg and Seckl 2001; Meaney 2001; Welberg et al. 2000; Plotsky and Meaney 1993; Plotsky et al. 2005). By increasing the level of CRF, through glucocorticoid agonists for instance, and altering the mineralocorticoid-receptor and glucocorticoid-receptor sites in the central nucleus of the amygdala of the mother during gestation, there is a greater likelihood in the offspring experiencing cautious avoidance of unfamiliar events (Welberg et al. 2000; Weinstock 2005; Bingham et al. 2013).

During development in cross-fostering experiments, changes in DNA demethylation are linked to the transmission of various social behaviors. They are likewise linked to CRF (Elliot et al. 2010). Depicted below are epigenetic changes in CRF under conditions of adversity. Addition of DNA-methyltransferase inhibitor 5-aza-2′-deoxycytidine (5-Aza) to hypothalamic cells increased CRF mRNA expression, as did the co-incubation of cells with 5-Aza and N6,O2’-dibutyryl-cAMP (Elliot et al. 2010; Figure 6.10).

Animal studies suggest that excessive prenatal stress and fear can have long-term effects on offspring (Weinstock 1997; Chen et al. 2013; Leshem and

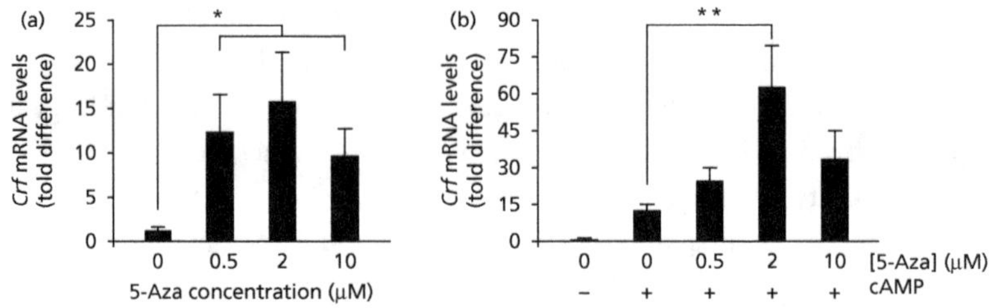


**Figure 6.10** Methylation of the CRF promoter regulates gene expression. (a) Incubation of N42 immortalized mouse hypothalamic cells with 5-Aza increased CRF mRNA expression (n = 4). Bars represent mean ± s.e.m. *P = 0.029. (b) Co-incubation of 5-Aza and N6,O2′-dibutyryl-cAMP increased CRF mRNA expression (n = 4). **P = 0.0021.



Schulkin 2012). One study, for example, has shown that aversive stress results in elevated cortisol levels in both the mother and the fetus(es). When sacrificed as young adults, the offspring demonstrate increased levels of CRF (49%) in the central nucleus of the amygdala compared with control rats; prenatal stress experiences can increase fear responses when these animals are provoked as adults (Cratty et al. 1995; Bingham et al. 2013). Consistent with this picture are the experiments demonstrating that CRF injections during pregnancy in rats increase the vocalization of the offspring when tested in isolation chambers during ontogeny (Williams et al. 1995).

One mechanism that may underlie epigenetic gene regulation is methylation or demethylation (e.g., silencing or enhancing the expression of genes; Holliday and Ho 1998; Holliday 2002); demethylation prevents transcriptional expression, and one result is the silencing of gene expression. Variations in parental behaviors, for instance, are linked to oxytocin (or prolactin) expression (Francis et al. 1999, 2000, 2002; Dixson and George 1982; Bridges et al. 2001; Carter et al. 1997, 1999; Keverne and Curley 2008). The high levels of grooming in cross-fostering experiments are knotted to estrogen and oxytocin expression (Meaney 2001; Cameron et al. 2008), as well as glucocorticoid receptors and diverse neurotransmitter and neuropeptide expression such as CRF, 5HT, dopamine, and DNA methylation (Weaver et al. 2004, 2005).

One notable lesson learned with molecular science is the danger of deleting one gene, or overemphasizing one level of analysis or one particular information molecule. Take out a particular CRF gene and one can still note fear-related behavioral responses (Majzoub et al. 1999). Uncertain comforting social contact and lower levels of social grooming results, for example, in higher levels of CRF in several species that have been studied to date (macaques, rats, and mice [Coplan et al. 1996, 1998]; Erickson et al. 2005; Habib et al. 2000; Plotsky et al. 2005). Developmental placation by robust social grooming and social comfort places the activation of the HPA axis under basal conditions (e.g., Levine 1975; Meaney 2001; Liu et al. 1997).

What we know is that the postnatal period, in addition to the neonatal, can have long-term effects in terms of CRF levels and other neuropeptide systems in the brain (Welberg and Seckl 2001). One feature affecting these systems is social contact (Carter et al. 1997, 1999; Carter 2007). CRF, a neuropeptide linked to social withdrawal, and social learning affects long-term behavioral consequences on behavior and brain function (Coplan et al. 1996, 1998; Kalin et al. 1998; Habib et al. 2000; Takahashi et al. 1993; Heinrichs 2003). We are more fearful of unfamiliar, uncertain events, many of which are social events (Kalin et al. 1998; Schwartz et al. 2003; Beaton et al. 2008, 2015).

Neural events that may reflect epigenetic changes are greater amygdala activation, higher cortisol levels, and greater CRF expression (Kalin et al. 1998). Chronic adversity has differential effects of methylation and CRF expression in male and female rat brains (Sterrenburg et al. 2011), in which there may already be sex differences in CRF expression and regulation. It has been shown that there may be significant differences between gender and between stressed and nonstressed levels of corticosterone in male and female rats. There seems consistently to be a significant difference in the weight of stressed and unstressed male rats (at days 10 and 14), and stressed and unstressed female rats (at day 14; Sterrenburg et al. 2011).

In other words, epigenetic regulation of CRF and other information molecules has also been discerned in placental tissue; namely changes in chromatin structure and histone deacetylation (Abou-Seif et al. 2012; Nicholson et al. 2004; Di Stefano et al. 2015), pre- and postnatally and across the life cycle are diverse regulation of genes.

Interestingly, glucocorticoids facilitate CRF gene expression that can elicit cautious avoidance behaviors through epigenetic mechanisms, such as histone modification (e.g., deacetylation). Such epigenetic mechanisms may underlie changes in behavior, in this case cautious avoidance behaviors by the elevation of CRF. These mechanisms may also underlie longer term changes in CRF expression (Tran et al. 2014). When corticosteroid-implanted animals were infused with either a vehicle or trichostatin A (TSA, a histone deacetylase inhibitor) into the central amygdala and were compared using fluorescence to visualize CRF mRNA gene expression, CRF mRNA showed a significant decrease in those infused with TSA.

## Conclusion

CRF underlies both approach and avoidance behaviors. Social contact, as in parental and developmental behaviors, is essential for long-term viability in many species. Diverse steroids facilitate CRF gene expression. For instance, the data provide evidence that estrogen, progesterone, and testosterone can influence fear. In addition, the present data demonstrate that estrogens increase CRF mRNA expression within the CEA, a region that is important for regulating fear responses

Regions of the BNST and CRF expression are differentially impacted by, in this case, paternal separation from their young (Braun et al. 2011), as there are a number of cases in which diverse neuropeptides are affected by maternal separation. In other words and not surprisingly, distress is emitted from the Octodon degus pups. This is a biparental care setting; the pups brains are affected, and CRF expression is altered in several regions of the brain that include the BNST, amygdala, and hippocampus in the pups who were paternally and maternally deprived (Seidel et al. 2011; Gos et al. 2014). Interestingly and not surprisingly, the BNST is also a sexually dimorphic brain region that is tied to parental-related behaviors (Becker et al. 2007).

One form of regulation of feedforward systems in the brain is steroid induced, and it regulates diverse information molecules, including CRF (Herbert 1993; Fink 1997; Schulkin et al. 1998; Pfaff 1980). Indeed such feedforward regulatory systems, namely the positive induction of CRF by diverse steroids, and not just glucocorticoids, underlie appetitive/consummatory behavioral expression. While an exaggerated emphasis has been on negative restraint as a primary model for steroid peptide interactions (though see Fink 1988; 1997; Kalra 1993) on the model of the HPA axis, the behavioral model is less in conformity with this view (see Herbert 1993; Schulkin et al. 1998).

As this chapter and the previous suggest, diverse steroids (both adrenal glucocorticoids and mineralocorticoids and estrogenic and anabolic steroids) can also facilitate feedforward CRF gene expression (Oberlander and Henderson 2012a). And there is again differential regulation of CRF gene expression across brain regions.

Sex differences may reflect social contact capability and necessity (Bale 2009), and are rooted in development (e.g., brain and placental development; Rossant and Cross 2001; Howerton et al. 2013). The effects of these differences may occur during both prenatal and postnatal periods (Bale 2015). Gender-specific CRF regulation in development and puberty (Weathington and Cooke 2012; Weathington et al. 2014) and anxiety-related behaviors are perhaps core themes reflective of successful evolution (Barna et al. 2003).

Gender-related differences are tied to CRF in core neurotransmitter systems (norepinephrine in the locus, Valentino et al. 2013; Gilman et al. 2015). Serotonin, one very basic transmitter, is linked to CRF-R1 receptor regulation in a sex-specific manner in response to adversity in the dorsal raphe nucleus (Howerton et al. 2014). However, in diverse experiments, both male and female rats display sex-specific responses; the males in this case are more vulnerable to devolution of CRF regulation during adversity (Mueller and Bale 2008).

One feature of elevated CRF in the brain is the arousal of behaviors. Arousal systems (Pfaff 1999) are both specific to a context and knotted to general activation that underlies diverse behavior by increasing CRF gene expression in the central amygdala, and estrogen or testosterone, for instance, facilitate one role of CRF. CRF is responsive and it enhances attention to changing circumstances in environmental events (Merali et al. 1998, 2004, 2008).

CRF is not merely a signal of anticipated danger: it is differentially expressed in female and male brains. In males, the forebrain sites are clearly larger than those in the female brain are, but in the brainstem, sites are larger in females. One striking result is the vulnerability in development for CRF changes in sexually dimorphic regions (Funabashi et al. 2004) affecting expression, which in females is more pronounced on the surface. We now turn to a deeper consideration in the context of adversity, survival, memory, and aging.

Chapter 7

# Surviving, Remembering, Adversity

## Introduction

The role of information molecules is to promote survival systems across life cycles. On the adaptive side, CRF promotes change and attention to change; on the nonadaptive side, CRF promotes decreased tissue capability and the acceleration of an aging process in end organ systems, as this chapter will discuss.

The effects on the adaptive side can be quick or not; in adapting to change, the expression and regulation has to be both. The world is living with social duress and uncertainty, placated with bouts of stability setting the conditions for longer term viability. The precarious and the stable are common features of our world (Dewey 1925).

In this chapter, fast and slow regulation of CRF: CRF in memory formation, in brain deterioration, and in adapting to adversity are discussed. Information molecules such as cortisol and CRF figure importantly across these events. One model of CRF in the context of neurotransmitters is depicted below (Gallagher et al. 2008; Figure 7.1). CRF and its receptors play many roles within the CNS. CRF primes the principal transmitter and/or neuromodulator's postsynaptic action. CRF enhances or depresses the primary transmitter's action.

Context, steroids, and CRF: Events can influence, for instance, levels of gonadal steroids and adrenal steroids that vary with context, need, and social capabilities. Chronic duress can result in longer term elevated levels of adrenal steroids or not (Sapolsky 1992; Albeck et al. 1997). Dominance with high or low cortisol is not one-dimensional and it varies with species and capability. The issue is always regulation; capable and viable end organ regulation.

One feature on the behavioral side is the prediction and reliability of events. Of course, many things are just not like that; chronic duress is one example as it affects the internal milieu. Chronic duress coupled with uncertainty is a recipe for disaster in the maintenance of the internal milieu. End organ systems are all finite. One can show high or low cortisol levels, as in PTSD or anxious depression, respectively, and still have elevated CRF in the brain (Bremner et al. 1997).

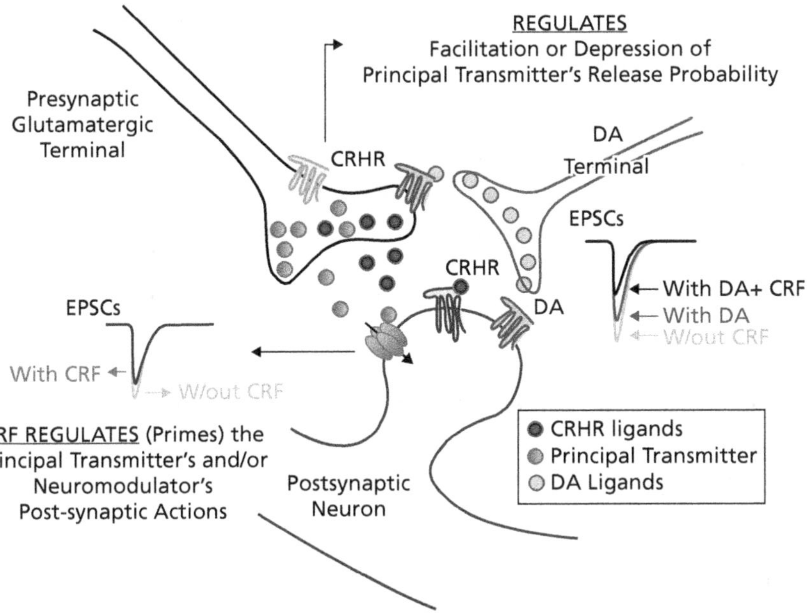


**Figure 7.1** (See color plate section) Roles for CRF, CRF-receptor (CRFR) ligands as neuroregulators ("primers") within CNS synapses tonic role of endogenous CRF as demonstrated following application of a CRF-receptor antagonist results in an enhancement or depression of the primary transmitter's action (e.g., upon glutamate transmission at the central amygdala nucleus or lateral septum mediolateral nucleus synapses); OR, phasic role of CRF acting either by evoked or volume transmission to modulate the action of a principal transmitter (e.g., glutamate; or, a modulator, e.g., enhance dopamine's affects on basolateral amygdala to medial prefrontal cortex glutamatergic synaptic transmission).

Reprinted from *European Journal of Pharmacology*, 583 (2–3), Joel P. Gallagher, Luis F. Orozco-Cabal, Jie Liu, and Patricia Shinnick-Gallagher, Synaptic physiology of central CRH system, pp. 215–225, doi:10.1016/j.ejphar.2007.11.075, Copyright © 2008 Elsevier B.V. with permission from Elsevier.

In other words, chronic social duress leads to high CRF under diverse conditions. One condition is dominant and subordinate relationships; levels of CRF are greater with both (with or without high cortisol levels) particularly in the central nucleus of the amygdala. Dark-field photographs show that CRF mRNA expression in the central amygdala is much higher in non-responder subordinate and subordinate responder rats than in control rats. Subordinate responder rats displayed higher levels of CRF mRNA expression compared with dominant rats (Albeck et al. 1997).

Survival systems, of which diverse emotions are one kind (Darwin 1958; LeDoux 2014, 2015), underlie brain generated and suitable contexts of expression. CRF is one information molecule tied to these events.

CRF can elicit fear-related avoidance behaviors or aggressiveness toward conspecifics (Wiersma et al. 1998). Levels of CRF and the duration of CRF expression can elicit diverse behaviors, depending on the behavioral options and biological needs. For instance, CRF within one part of the accumbens might elicit approach behaviors or avoidance behaviors. GABA infusion can elicit defensive behaviors in the caudal end of the accumbens, while infusion elsewhere may elicit appetitive behaviors (Reynolds and Berridge 2008). Similarly, CRF in the amygdala can elicit either approach or avoidance depending upon context, needs, and capability (Pecina et al. 2006). Peptides typically are fast acting; events in nature are often rapid. Mobilization of these peptides is a key feature in the execution of behavioral options. Evolution is about the expansion of these preadaptive capabilities. One route is through diverse information molecules whose functional expression varies with context.

The issue of flexibility is a critical feature in evolutionary success along with suitable capabilities and context. Earlier in the text, I alluded to both fast and slow genomic changes by glucocorticoids on the regulation of CRF. Let us consider this more and consider CRF generally as a signal of danger in tissue, in the arousal of behavior and memory, the blood brain barrier, mast and glial cells, and further CRF expression. Finally, we will come back to the core theme of CRF signaling in adversity.

**Fast and slow regulation of CRF: fast membrane-related events**: Adapting to change across the life cycle, as I indicated at the onset, is directly controlled by glucocorticoids, which in turn control neuronal excitability (Joëls 1997; Groeneweg et al. 2011; McEwen 1995; Hill and McEwen 2009). This is in part due to changes in CRF regulation. Some of the glucocorticoid effects on the brain are quite rapid (Groeneweg et al. 2011), suggesting that corticosterone has nongenomic membrane effects via GABAergic mechanisms (Orchinik et al. 1994; Tasker et al. 2006), or perhaps via endocannabinoid receptors (Gray et al. 2015; Laskey and Polosa 1988; Coddington et al. 2007; Tasker et al. 2015).

Orchinik, Murray, and Moore (1994) demonstrated fast-acting effects of cortisol on glucocorticoid receptors on clasping behaviors in amphibians, perhaps via the induction of CRF. Through an endocannabinoid-dependent pathway, corticosterone inhibits a clasping behavior seen in rough-skinned newts. This behavior is inhibited by stress or corticosterone injections. On the membranes of medullary neurons, corticosterone binds to G-protein coupled receptors, causing endocannabinoid release, which act on presynaptic neurons, inhibiting the release of glutamate and/or to stimulate GABA release. This ultimately leads to the inhibition of clasping (Denver 2007). These fast-acting membrane–related events have been suggested as a more phylogenetic, ancient signal system (Dallman 2003) that allows for fast forms of behavioral adaptation, quickly switching behavior in a changing context.

In fact, both negative- and positive-feedback systems in the interactions between steroid hormones and neuropeptides underlie behavioral responses (Fink 1997; Kalra 1993). These responses range across vertebrates and include appetitive behaviors in newts (Moore et al. 1992; Moore and Rose 2002), and are membrane-related effects (Tasker et al. 2006). Steroid-related events have long been noted to be fast, as noted in the introduction to the book.

Others used rats to demonstrate, for instance, that the consolidation of avoidance learning was both enhanced and inhibited by endocannabinoids (Hill and McEwen 2009). In particular, it was enhanced within the lateral amygdala, an important site for the interaction between glucocorticoids, neuroadrenergic activation, and CRF. Indeed, endocannabinoids may underlie the acute sense of fear via both glutamatergic, as well as CRF systems (Kamprath et al. 2009; Gray et al. 2016). Additionally, interference with cannabinoid 1 receptors reduces the anxiety-related behaviors tied to CRF (Kupferschmidt et al. 2012).

Nonnuclear activation is one of several mechanisms by which steroids interact with transmitters and perhaps peptides, such as CRF, in rapid activation (Hill et al. 2009). Intraventricular injections of CRF, for instance, result in increased levels of the endocannabinoid ligand 2-arachidonoylgycerol (2-AG) and anandamine (AEA) in several regions of the brain (amygdala and frontal cortex; Gray et al. 2015; Kim et al. 2015). Regions of both the amygdala and prefrontal endocannabinoid signaling are tied to adaptive responses to adversity conferring some adaptive capability (McLaughlin et al. 2014). Corticosterone, which increased CRF in both the amygdala and frontal cortex, resulted in reductions of AEA in both regions, but increases in 2-AG in the frontal cortex. Both glucocorticoids and CRF interact with endocannabinoid signaling systems via in part CRF type 1 receptors (Gray et al. 2015; Gray et al. 2013).

Endocannabinoid signaling is tied to critical neurotransmitters as is seen in GABAergic function and in CORT reduction in PTSD patients (Trezza and Compolongo 2013; Hill et al. 2013; Yehuda 2002); this appears to modulate the effects of adversity (Coddington et al. 2007). To induce endocannabinoid synthesis, corticosterone binds to an uncharacterized membrane-bound glucocorticoid receptor that activates the G-protein signaling cAMP/PKA pathway. When endocannabinoids are released into the synapse, they bind to CB1 receptors on GABAergic terminals and ultimately inhibit GABA release and disinhibit norepinephrine release. The binding also increases activation of postsynaptic-adrenoceptors, which helps to increase memory consolidation of emotionally aversive experiences (Hill and McEwen 2009).

One of the other fundamental neurotransmitters that CRF interacts with in regulation is norepinephrine. Moreover, norepinephrine and CRF are fundamental units in arousal, essential for attending and foraging, for coherence, and in adapting to events, particularly under conditions of adversity (Koob 2009; Craig 1918; Berridge 2004). A possible model for psychopathology is described as a feedforward system, where CRF activates brainstem noradrenergic activity through norepinephrine. Consequently, forebrain CRF activity is activated, creating a closed loop of stress response potentiation when faced with repeated exposure to a stressful stimulus. Repeated exposure could ultimately result in psychopathology. From the forebrain (CEA, BNST, and PVN), CRF is released and sent to the locus coeruleus, which sends norepinephrine back to the forebrain. Both brain regions contribute to a behavioral and autonomic response to stress (Koob 1999, 2015; McCall et al. 2015).

Indeed, CRF and noradrenergic activity are involved in most behaviors requiring the activation of the brain through the enhancement of attention and will reflect both experience and epigenetic-related events along with genetic conditions that set the internal expression of CRF; all of this has to be understood in the light of context and circumstance.

## Basal/Lateral Amygdala and Hippocampus: Glucocorticoids and CRF and the Facilitation or Degradation of Memory

Generally, glucocorticoids and CRF have been consistently linked to memory; both the facilitation and degradation of memory are related to the level and duration of these hormones (e.g., Lupien et al. 2005; Okuda et al. 2004; McGaugh 2000, 2003; Roozendaal et al. 2002, 2004; Kim et al. 2001; Blank et al. 2002, 2003). Thus, although aversive or fear conditioning is facilitated by glucocorticoids, glucocorticoids can interfere with retrieval of memory in both rats and humans via adrenoceptor activation (e.g., Hui et al. 2004; Conrad et al. 2004; Roozendaal et al. 2002, 2008; Cordero and Sandi 1998).

CRF effects on learning and memory have also been reported within the basolateral amygdala. For example, local infusion of a CRF antagonist has been shown to reduce memory consolidation of an inhibitory avoidance task (Roozendaal et al. 2002, 2004). For instance, one study reported that injection of a CRF-R1 antagonist into the basolateral complex of the amygdala immediately following foot shock diminished retention of aversive conditioning in an inhibitory avoidance task. Rats given higher levels of alpha-helical CRF exhibited lower levels of retention latencies when undergoing foot-shock experiments post training. It was also shown that expression of CRF in the amygdala increased 30 minutes following foot shock (Roozendaal et al. 2002). It has been suggested that the source of the CRF acting on the basolateral amygdala originates in the CEA (Roozendaal et al. 2002). One hypothesis is that glucocorticoids, by increasing CRF availability, facilitate the consolidation of the aversive memory (Rosen and Schulkin 1998; Isogawa et al. 2013).

Importantly, injections of the glucocorticoid receptor antagonist, RU-486, into the basolateral complex of the amygdala will block the consolidation of aversive conditioning. Other experiments have shown that glucocorticoid injections into this region of the amygdala can facilitate aversive conditioning (Moriceau et al. 2004) perhaps via the induction of CRF (Roozendaal et al. 2002: Moriceau et al. 2004). Glucocorticoid activation (or deactivation) aimed at several regions of the brain, including the lateral amygdala, can either facilitate or disrupt the expression of fear via CRF, at least in part (Roozendaal et al. 2000, 2002; Thompson et al. 2004; Donley et al. 2005). Moreover, it has been shown through optogenetic manipulation that CRF influences memory and learning via the lateral amygdala (Asok et al. 2016a).

Diverse brain regions underlie coping with adversity (Joels and Baram 2009). One key area in which there are both glucocorticoids and CRF-receptor sites is the lateral region of the amygdala (e.g., Roozendaal et al. 2002; Makino et al. 1995), an area known to be influenced by both glucocorticoids and CRF and tied to memory (e.g., Donley et al. 2005; Moriceau et al. 2004; Roozendaal et al. 2002; Roozendaal 2000; Bakshi et al. 2002; Kavushansky and Richter-Levin 2006; Thoeringer et al. 2012). Moderate levels are essential not only in coping with adversity but in learning that may be necessary (Lucas et al. 2013).

CRF-related activity within the amygdala has been consistently linked to distress (e.g., Rainnie et al. 2004; Dabrowska et al. 2013; McGill et al. 2006). Indeed the development of fear is tied to synaptic plasticity in, for instance, the basal lateral amygdala (Thompson et al. 2004; LeDoux et al. 1988; Cohen et al. 2014) and CRF CEA (Sanford et al. 2016; Pomrenze et al. 2015), but also linked to diverse regions of the brain and CRF synaptic plasticity physiology (Gallagher et al. 2008; Orozco-Cabal et al. 2006a, b; Chen et al. 2013). For instance, CRF is located in the axonal terminals of GABAergic neurons (Chen et al. 2012). CEA CRF neurons release GABA (Pomrenze et al. 2015).

CRF infusions into the lateral regions can either enhance or degrade memory consolidation via interactions with adrenoceptor activation and glucocorticoids (Roozendaal et al. 2008), and modulates contextual fear (Hubbard et al. 2007). The CRF type 1 receptor in particular is tied to modulating dendritic growth in the hippocampus (Sheng et al. 2012). Indeed atrophy in hippocampal dendrites is tied to CRF over expression and pathology (Chen et al. 2004, 2010,

2012) and corticosterone induced deterioration (Sapolsky 1992), and that can affect memory (McEwen et al. 2015).

These events in development can either create vulnerability or enhance neurogenesis in adulthood, creating a condition of less adaptive capabilities (e.g., Korosi et al. 2010). Long-term priming activation affects hippocampus and hippocampal CRF (Blank et al. 2002). CRF expression in the dorsal hippocampus affects dendritic spines under conditions of duress, especially and

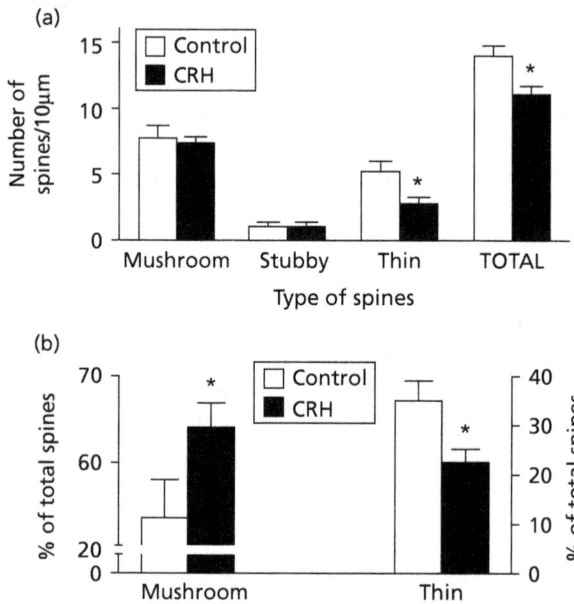


**Figure 7.2** Corticotrophin-releasing factor (CRF) preferentially affects the integrity of thin spines. (a) Thin spines on CA1 pyramidal neurons were selectively eliminated in response to CRF infusion (100 nM, 60 min), whereas the density of mushroom-type spines was unaffected after CRF application. Quantitative analysis of spine density and spine types is based on 9–12 CA1 pyramidal neurons that were imaged and analyzed without knowledge of the treatment group (*P < 0.05; analysis of variance (ANOVA) with the Bonferroni post hoc tests). (b) Analysis of the relative contributions of different spine types to the overall population before and following exposure to CRF. Whereas mushroom-type spines were minimally affected by CRF (A), there is a significant increase in the contribution of these spines, which are not considered to undergo further changes upon long-term potentiation (LTP) induction, to the overall spine population. Commensurately, there is a major reduction in the availability of thin spines (*P < 0.05). Scale bar = 5 µm.

significantly, thin spines (Chen et al. 2013; Figure 7.2). CRF antagonists for the type 1 receptor can ameliorate the deficits in learning as a function of these events on dendrites (Blank et al. 2003). The CRF type 1 receptor expression is tied to both remodeling of dendritic branches and the decline in function (Wang et al. 2011).

CRF tends to be over expressed in hippocampal decline, and hippocampal degeneration is tied to both the overexpression of CRF and early life adversity (Ivy et al. 2010). Analysis of both apical and basal dendrites displayed significant differences in dendritic complexity when treated by CRF. Overall, there was a significant decrease in the number of intersections when compared with the control. This is pictured in fluorescent images of CA1 pyramidal neurons expressing YFP, which show a decrease in intersections and overall dendritic complexity when induced by CRF exposure (Ivy et al. 2010; Figure 7.3). But CRF is also important in promoting neurogenesis of and synaptogenesis in hippocampal neurons (Garcia et al. 2014). The question is one of degree and duration: from adaptation to pathology.

We know in several species including macaques that early adverse experiences can result in long-term expression of CRF, as shown by experimentation with nonpredictive events in the laboratory (Coplan et al. 1996). CRF elevation can promote damage that can occur overtime (Chen et al. 2010). It might be the case that the immediate induction of CRF gene expression in hippocampal tissue is an adaptation to new events, to remembered events that over time begin to degrade, facilitated by glucocorticoids neuronal expression (Sapolsky 1992).

CRF is known to influence dendritic differentiation, cell loss, and neurogenesis (Chen et al. 2005a, 2005b; Brunson et al. 2001, 2002; Gibson et al. 2010). Disruption of dendritic spines is tied to NMDA-receptor activation (Andres et al. 2013). Following social defeat there can be a loss of dendrites linked to CRF expression, and this is ameliorated by infusions of CRF-R1-receptor antagonists (Chen et al. 2008), along with other changes in brain function such as increased dopamine release modified by mesolimbic regions tied to CRF regulation (Holly et al. 2015).

The point about CRF is its involvement in diverse forms of behavioral adaptation; the type 1 receptor's widespread functionality is a primary feature after activation of CRF secretion (Chen et al. 2012). CRF-R1 receptor is located at the dendritic spine heads of CA3 pyramidal neuronal dendrites. Electron micrographs show regions of CRF-R1-immunogold particles at the postsynaptic density of an excitatory synapse in stratum oriens of hippocampal area CA3. These experiments localize the CRF-receptor's postsynaptic location to the head of the dendritic spine (Chen et al. 2012).

Finally, amygdala stimulation influences the hippocampus (Roozendaal et al. 2000, 2002; McGaugh 2000), which is linked to both memory and CRF (Cohen et al. 2014; Rosen and Schulkin 1998).

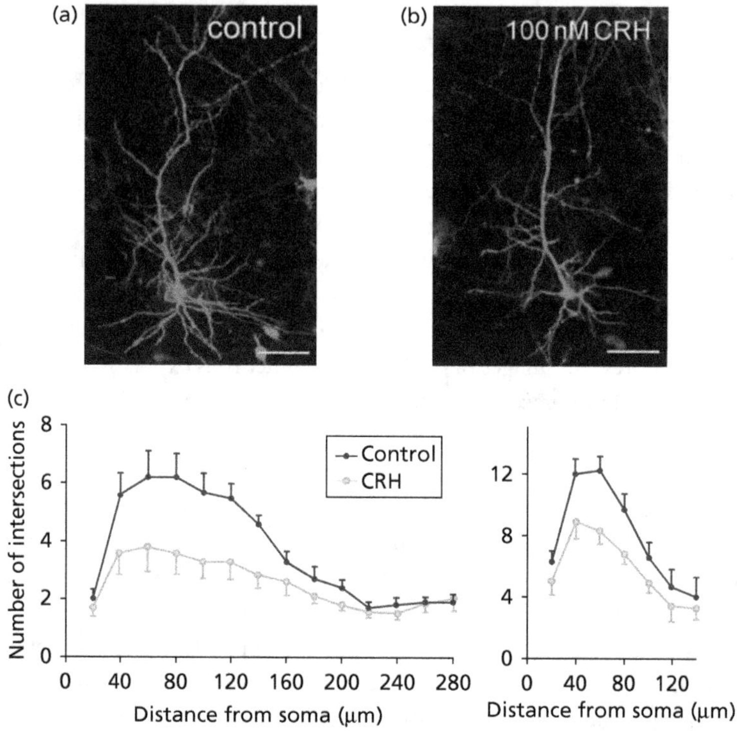


**Figure 7.3** (See color plate section) CRF application onto hippocampal organotypic-slice cultures reduced dendritic complexity. Cultures were prepared from postnatal day 1 mice and grown in the presence of CRF (100nM) for 2 weeks. CA1 pyramidal neurons expressing YFP (n = 10 per group) are shown in a, control condition, and b, in cultures grown with CRF. C, Sholl analysis of both apical dendrites (left panel; $F_{1,16} = 26.99$; $p < 0.0001$) and basal dendrites (right panel; $F_{1,6} = 18.35$; $p < 0.0001$) revealed significant reductions in dendritic complexity induced by CRF exposure.

Reproduced from Autumn S. Ivy, Christopher S. Rex, Yuncai Chen, Céline Dubé, Pamela M. Maras, Dimitri E. Grigoriadis, Christine M. Gall, Gary Lynch, and Tallie Z. Baram, Hippocampal Dysfunction and Cognitive Impairments Provoked by Chronic Early-Life Stress Involve Excessive Activation of CRH Receptors, *The Journal of Neuroscience*, 30 (39), pp. 13005–13015, Figure 2, doi: 10.1523/JNEUROSCI.1784-10.2010 

## Visceral Peripheral Nervous System: CRF and the Blood Brain Barrier

CRF, as has been indicated throughout this book, is very much expressed outside the central nervous system. One function of CRF is signaling adversity; however, does peripherally produced CRF get into the central nervous system? Consider another context in which CRF might affect the brain and perhaps pass into the brain—the gut.

The brain and the gut are linked through the nervous system and through a multitude of gut-brain peptides. Many of these peptides have been shown to be produced in both the gut and brain; some appear to be produced only in the gut, but are transported to brain areas where they bind to receptors. Some of these peptides act primarily on vagal afferents.

For instance, CRF (or urocortin) is expressed in the stomach along with the CRF type 2 receptor; urocortins and CCK act synergistically to delay gastric emptying and reduce food intake in mice resistant to dietary-induced obesity (Gourcerol et al. 2007). CRF, urocortin and both CRF type 1 and type 2 receptors are all expressed in the colon (Tache and Perdue 2004). Exogenous urocortin and CRF delay gastric emptying but enhance colonic motility (Martinez et al. 2002; Sweetser et al. 2009). Interestingly, CRF-receptor 2 deficient mice have a reduced intestinal inflammatory response (Kokkotou et al. 2006; Novembri et al. 2015).

As I have indicated, diverse forms of tissue vulnerability increase CRF expression, for instance, in the colon. RT-PCR shows that CRF is present in both the proximal and distal colon, as well as in the mucosa, submucosa, and muscle layers of the proximal colon layers. Quantitatively, it has been shown that the distal colon has more CRF than the proximal colon, as does the proximal colonic submucosa (plus muscle layers) compared with the proximal colonic mucosa. CRF immunostaining in the proximal and distal colon of a saline-treated rat showed that CRF immunoreactivity was localized in nerve terminals of the myenteric plexus (Yuan et al. 2010).

Thus distressful stimuli would result in digesta being excluded from the intestinal tract (where digesta stimulates blood flow and energy expenditure necessary for digestion), allowing blood flow and energy expenditure to be prioritized to brain and muscle tissue to respond to the perceived threat (Power and Schulkin 2006). Fear really does empty the bowels.

Importantly, CRF is elevated in skin following psoriasis (Loite et al. 2013) and CRF is tied to breast cancer motility and expansion, via CRF-R1 (Androulidaki et al. 2009). Elevated levels of CRF family ligands acting through different receptors have opposite actions on the flow of digesta from the stomach or colon. The result may explain why, under circumstances of fear or distress, gastric emptying is delayed, while colonic emptying is stimulated (Kawahito et al. 1995). The net result is to exclude digesta from the intestinal tract, where it would stimulate blood flow and energy expenditure; thus, blood flow and energy expenditure can be prioritized to brain and muscle tissue to meet the challenge presented to the animal (e.g., a predator). CRF was significantly more present and extensive in the colonic tissues of ulcerative colitis patients than in those of normal patients (Kawahito et al. 1995; Figure 7.4).

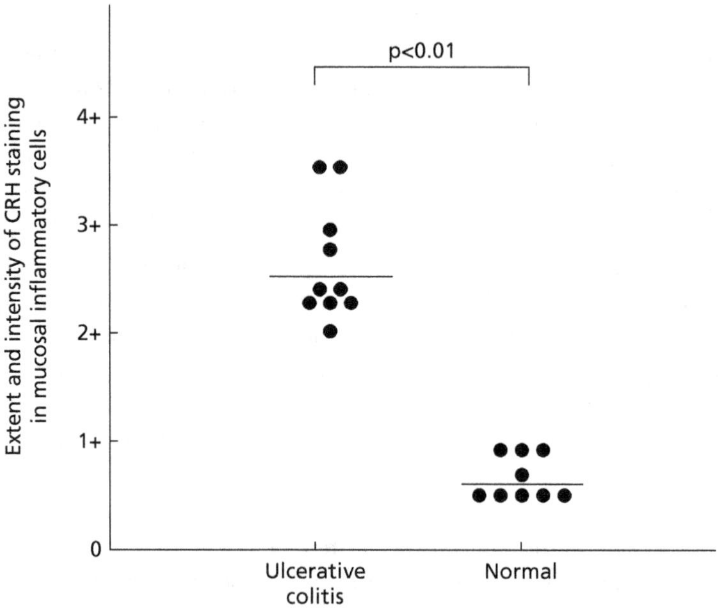


**Figure 7.4** CRF immunostaining of colonic mucosal inflammatory cells from patients with ulcerative colitis. The extent and intensity of immunostaining with anti-CRF antibody in mucosal inflammatory cells of colonic tissues with normal subjects and ulcerative colitis patients were graded 0 to 4 + by two blinded observers. CRF immunostaining was significantly more extensive and intensive in colonic tissues of ulcerative colitis patients (n = 10) than normal colonic tissues (n = 10), by Mann-Whitney U test ($p < 0.01$).

Reproduced from *Gut*, Y Kawahito, H Sano, S Mukai, K Asai, S Kimura, Y Yamamura, H Kato, G P Chrousos, R L Wilder, and M Kondo, 37 (4), pp. 544–551, doi:10.1136/gut.37.4.544, Copyright © 1995, BMJ Publishing Group Ltd and the British Society of Gastroenterology, with permission from BMJ Publishing Group Ltd.

Sepsis induced activation of CRF, and more generally the hypothalamic pituitary adrenal (HPA) activation, is just one of many instances in which insult and danger activates this very general activating system (Carlson et al. 2007; Gonzales et al. 2008).

**Adaptation and adversity:** One issue is whether such events increase, for instance, brain regions that both produce CRF and or have CRF receptors that are impacted by such events. Researchers do not know.

One view is that peptides are not typically accessible to the brain from peripheral sites. Indeed, several regions of the brain are outside the easily detectable entry; they are called circumventricular organs. Several prominent examples include the subfornical organ, organum vasculosum of the lamina terminalis, and the area postrema in the brainstem.

For instance, we know that placenta pumps out CRF, but we do not know if this is getting into the maternal or fetal brain; we do now know that the barrier is less formidable and that diverse peptides can get in; the issue about CRF is unresolved.

One study showed that following intracerebral infusions into the brain (Martins et al. 1997b), CRF was not readily apparent in the blood. The blood brain barrier is thought to be more permeable under conditions of adversity. One prominent way in which it is expressed is during skin disorders; CRF induces increased vascularization (Donelan et al. 2006). Recall that the skin is a rather large repository of information molecules, including putative HPA-related events, as noted previously. Another condition of potentially greater blood brain barrier permeability is during pregnancy, when circulating CRF levels are increased.

Mast cells come in with the regulation of the blood brain barrier particularly under conditions of adversity; the blood brain barrier is more like a regulatory organ than a fixed pillar of defense against outside influences of the brain. An important point, and relevant to CRF as a peptide and other peptides produced in the periphery, is that the blood brain barrier is like a regulatory organ subject to context and biological variability (Banks 2015; Banks and Erickson 2010).

Interestingly, both CRF and mast cells help regulate blood brain barrier permeability under duress (Esposito et al. 2001, 2002; Singh et al. 1999). Indeed mast cells themselves in dermal tissue synthesize CRF and a number of other information molecules (Theoharides and Cochrane 2004; Theoharides et al. 2004). A hypothesized mechanism for the effect of stress on skin mast cells has been suggested by Theoharides et al., as shown in Figure 6.5. CRF and Ucn (as well as UV-activated keratinocytes) stimulate mast cells, with the help of additional neuropeptides. This prompts a secretory response from mast cells, releasing signals (including CRF) that result in inflammation, leukocyte extravasation, and vascular permeability (Theoharides and Cochrane 2004; Figure 7.5). A study of patients suffering with Fibromyalgia suggests that elevated levels of some neuropeptides, like CRF, might stimulate mast cells to secrete inflammatory cytokines, worsening symptoms (Tsilioni et al. 2016).

**Glial cells and CRF:** Glia or microglia are so much more abundant in the brain than neurons are.

Stevens and colleagues have been able to visualize CRF receptors through co-localization, resulting in localization of the receptors in astrocytes and

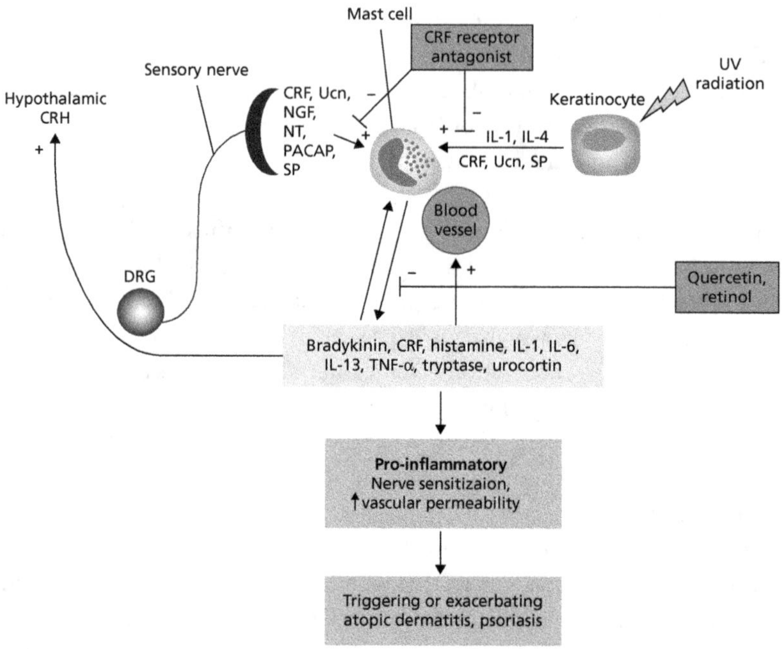


**Figure 7.5** (See color plate section) The hypothesized effect of stress on skin mast cells. Corticotrophin-releasing factor (CRF) and urocortin (Ucn) secreted from dorsal root ganglia (DRG) or UV-activated keratinocytes stimulate mast cells alone or in association with other neuropeptides such as nerve growth factor (NGF), neurotensin (NT), pituitary adenylate cyclase activating polypeptide (PACAP), and substance P (SP). Mast cells then secrete CRF, histamine, interleukin 6 (IL-6), tumor necrosis factor a (TNF-a), tryptase, and Ucn (along with other mediators), which induce vascular permeability, leukocyte extravasation, and inflammation. Histamine and cytokines can stimulate further local CRF and Ucn release by acting on the hypothalamus and on mast cells themselves. In addition, histamine, bradykinin, and tryptase might further activate DRG. Pro-inflammatory mediators could then participate in skin physiology and pathology. CRF-receptor antagonists could interfere with the pathological processes by blocking the action of CRF and related peptides, released from sensory neurons and activated keratinocytes on mast cells. Natural molecules such as quercetin and retinol could inhibit mediator secretion from stimulated mast cells, preventing or limiting local inflammation.

Reprinted from *Trends in Pharmacological Sciences*, 25 (11), Theoharis C. Theoharides, Jill M. Donelan, Nikoletta Papadopoulou, Jing Cao, Duraisamy Kempuraj, and Pio Conti, Mast cells as targets of corticotropin-releasing factor and related peptides, pp. 563–568, doi:10.1016/j.tips.2004.09.007, 

microglia (2003). CRF is one of a number of information molecules (e.g., neuropeptide y) found in microglial cells (Yang et al. 2005; Wang et al. 2004). CRF has both a facilitation and degradation role in the induction of apoptosis in microglia (Ock et al. 2006). CRF affects apoptosis and proliferation, as shown in dexamethasone-treated animals. There was a significant difference between animals treated with CRF and those untreated, after dexamethasone treatment, for both levels of cell proliferation and apoptosis (Koutmani et al. 2013).

While little is still known about glial function, they are linked to restorative functions in response to tissue damage and to inflammatory events. They are also tied to cell proliferation. CRF infusions activate cytokine expression in mouse glial cells (Yang et al. 2005).

CRF is known to facilitate the expression of microglia, which appears to be mediated by CRF type 1 receptors (Wang et al. 2003). In the absence of microglia, the effect of CRF on neuronal apoptosis does not seem to be as effective as it is in the presence of microglia (Wang et al. 2004; Figure 7.6). Indeed cerebral injury is reduced in CRF-receptor-deficient mice (Stevens et al. 2003).

Adversity impacts CRF neurons, which in turn play a role in affecting various output systems, microglia, and cytokine action (Knapp et al. 2011b).

The incredible diversity of function that is created by these information molecules is due in part to these potent molecules having diverse functions in diverse tissues, and tissue- and context-specific regulation and regulatory function. Many of the original molecules have "split" via gene duplication to form families of signaling molecules and pathways. Information molecules have diverse functions in diverse tissues, and their effects depend upon complex interactions with other signaling molecules.

**Effect of adversity and CRH on hippocampus and amygdala:** Again, two critical areas of the brain have long dominated the intellectual landscape; one is a seahorse-shaped structure called the hippocampus, the other is an almond-shaped structure called the amygdala. Both areas are fundamentally affected by adrenal glucocorticoids (corticosterone in rodents, cortisol in primates), but in very different manners.

Sustained and high levels of cortisol under conditions of adversity decrease dendritic spines in regions of the hippocampus (Sapolsky 1992); the regulation is tied to natural rhythmic expression of both progesterone expression and corticosterone levels (Gilhooley et al. 2014; AlAhmed and Herbert 2010). Importantly neurogenesis is promoted by decreasing cortisol and CRF levels and by physical activity and fluoxetine (Pinnock et al. 2012; Pinnock and Herbert 2009).

For example, in rats under conditions of adversity, corticosterone decreases dendritic branching in the hippocampus while expanding it in the amygdala. Glucocorticoids and the hippocampus are associated with negative-feedback mechanisms for the HPA axis (i.e., suppression of function). However, glucocorticoids sustain diverse forms of adaptive behaviors by activation of amygdala

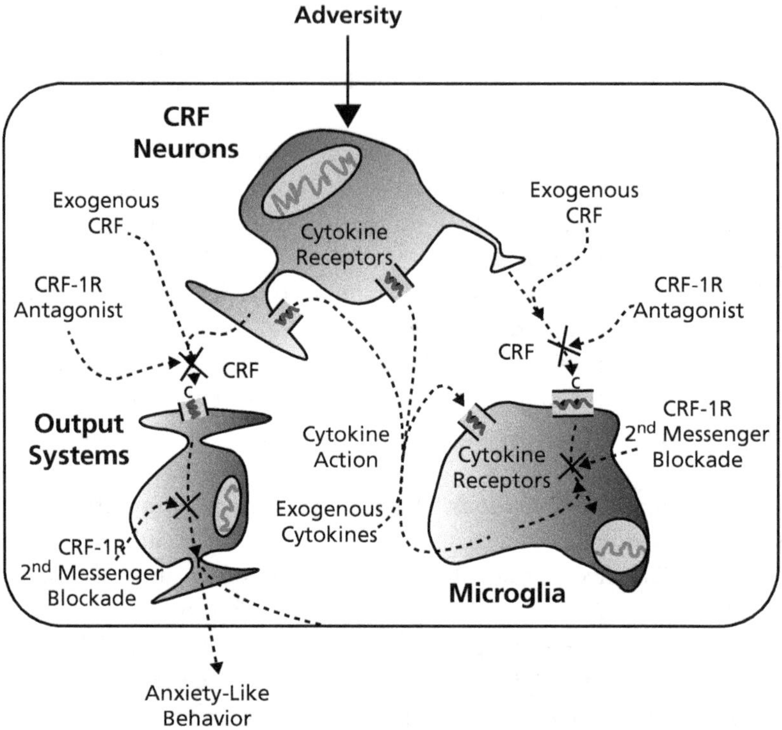


**Figure 7.6** Adversity affects CRF neurons, affecting receptors, cytokines, microglia, and various output systems, resulting in anxiety-like behavior in some scenarios.
Reprinted from *Brain, Behavior, and Immunity*, 25, Darin J. Knapp, Buddy A. Whitman, Tiffany A. Wills, Robert A. Angel, David H. Overstreet, Hugh E. Criswell, Zhen Ming, and George R. Breese, Cytokine involvement in stress may depend on corticotrophin releasing factor to sensitize ethanol withdrawal anxiety, pp. S146–S154, doi:10.1016/j.bbi.2011.02.018, Copyright © 2011 Elsevier Inc., with permission from Elsevier.

function. Thus, while glucocorticoids like cortisol are known to suppress the activation of the HPA axis, there are also regions in which cortisol promotes growth and excitation and not suppression. And both use and exaggerated use with age facilitate decline in the HPA regulation (Sapolsky 1992; Romeo et al. 2007).

CRF is known to influence dendritic differentiation, cell loss, and neurogenesis (Chen et al. 2004, 2005a, 2005b; Brunson et al. 2001; Gibson et al. 2010). Importantly, parental attention decreases distress-related events tied to CRF expression and adverse conditions (Hennessy et al. 2013). Interestingly, dendritic expansion can occur in the amygdala, while dendrites decrease in the hippocampus (Mitra et al. 2005).

These effects may be mediated by glucocorticoids and the induction of CRF and by type 1 CRF receptors. Repeated immobilization stress increased the dendritic

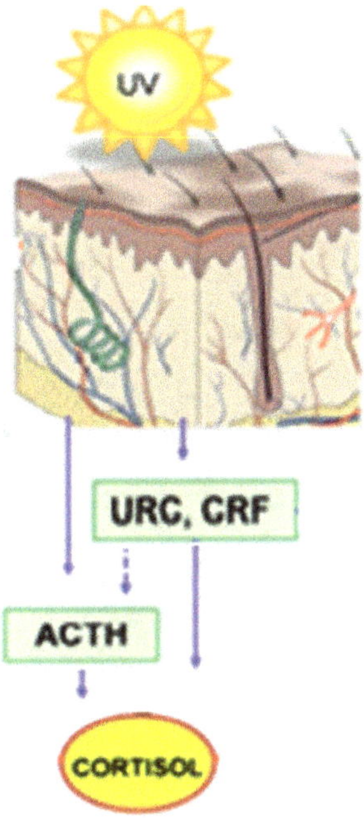


**Figure 2.2** *The* skin stress-response system can activate the central HPA with its direct regulatory, metabolic, and phenotypic consequences.
*We hypothesized that global responses (on the organism level) to UVR initiated in the skin involved simultaneous activation of sensory receptors and local production of humoral messages. The net effect of all of these processes is release of cortisol/corticosterone and induction of steroidogenesis with subsequent metabolic and regulatory effects.*
Reproduced from Andrzej T. Slominski, Michal A. Zmijewski, Cezary Skobowiat, Blazej Zbytek, Radomir M. Slominski, and Jeffery D. Steketee, Sensing the Environment: Regulation of Local and Global Homeostasis by the Skin's Neuroendocrine System, *Advances in Anatomy, Embryology and Cell Biology*, Volume 212, pp. 1–115, Figure 17, Copyright © 2012, Springer-Verlag Berlin Heidelberg.

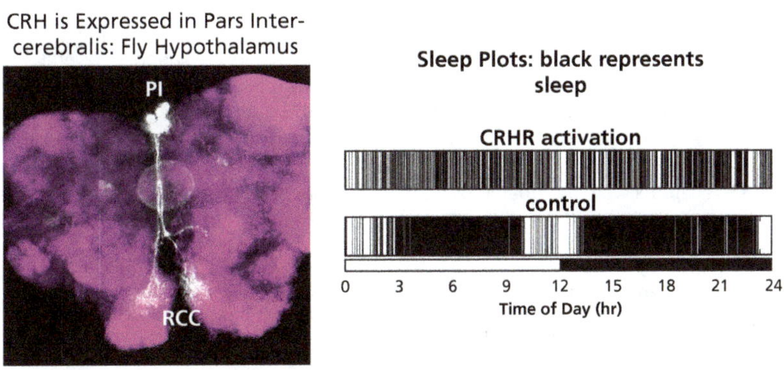


**Figure 2.4** (Left) Corticotrophin-releasing hormone (CRH) peptide is secreted by large neurons of the fly pars intercerebralis, a brain region homologous to the vertebrate hypothalamus. (Right) Activation of central nervous system (CNS) CRH receptors using a genetically targeted membrane-tethered CRH ligand strongly suppresses and fragments sleep.

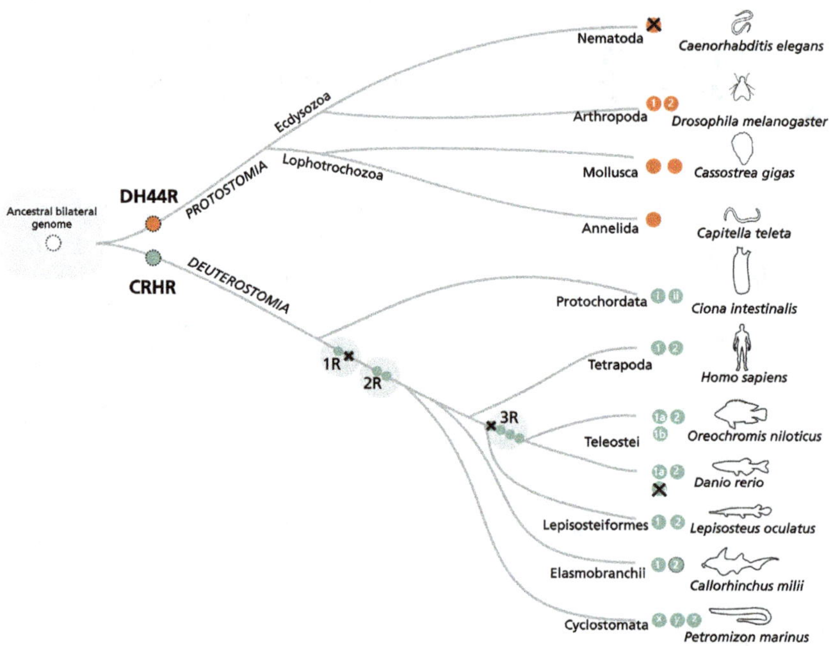


**Figure 2.5** Outline of the evolution for the CRF/DH44 receptors in relation to several important evolutionary lineage divergences. The deuterostome CRF receptors and protostome DH44 receptors are represented by green and orange filled circles, respectively. The genome doubling events in the vertebrates are marked as 1R, 2R, and 3R, respectively. Lost genes are marked with black crosses. To facilitate interpretation, the Ciona duplicate CRFR-like receptors were designated by i and ii, the duplicate receptors in lamprey by x, y, and z, and the teleost CRFR1 duplicates by a and b. The timeline is not drawn to scale having in consideration the species evolutionary events.

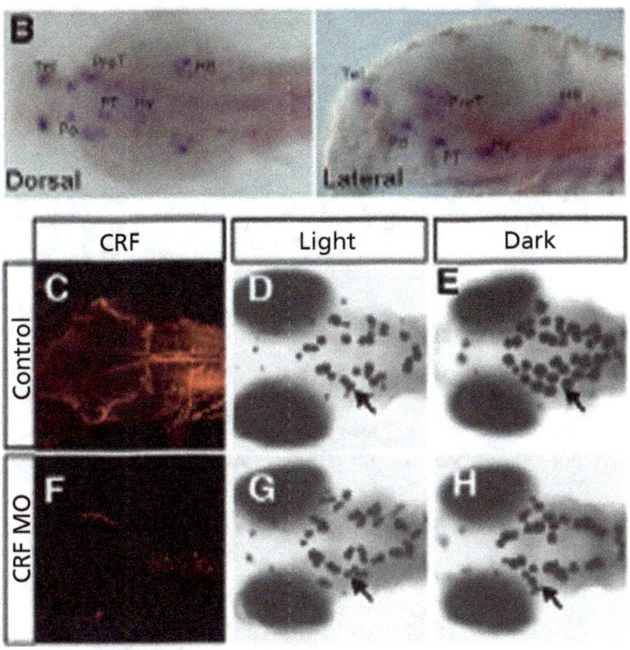


**Figure 3.3** Corticotrophin-releasing factor (CRF) is necessary for the camouflage behavior. (b) Dorsal and lateral views of 5dpf zebrafish larvae showing CRF expression in the telencephalon (Tel), pretectum (PreT), preoptic (Po), posterior tubercle (PT), hypothalamus (H), and hindbrain (HB). (c-h) The CRF immunostaining is detected in axonal processes of 5-day old control morphant (c) and is greatly reduced in the crf morphant (F). Images of 5 dpf control (d, e) or the crf morphant (g, h) exposed to light (d, g) or darkness (e, h). Adapted from Mahendra Wagle, Priya Mathur, and Su Guo, Corticotropin-Releasing Factor Critical for Zebrafish Camouflage Behavior Is Regulated by Light and Sensitive to Ethanol, *The Journal of Neuroscience*, 31 (1), pp. 214–224, Figure 1, doi: 10.1523/JNEUROSCI.3339-10.2011, Copyright © 2011 by The Society for Neuroscience.

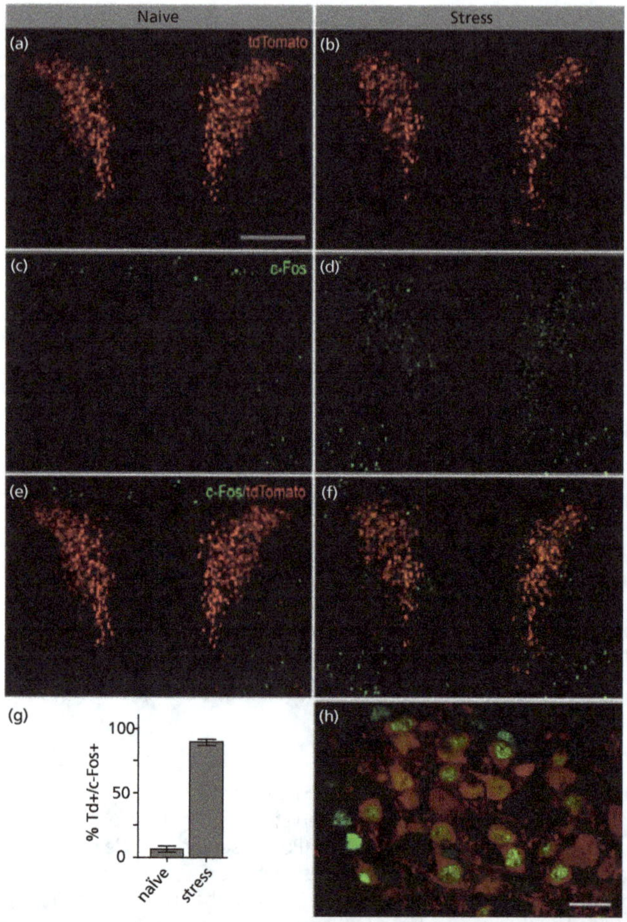


**Figure 3.6** Induction of cFos protein in PVN tdTomato neurons following stress. (a and b) tdTomato expression in PVN of naïve CRF-IRES-Cre; Ai14, and one exposed to a 15 min forced swim stress, 90 min before sacrifice. Confocal image at 15x. (c and d) cFos immunoreactivity in these naïve or stress brain sections, showing an increased number of cFos containing nuclei following stress. (e and f) Merged image showing high neurons co-expressing cFos in naïve (n = 5) versus stressed (n = 6) mice. (h) Higher magnification (100x) of cFos immunoreactivity in tdTomato cells from a stressed mouse. Data summarized in (g) are mean + / - SEM. Scale bars are 200 um (a–f) and 20 um (h).

Reproduced from Wamsteeker Cusulin JI, Füzesi T, Watts AG, Bains JS, Characterization of Corticotropin-Releasing Hormone neurons in the Paraventricular Nucleus of the Hypothalamus of *Crh-IRES-Cre* Mutant Mice, *PLoS ONE*, 8(5): e64943. doi:10.1371/journal.pone.0064943, © 2013, Wamsteeker Cusulin et al.

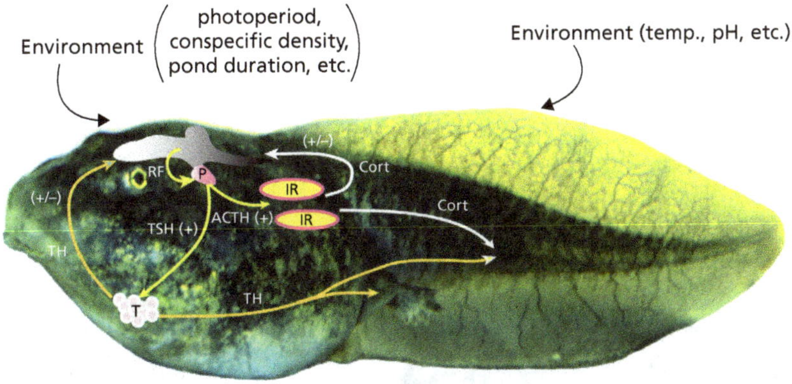


**Figure 4.2** Endocrine systems controlling tadpole metamorphosis. P, pituitary gland; RF, releasing factor (CRF regulates both TSH and ACTH secretion in tadpoles); IR, interrenal gland; ACTH, adrenocorticotropic hormone; TSH, thyroid-stimulating hormone; T, thyroid gland; TH, thyroid hormone; CORT, corticoids. Plus signs indicate a stimulatory effect and minus signs a negative feedback. In the case of TH and Cort effects on the brain, (+/–) indicates that these hormones promote differentiation of neurosecretory centers (and other brain regions) in addition to their negative feedback effects on neurohormone and pituitary hormone secretion.



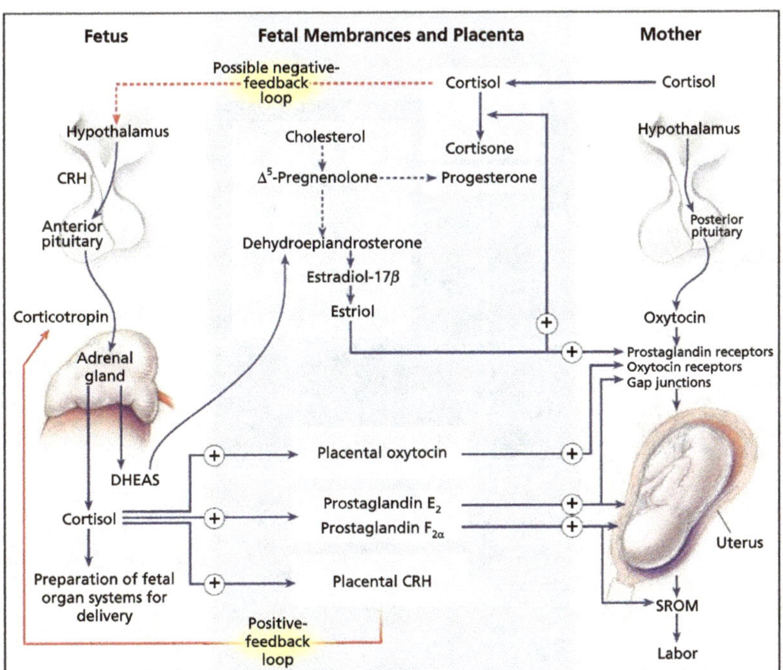


**Figure 4.6** The maternal and fetal HPA axes are interconnected via the placenta, fetal membranes, and circulating information molecules. Some of the key events involved in gestation include progesterone, estrogen, cortisol, oxytocin, and CRF.

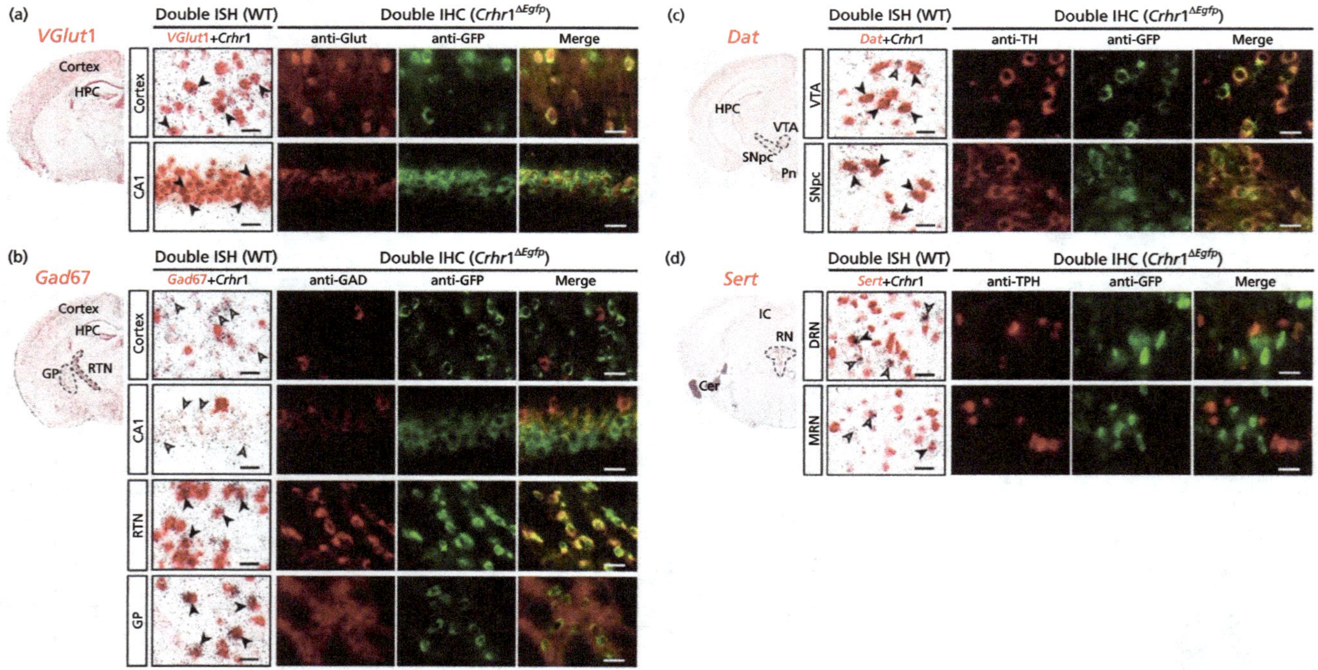


**Figure 5.1** CRFR1 is expressed in neurons of different neurotransmitter systems. CRFR1 was co-localized with neurotransmitter-specific markers by double ISH and double immunohistochemistry (IHC), using wild-type and CRFr1DEGFP mice, respectively. (a) In the cortex and hippocampus (HPC), CRFR1 is expressed in glutamatergic (VGlut1, glutamate) neurons. (b) CRFR1 is not expressed in GABAergic (GAD67) neurons of the cortex or of the hippocampus but in GABAergic neurons of the RTN and the GP. (c) CRFR1 is expressed in dopaminergic (Dat, TH) neurons of the ventral tegmental area (VTA) and the SNpc. (d) CRFR1 is scarcely expressed in serotonergic (Sert, TPH) neurons of the dorsal (DRN) and median raphe (MRN). Gray arrowheads indicate cells expressing only CRFR1 (silver grains). Black arrowheads indicate cells co-expressing CRFR1 and the respective markers (red staining).

Reproduced from Damian Refojo, Martin Schweizer, Claudia Kuehne, Stefanie Ehrenberg, Christoph Thoeringer, Annette M. Vogl, Nina Dedic, Marion Schumacher, Gregor von Wolff, Charilaos Avrabos, Chadi Touma, David Engblom, Günther Schütz, Klaus-Armin Nave, Matthias Eder, Carsten T. Wotjak, Inge Sillaber, Florian Holsboer, Wolfgang Wurst, and Jan M. Deussing, Glutamatergic and Dopaminergic Neurons Mediate Anxiogenic and Anxiolytic Effects of CRHR1, *Science*, 333 (6051), DOI: 10.1126/science.1202107, Copyright © 2011, American Association for the Advancement of Science. Reprinted with permission from AAAS.

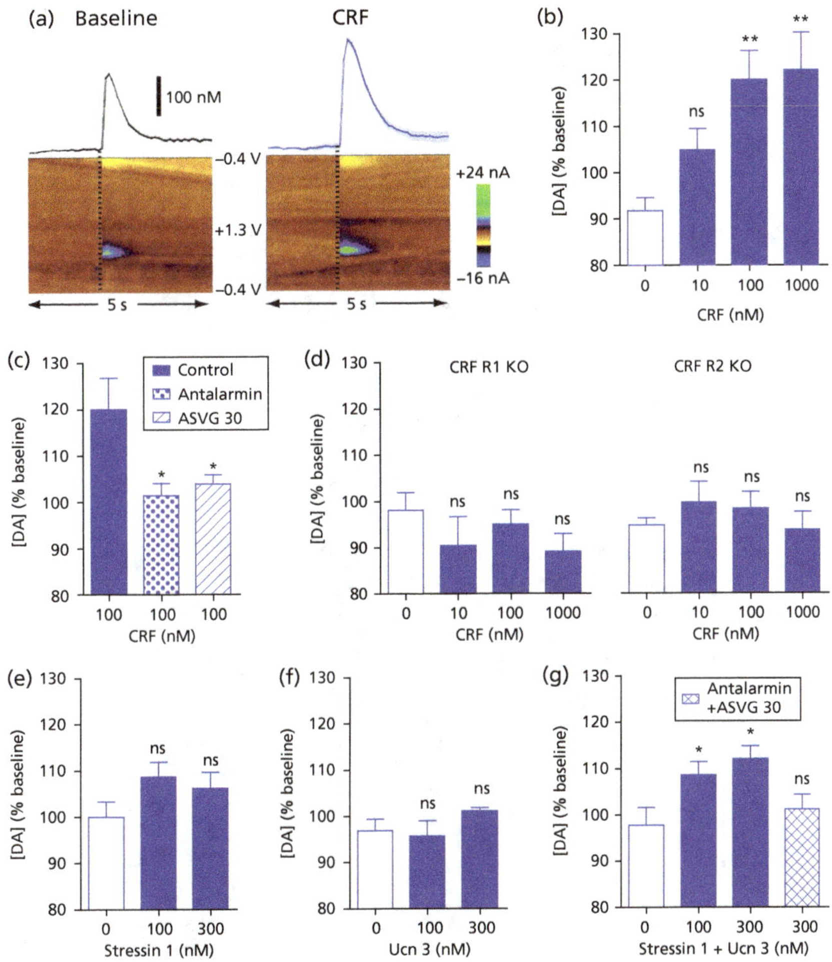


**Figure 5.2** CRF increases dopamine release in the nucleus accumbens through coactivation of CRFR1 and CRFR2. (a) Representative dopamine release evoked by electrical stimulation (dashed lines) before (left) and after (right) application of 100 nM CRF (mean ± s.e.m. for 5 consecutive stimulations, top) and corresponding two-dimensional plots depicting changes in faradaic current (pseudocolor) with time as the abscissa and applied potential as the ordinate (bottom). (b) Concentration response to CRF, n = 11–18. (c) Effect of antagonists for CRFR1 (antalarmin, 1000 nM) or CRFR2 (anti-sauvagine 30 (ASVG 30), 250 nM), n = 18–20. (d) CRF in mice lacking the gene encoding the CRFR1 (left) or CRFR2 (right) receptors, n = 7–13. (e–g) Effect of the CRFR1 agonist, stressin 1, n = 9–15 (e), the CRFR2 agonist urocortin 3 (100 or 300 nM), n = 5–8 (f) or their co-application, n = 8–15 (g). Error bars, s.e.m. DA, dopamine; NS, not significant (with $P > 0.05$); *$P < 0.05$; **$P < 0.01$ versus vehicle.

**Figure 6.1** There are examples of different parental care strategies across vertebrates (depicted here). Male uniparental care is lacking only in the mammalian and reptilian lineages, although there are male-biased parental care systems in some canids and primates. (a) Common deer mouse (Peromyscus maniculatus). (b) Oldfield mouse (Peromyscus polionotus). (c) Kori bustard (Ardeotis kori). (d) Adélie penguin (Pygoscelis adeliae). (e) Pheasant-tailed jacana (Hydrophasianus chirurgus). (f) Water python (Liasis fuscus). (g) Black rock skink (Egernia saxatilis). (h) Diablito frog (Oophaga sylvatica). (i) Mimic poison frog (Ranitomeya imitator). (j) Dyeing poison frog (Dendrobates tinctorius). (k) Burton's mouthbrooder (Astatotilapia burtoni). (l) Convict cichlid (Amatitlania nigrofasciata). (m) Three-spined stickleback (Gasterosteus aculeatus). (n) Golden brown stink bug (Anchises parvulus). (o) Burying beetle (Nicrophorus vespilloides). (p) Giant water bug (Abedus herberti).
From Catherine Dulac, Lauren A. O'Connell, and Zheng Wu, Neural control of maternal and paternal behaviors, *Science*, 345 (6198), pp. 765–770, DOI: 10.1126/science.1253291 

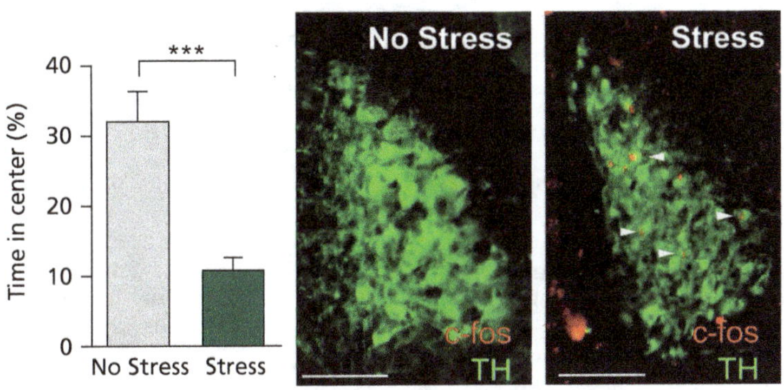


**Figure 6.6** Stressed animals spend significantly less time exploring the center of the open-field test than nonstressed control (left). Representative immunohistochemistry (IHC) and quantification of CRH expression (right). TH, tyrosine hydroxylase.

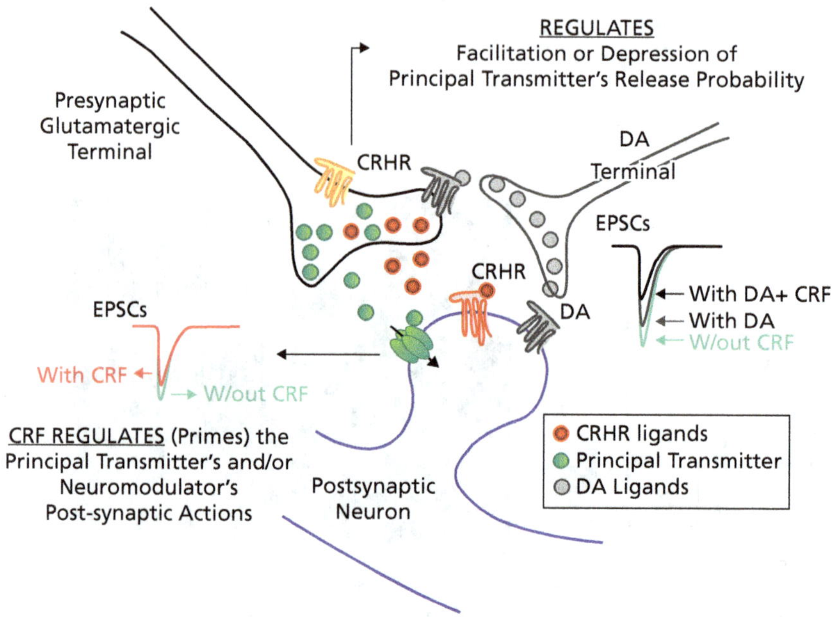


**Figure 7.1** Roles for CRF, CRF-receptor (CRFR) ligands as neuroregulators ("primers") within CNS synapses tonic role of endogenous CRF as demonstrated following application of a CRF-receptor antagonist results in an enhancement or depression of the primary transmitter's action (e.g., upon glutamate transmission at the central amygdala nucleus or lateral septum mediolateral nucleus synapses); OR, phasic role of CRF acting either by evoked or volume transmission to modulate the action of a principal transmitter (e.g., glutamate; or, a modulator, e.g., enhance dopamine's affects on basolateral amygdala to medial prefrontal cortex glutamatergic synaptic transmission).

Reprinted from *European Journal of Pharmacology*, 583 (2–3), Joel P. Gallagher, Luis F. Orozco-Cabal, Jie Liu, and Patricia Shinnick-Gallagher, Synaptic physiology of central CRH system, pp. 215–225, doi:10.1016/j.ejphar.2007.11.075, Copyright © 2008 Elsevier B.V. with permission from Elsevier.

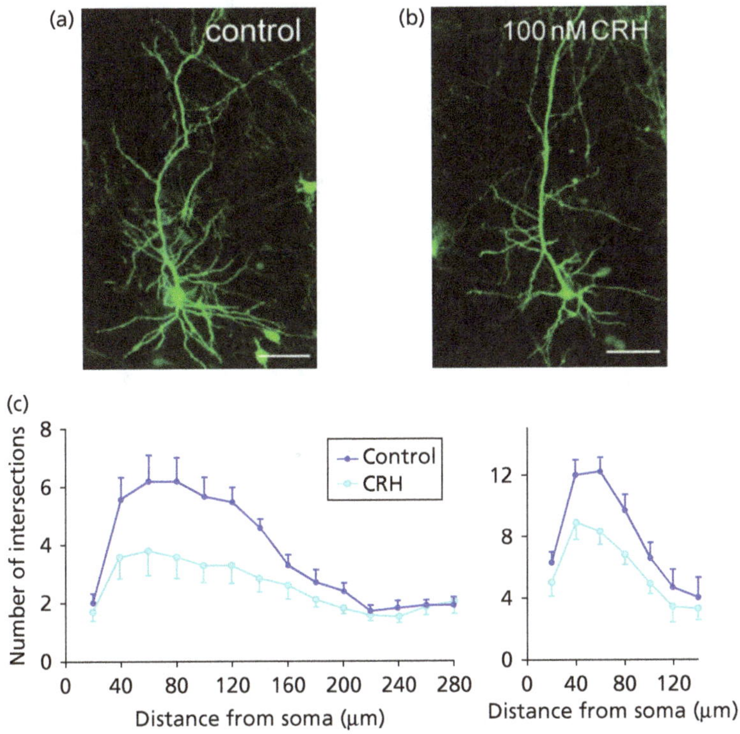


**Figure 7.3** CRF application onto hippocampal organotypic-slice cultures reduced dendritic complexity. Cultures were prepared from postnatal day 1 mice and grown in the presence of CRF (100nM) for 2 weeks. CA1 pyramidal neurons expressing YFP (n = 10 per group) are shown in a, control condition, and b, in cultures grown with CRF. C, Sholl analysis of both apical dendrites (left panel; $F_{1,16} = 26.99$; $p < 0.0001$) and basal dendrites (right panel; $F_{1,6} = 18.35$; $p < 0.0001$) revealed significant reductions in dendritic complexity induced by CRF exposure.

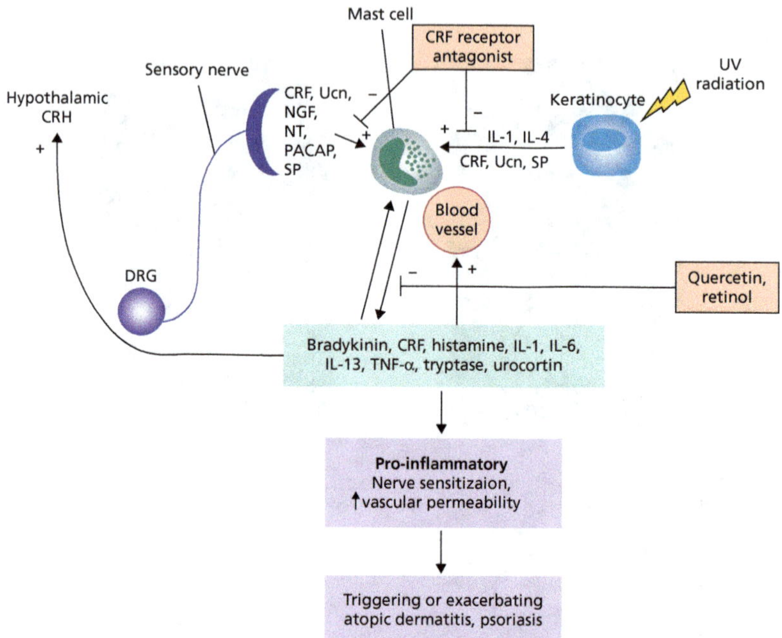


**Figure 7.5** The hypothesized effect of stress on skin mast cells. Corticotrophin-releasing factor (CRF) and urocortin (Ucn) secreted from dorsal root ganglia (DRG) or UV-activated keratinocytes stimulate mast cells alone or in association with other neuropeptides such as nerve growth factor (NGF), neurotensin (NT), pituitary adenylate cyclase activating polypeptide (PACAP), and substance P (SP). Mast cells then secrete CRF, histamine, interleukin 6 (IL-6), tumor necrosis factor a (TNF-a), tryptase, and Ucn (along with other mediators), which induce vascular permeability, leukocyte extravasation, and inflammation. Histamine and cytokines can stimulate further local CRF and Ucn release by acting on the hypothalamus and on mast cells themselves. In addition, histamine, bradykinin, and tryptase might further activate DRG. Pro-inflammatory mediators could then participate in skin physiology and pathology. CRF-receptor antagonists could interfere with the pathological processes by blocking the action of CRF and related peptides, released from sensory neurons and activated keratinocytes on mast cells. Natural molecules such as quercetin and retinol could inhibit mediator secretion from stimulated mast cells, preventing or limiting local inflammation.

Reprinted from *Trends in Pharmacological Sciences*, 25 (11), Theoharis C. Theoharides, Jill M. Donelan, Nikoletta Papadopoulou, Jing Cao, Duraisamy Kempuraj, and Pio Conti, Mast cells as targets of corticotropin-releasing factor and related peptides, pp. 563–568, doi:10.1016/j.tips.2004.09.007, 

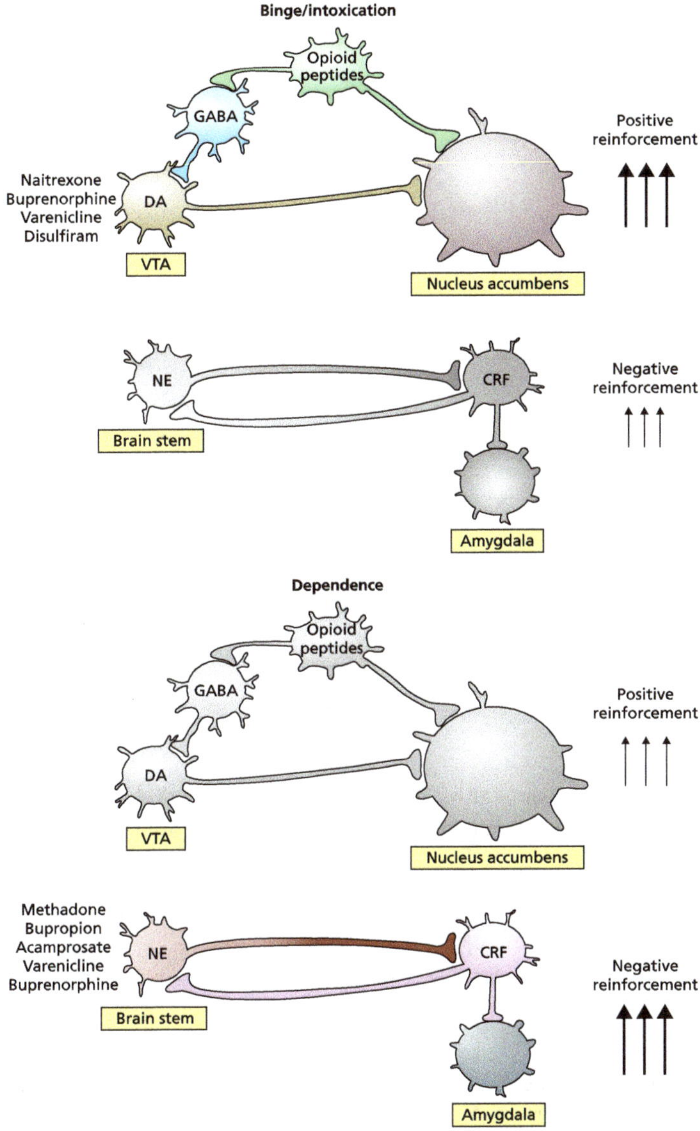


**Figure 8.3** Neurocircuitry associated with the acute positive reinforcing effects of drugs of abuse and the negative reinforcement of dependence, and changes in the transition from nondependent to dependent drug taking. (a) Key elements of the reward circuit include dopamine and opioid peptide neurons that intersect at both the VTA (ventral tegmental area) and the nucleus accumbens that are activated during initial drug use and the early binge/intoxication stage of the addiction cycle. Drugs approved for the treatment of this component of the addiction cycle include naltrexone, buprenorphine, varenicline, and disulfiram. (b) Key elements of the stress circuit include CRF (corticotrophin-releasing factor) and noradrenergic neurons that converge on GABA interneurons in the central nucleus of the amygdala that are activated during the development. Drugs approved for the treatment of this component of the addiction cycle include methadone, bupropion, acamprosate, varenicline, and buprenorphine. DA, dopamine; NE, norepinephrine.

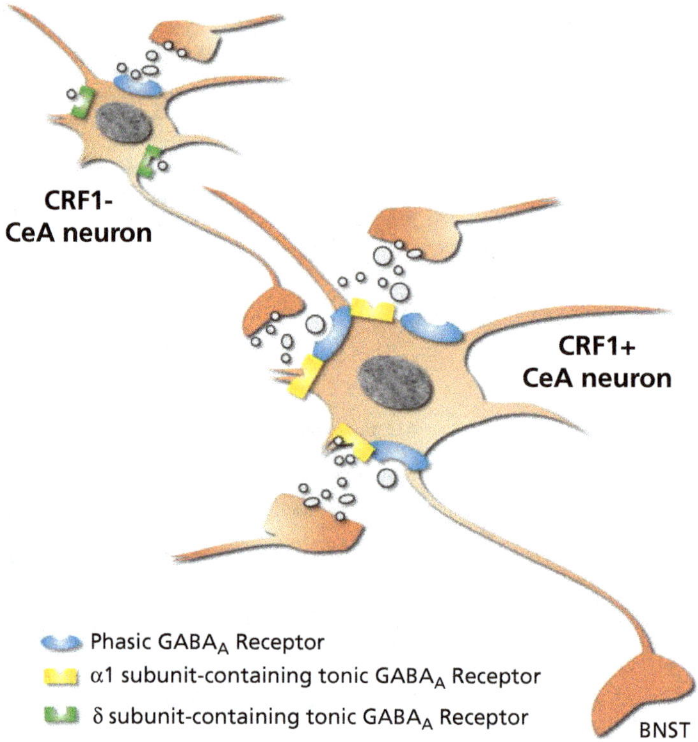


**Figure 8.5** Proposed microcircuitry of the CEA illustrating local phasic and tonic GABA$_A$ receptor transmission in CRF1 + and CRF1− neurons. Phasic GABA$_A$ receptor transmission is depicted on both CRF1 + and CRF1− neurons, although CRF1 + neurons are depicted with more baseline GABA activity. The CRF1 + neuron possesses α1 subunit-containing tonic GABA$_A$ receptors located in or near the synaptic cleft to reflect the dependence on action potential-dependent GABA release and projects out of the CEA into the BNST. The CRF1− neuron possesses δ subunit-containing GABA$_A$ receptors located outside the synaptic cleft to reflect activation by ambient GABA. We hypothesize that ethanol acts on this CEA circuitry by enhancing tonic conductance in CRF1− neurons through actions at δ-containing GABA$_A$ receptors, resulting in decreased firing of this subpopulation. The ongoing tonic inhibition of CRF1 + neurons is subsequently diminished, resulting in disinhibition and increased firing in this subpopulation. This increased firing then results in increased GABA release in target brain regions such as the BNST.

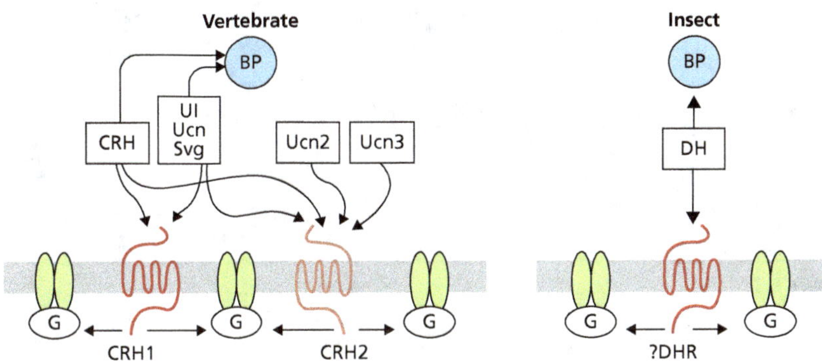


**Figure 9.1** Scheme of the molecular interaction among CRF ligands, receptors, and the binding protein in chordates and insects.

Reproduced from David A. Lovejoy, Belinda SW Chang, Nathan R. Lovejoy, and Jon del Castillo, Molecular Evolution of GPCRS: CRH/CRH Receptors, *Journal of Molecular Endocrinology*, 52 (3), pp. T43–T60, Figure 2, doi: 10.1530/JME-13-0238, Copyright © 2014, Society for Endocrinology.

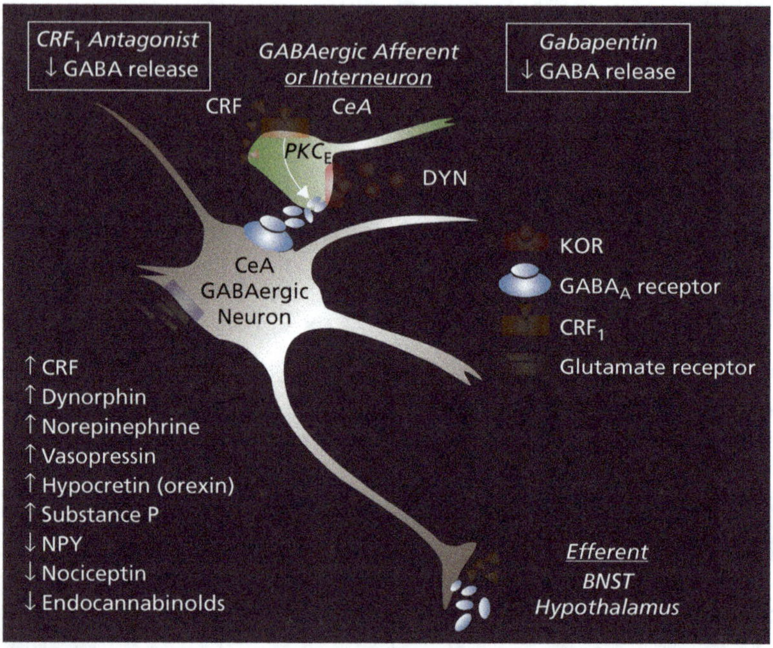


**Figure 9.3** Cellular neuroadaptive mechanisms in the central nucleus of the amygdala in drug addiction. Simplified schematic of rodent central nucleus of the amygdala circuitry and hypothetical sites of ethanol and CRF action on GABAergic synapses. Most neurons in the CEA are GABAergic inhibitory projection neurons or interneurons that contain CRF or other neuropeptides as co-transmitters. (Upper synapse) Ethanol may enhance the release of GABA (filled ellipsoids) from GABAergic afferents or interneurons either via release from the same terminal as CRF (triangles), which then acts on CRF1 receptors on the terminal to elicit (arrow) release of more GABA via a PKCε-mediated mechanism, or direct activation of CRF1 receptors to elicit the release of more GABA (Bajo et al. 2008). CRF1 antagonists and the drug gabapentin decrease presynaptic GABA release in dependent animals (Roberto et al. 2008, 2010). κ-Opioid antagonists have similar effects as CRF1 antagonists in rats that present an escalation in cocaine intake (Kallupi et al. 2013). Thus, CRF, dynorphin, and ethanol augment the inhibition of CEA projection interneurons (co-containing CRF, opioids, or NPY), leading to the excitation of downstream (e.g., BNST) neurons through disinhibition. The activation of presynaptic cannabinoid CB1 or NPY receptors (not shown) may reduce GABA release onto CEA inhibitory projection neurons, increasing their excitability and release of GABA onto downstream targets, such as in the BNST. CRF, corticotrophin-releasing factor; GABA, γ-aminobutyric acid; CEA, central nucleus of the amygdala; DYN, dynorphi; KOR, κ opioid receptor; NPY, neuropeptide Y; BNST, bed nucleus of the stria terminalis.

length and arborization of pyramidal and stellate cells of basolateral amygdala neurons (Vyas et al. 2002; Figure 7.7). This finding is even more remarkable because hippocampal pyramidal neurons in the same rats had decreased dendritic length and branching. The immobilized rats had adrenal hypertrophy suggesting that these rats had increased synthesis and release of glucocorticoids.

The increased glucocorticoid production results in changes in neuronal morphology in the basolateral complex of the amygdala following repeated stress (Vyas et al. 2002; Mitra et al. 2005). The elevated glucocorticoids could then facilitate changes in CRF expression within the amygdala (Matys et al. 2004; Chattarji et al. 2015). Upregulation of tissue plasminogen activator (tPA) activity in the central and medial amygdala is facilitated by CRF and is inhibited by a CRF-R1 antagonist, antalarmin, but not a CRF-R2 antagonist, antisauvagine (Matys et al. 2004).

Interestingly, a single injection of corticosterone was sufficient to induce dendritic expansion in the lateral amygdala (after 1 day; Mitra and Sapolsky 2008). The growth in dendritic branching would facilitate processing of information in the amygdala and possibly increase vigilance and interpretation of stimuli as being unfamiliar, discrepant, or threatening (Rosen and Schulkin 1998).

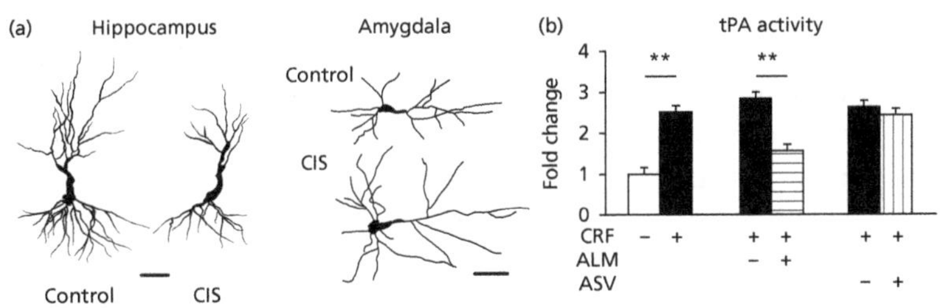


**Figure 7.7** (a) Dendritic reduction of the hippocampus in animals and expansion of the amygdala during duress. (b) CRF increases tissue plasminogen activator (tPA) activity in the amygdala in brain slices by means of the CRF-R1 receptor. CRF caused a significant upregulation of tPA in the central and medial amygdala, in comparison with the slice perfused with RS only. This upregulation was inhibited by the CRF-R1 antagonist, antalarmin (ALM), but not the CRF-R2 antagonist antisauvagine 30 (ASV). **, $P < 0.05$.

(a) Reproduced from Ajai Vyas, Rupshi Mitra, B. S. Shankaranarayana Rao, and Sumantra Chattarji, Chronic Stress Induces Contrasting Patterns of Dendritic Remodeling in Hippocampal and Amygdaloid Neurons, *The Journal of Neuroscience*, 22 (15), pp. 6810–6818, Figure 1b and 2b, Copyright © 2002 Society for Neuroscience.

(b) Reproduced from Tomasz Matys, Robert Pawlak, Elzbieta Matys, Constantine Pavlides, Bruce McEwen, and Sidney Strickland, Tissue plasminogen activator promotes the effects of corticotropin-releasing factor on the amygdala and anxiety-like behavior, *Proceedings of the National Academy of Sciences of the United States of America*, 101 (46), pp. 16345–16350, Figure 2e, DOI: 10.1073/pnas.0407355101, Copyright © 2004, The National Academy of Sciences.

Indeed, differential effects of decreases in hippocampus and increases in amygdala are fairly well established in animal models (Leuner and Shors 2013; Chen et al. 2014) and may reflect affective disorders; the medial prefrontal cortex is also affected (Chattarji et al. 2015). High levels of glucocorticoids are known to produce morphological changes in several regions of the brain; they typically decrease hippocampal and prefrontal dendritic branching (e.g., Wellman 2001; Radley et al. 2006a, b; Gould et al. 2001; Wang et al. 2011). Glucocorticoids can exacerbate neural deterioration (Sapolsky 1992) and decrease neurogenesis (e.g., Gould et al. 2001; Wong and Herbert 2005).

## Conclusion

CRF is tied to diverse forms of adaptive responses signaling danger and is expressed in diverse peripheral and central nervous system tissue. For instance, skin is a repository of information sensors of which, along with neurotensin, participate with regard to vascular permeability and inflammatory events (Donelan et al. 2006). Neurotensin, like CRF, is affected in brain tissue by corticosteroids (Watts and Sanchez-Watts 1995). Activation of tissue may affect blood brain barrier permeability and CRF access to central tissue, perhaps exacerbating tissue vulnerability.

Diverse informational contexts affect CRF expression that underlies responding and then coping with adversity in different end organ systems (Haramati et al. 2011; Gunduz-Cinar et al. 2013). CRF can facilitate the degradation of dendritic spines (Chen et al. 2013), but can also facilitate long-term potentiation, which is important for learning (Blank et al. 2002), regardless of its impact on synaptogenesis (Garcia et al. 2014).

Most tissues seem to contain CRF and can be induced into expressing CRF. Diverse mechanisms underlie the adaptive responses such as in the instance of inflammation (e.g., cannabinoid receptors; Marchalant et al. 2007; Bardou et al. 2012; Venihaki et al 2001; Webster et al. 1998; Wei et al. 2002), where tissue integrity can be influenced by the regulation of CRF and perhaps microglial cells and blood brain barrier permeability.

Fast-acting endocannabinoid mechanisms underlie CRF (Hill et al. 2005). Cell-specific manipulations of CRF neurons and CRF indeed enhanced anxious or fear-related behaviors and related memories (Gafford et al. 2012). Localization of diverse neurotransmitters or receptors underlie the utilization of CRF in diverse functional contexts (e.g., GABA and NMDA receptors; Delaney and Sah 2001; Partridge et al. 2014; Gafford and Ressler 2015) that include fear or vulnerability to tissue damage.

It is in this context of duress that systems like the blood brain barrier are vulnerable for peripheral peptide production and entry into the brain in greater amounts than normal; but we do not really know that yet. We do know that CRF underlies behavioral, genomic, and broadly conceived endocrine responses in coping with adversity (Campbell et al. 2004).

One must take into account a broadening array of events including re-envisioning the blood brain barrier, glia contribution, and an expansion of traditional motor areas of the brain that have a much larger contribution to the organization of behavior and CRF, such as in the case of the globus pallidus (Sztainberg et al. 2010). It is important to understand the role of CRF in diverse end organ systems and different regions of the brain including the globus pallidus. For example, in mice exposed to stress and innervated by CRF terminals, a reduction in CRF-R1 mRNA expression in the external globus pallidus (GPe) was shown through in situ hybridization, immunostaining, and RT-PCR. And cells co-expressing parvalbumin and CRF-R1 have been found in the external globus pallidus (Sztainberg et al. 2010).

The next chapter looks at the role of CRF in magnifying rewards to addictive behaviors: from adaptation to pathology.

Chapter 8

# Cravings and Addictions: The Contributions of CRF to Approach, Ingestion, and Withdrawal

## Introduction

The allure of afflictions and appetites gone awry are endless in the modern era. They range from the endless junk food we eat, to the computer games that lock our children to distraction, compulsion, and fixation on a screen. A sense of compulsion pervades addiction (Koob 2009). For both appetite and addiction, incentives are mediated by diverse information molecules, which include CRF and dopamine (Berridge et al. 2004; Peciña et al. 2006).

Take smoking: both CRF and dopamine are tied to smoking behaviors such as incentives, ingestion, and withdrawal (Koob 2013). Indeed, both CRF and central dopamine are elevated before and following ingestion or the anticipation of ingestion. But CRF is also tied to the withdrawal of events (Koob and LeMoal 2005). CRF type 1 receptors are tied both to vulnerability to affective distress and to self-medication (Ducottet et al. 2003). Enhanced vulnerability for both substance abuse and affective disorders may therefore reflect aberration in CRF type 1 activity (Contoreggi et al. 2003; Hsu et al. 2012; Pintor et al. 2013). We will see that the whole array of addictive behaviors is knotted to CRF.

Cigarettes, alcohol, and other addictive substances are associated with a quarter of all deaths in the United States (McGinnis and Foege 1999). Smoking can be traced back as early as the fourteenth century, as shown in images of Native Americans smoking from long pipes. Substance use also accounts for a large amount of economic burden and social costs to families and communities, including crime, birth defects, family violence, and divorce. The answers to addictive behaviors are not simple. Addictive objects vary considerably; some have properties in common, others do not. Consider the range of objects to which one could become addicted or obsessed with (See Table 8.1). These objects range from powerful drugs to everyday common behaviors and items.

**Table 8.1** List of Addictive Events

| |
|---|
| Heroin |
| Cocaine |
| Pain relievers |
| Sedatives |
| Stimulants (e.g., amphetamine and caffeine) |
| Hallucinogens |
| Alcohol and cigarettes |
| Gambling |
| Gaming |
| Sexual behaviors |
| Food |
| Work |
| Money |

CRF is involved in many features and behaviors that surround addiction, such as withdrawal and reward. One view of addiction, and my own, is that addiction is the dominance of one thing at the expense of all else, it totally consumes attention. Of course, these fixations may occur under healthy circumstances, such as in necessary food intake, the quenching of thirst, infant care, or other motivations under diverse conditions of survival. It is the maladaptive aspect of addiction that is also essential in its definition. Addiction is commonly broken down into three stages: preoccupation/anticipation, binge/intoxication, and withdrawal/negative affect. There is a component of drug addiction, a permeating mood, that exists both between and partly during the withdrawal and preoccupation stages: psychic pain (i.e., the dark side of addiction; Koob 2010). CRF plays a major role in contributing to the common elements of the psychic pain of addiction, such as malaise, dysphoria, and anxiety. This psychic pain, which permeates the stages of addiction in which one is not actively using their drug of choice further promotes addictive behavior.

Addiction is a good example of a central motive state (Koob 1998). Motivational states modulate emotional responses, inherently involve information-processing (Lang et al. 1998) and underlie our behavioral responses (Lashley 1938; Stellar 1954). Of course, the range of things for which one can form an obsession or addiction is endless (see Elster and Skog 1999). With this great array of candidates, social context matters. Let us start with a discussion of our attraction to food and then return to addictions for psychotropic drug use.

**Palatable food versus rat chow:** CRF magnifies the food ingestive behavior under conditions of intermittent access to food resources (Iemolo et al. 2013). High levels of CRF tend to inhibit food intake in diverse contexts (Stengel et al. 2009; Richardson et al. 2000; Grill et al. 2000 2002). Food-deprivation experiments in mice increased CRF mRNA expression in the hypothalamus but decreased immunofluorescent staining in the paraventricular nucleus, suggesting differential regulation of transcription and translation of CRF (Yadawa and Chaturvedi 2016).

Importantly, CRF in the midbrain modulates accumbens release of dopamine to reward (Wanat et al. 2013), such that CRF, like dopamine, magnifies incentive events (Dallman et al. 2005; Berridge et al. 2004). CRF, along with cortisol, is tied to energy balance. Insulin interacts with CRF and glucocorticoids in hunger-related ingestion (Richardson et al. 2002). Indeed, activity is tied to CRF expression; foraging recruits and requires glucocorticoids in addition to CRF and can contribute to appetitive behaviors (e.g., salt appetite; Tarjan et al. 1992).

A wide variety of objects (foods, people, and settings) either learned or not learned are powerful elicitors of approach behaviors. These approach behaviors are importantly linked to palatability judgments and appraisal mechanisms (i.e., what something looks like, what it tastes like, if it is sweet, if the texture is soft, if the body is inviting). Hedonic experiences linked to palatability judgments are also important features that motivate behavior (e.g., Berridge 2000; Pfaffmann and Norgren 1977; Stellar 1954).

The motivation to eat or to stop eating certain food types may have more to do with metabolic consequences and their effects than with energy balance per se. For example, Dallman and colleagues have proposed that eating high-fat, high-sugar foods (referred to as "comfort foods"; Dallman et al. 2005) results in physiological and metabolic responses (e.g., increases in insulin and glucocorticoids) that stimulate pleasure-associated areas of the nucleus accumbens. The motivation to eat "comfort foods" has been documented in humans and rats subjected to unpleasant, challenging, and stressful circumstances.

These behaviors are proposed, in part, to represent a form of self-medication, with the ingestion of food serving as more of a medicinal or therapeutic function than a nutritional one. In this way, eating comfort foods is one way to ameliorate hormonal signals that increase arousal (Tomiyama et al. 2011; Bell et al. 2000), an effect that has been shown to be stronger in human females than in males (Zellner et al. 2006). People who increase food consumption as a response to stress report both that the foods they overeat are the foods that they normally avoid for reasons of health or weight loss, and that they eat these foods to feel better (Zellner et al. 2006). Binge-eating was displayed in a recent study of female rats under food restriction, followed by inducement of frustration (i.e., exposure to the sight of food); the CRF gene showed upregulation in regions of the hypothalamus and ventral tegmental area. This upregulation of CRF may be due to epigenetic changes via loss of methylation at promoter sites (Pucci et al. 2015).

Adverse and challenging circumstances increase CRF in a number of brain regions. Adrenalectomy also changes the basal levels of CRF, increasing CRF in the PVN of the hypothalamus and decreasing CRF in the central nucleus of the amygdala (e.g., Swanson and Simmons 1989; Watts and Sanchez-Watts 1995; Palkovits et al. 1998). Interestingly, sucrose ingestion normalizes CRF in these brain regions in adrenalectomized rats (Dallman et al. 2003; Figure 8.1).

Rats can be trained to associate a sound with the availability of sucrose pellets if they press a bar (Peciña et al. 2006, Robinson and Berridge 1998). In these studies, injection of CRF into the nucleus accumbens increased cue-triggered bar pressing for the sucrose pellets (Figure 8.2b; Peciña et al. 2006). This can be interpreted as CRF increasing the salience of external cues and the motivation of the rats to respond (or diminished motivation when overly expressed; e.g., Bryce and Floresco 2016).

As neuropeptides are involved in attentional responses to both external and internal events (Peciña et al. 2006; Dallman et al. 2003), peptides such as CRF can increase incentive salience (Peciña et al. 2006; Merali et al. 2001 2003 2008; Dallman and Bhatnagar et al. 2000; Dallman et al. 2003). It can also

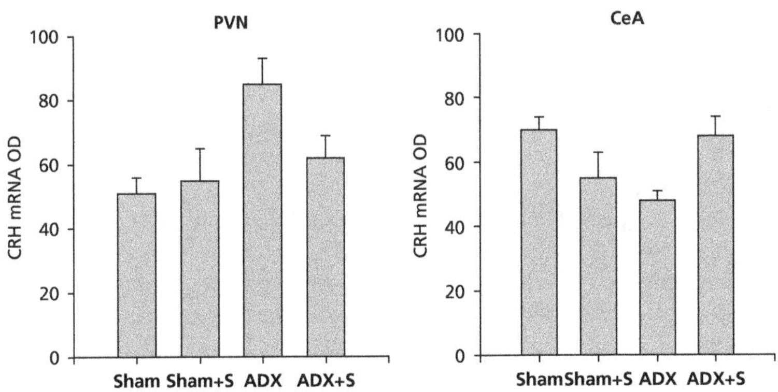


**Figure 8.1** Adrenalectomized rats had higher basal CRF in the paraventricular nucleus of the hypothalamus (PVN) and lower basal CRF in the central nucleus of the amygdala (CEA). Sucrose ingestion normalized CRF levels.

Reprinted from *Brain, Behavior, and Immunity*, 19 (4), Mary F. Dallman, Norman C. Pecoraro, and Susanne E. la Fleur, Chronic stress and comfort foods: self-medication and abdominal obesity, pp. 275–280, doi:10.1016/j.bbi.2004.11.004 Copyright © 2005 Elsevier Inc., with permission from Elsevier.

be interpreted as the rats reacting to the increased neural CRF with a higher motivation for sucrose because of its "comforting" attributes (i.e., the rats were self-medicating to alleviate the arousal owing to the artificially upregulated central CRF). Moreover, amphetamine injections (METH), which increase CRF expression in this region of the brain, produce the same results (Chu et al. 2016; Peciña et al. 2006; Cadet et al. 2014; Figure 8.2). Ultimately, injections of CRF and amphetamines into the caudal medial accumbens shell and nucleus

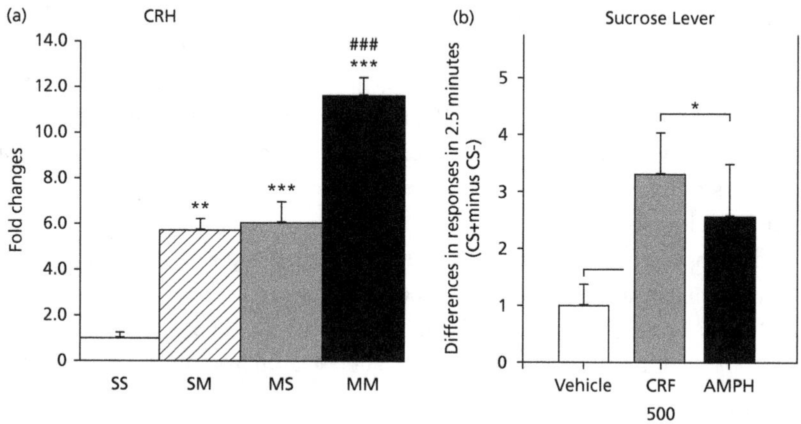


**Figure 8.2** (a) METH induced changes in the expression of neuropeptides in the rat NAc. The figure shows the acute and more delayed effects of METH injections on the mRNA levels of CRF. Rats were injected (4 rats in SS; 6 rats in SM; and 7 rats in each MS and MM groups, respectively) and total RNA was extracted from the NAc as described in the text. Key to statistics: * = $P < 0.05$, ** = $P < 0.01$; *** = $P < 0.001$, in comparison to the SS group; ### = $P< 0.001$, in comparison to the SM group and the MS group. (b) Enhancement of incentive motivation by CRF. Effects on cue-triggered lever pressing during extinction test caused by CRF (500 ng) and amphetamine (20 μg) microinjections in the caudal medial accumbens shell. Transform scores show a direct contrast between CS + and CS- effects of lever pressing. CS + effects on cue-triggered lever pressing are amplified by CRF (500 ng) and by amphetamine (20 μg) microinjections.

(a) Reproduced from Jean Lud Cadet, Christie Brannock, Bruce Ladenheim, Michael T. McCoy, Irina N. Krasnova, Elin Lehrmann, Kevin G. Becker, and Subramaniam Jayanthi, Enhanced Upregulation of CRH mRNA Expression in the Nucleus Accumbens of Male Rats after a Second Injection of Methamphetamine Given Thirty Days Later, *PLoS ONE*, 9(1): e84665, Figure 5a, doi:10.1371/journal.pone.0084665 (c) 2014, The Authors.

(b) Reproduced from Susana Peciña, Jay Schulkin and Kent C Berridge, Nucleus accumbens corticotropin-releasing factor increases cue-triggered motivation for sucrose reward: paradoxical positive incentive effects in stress?, *BMC Biology*, 4 (8): e84665, Figure 3b, DOI: 10.1186/1741-7007-4-8 (c) 2006, The Authors.

accumbens of rats both resulted in significantly increased efforts in auditory cue-triggered lever pressing after Pavlovian conditioning experiments.

As discussed later on in this chapter, levels of CRF are also knotted to amphetamine-related withdrawal, as seen in METH self-administered rats whose CRF levels significantly increased during withdrawal in the amygdala, but not significantly in the nucleus accumbens or the hypothalamus (Caberlotto et al. 2004; Nawata et al. 2012). Clearly, there is a differentiation in brain region location of CRF upregulation during these drug usage events (i.e. injection and withdrawal).

Regions of the nucleus accumbens are critical in appetitive behaviors (see Berridge 2004). These regions contain both glucocorticoids and CRF receptors (e.g., Lim et al. 2005). The intracranial infusion of CRF into the shell of nucleus accumbens can be visualized with a Fos plume map that localizes the magnification effect of the CRF microinjection (Peciña et al. 2006). In fact, one way to understand appetitive behaviors is that one function of glucocorticoids is to magnify the effects of CRF in increased attention to objects and their potential value, for example. This may be a particularly important aspect of the release of CRF when an object is unfamiliar or uncertain, as well as if it is dangerous (e.g., Habib et al. 2000; Kalin et al. 1998, 2016).

Recall that the nucleus accumbens was originally understood as the motor end of the limbic system (Rho and Swanson 1987; Nauta 1972). Gradually this region of the brain has been found to underlie a number of behavioral functions essential for adaptation, including higher level cognitive functions that include probability judgments about statistical likelihoods (Squire 2004), parsing of syntactic functions in language (Ullman 2001; 2004), and the organization of action and social responsiveness (Lieberman 2000). Not only is the rather large region of the brain not simply motor, but it is not strictly one dimensionally cognitive; it is adaptive in many dimensions.

Many brain regions are activated by the expectation of diverse rewards. For instance, expectancy for sucrose rewards results in activation of regions of the frontal cortex (Dallman et al. 2003; Merali et al. 2003; Peciña et al. 2006). In fact, both positive and negative events (good ingestion and drug withdrawal, e.g.) activate CRF expression in diverse regions of the brain, including the amygdala, frontal cortex, and PVN (Koob and Le Moal 2003 2005; Merali et al. 2003 2006 2008). Interestingly, using microdialysis to measure CRF release, corticosterone pretreatment potentiates the expression of CRF in both the amygdala and frontal cortex (Merali et al. 2008).

Importantly CRF activation under adversity helps drive appetitive behaviors to aversive behaviors in the nucleus accumbens. CRF and both receptor types 1 and 2 have been localized by immunoreactivity and transmission electron microscopy in the nucleus accumbens (Lemos et al. 2012). The link between

CRF and dopamine-related expression is broad and prominent (Lemos et al. 2012). It results in dopamine release through activation of both CRF-R1 and CRF-R2 in the mobilization of behavior. CRF evokes a dopamine release and appetitive behaviors that are then ended by exposure to stress. Following CRF infusions, animals were more likely to spend time in a CRF-paired chamber after conditioning. There was no difference in time spent in the center of an open field before and after the addition of a novel object for those given CRF infusions (Lemos et al. 2012; Hollon et al. 2015).

Under some conditions, CRF facilitates social preferences; when exposed to adverse conditions, CRF increases contact with familiar partners over contact with strangers (DeVries et al. 2002). Seeking comfort in the context of adversity is a nontrivial adaptation, although it is not always good (Dallman et al. 2005; Tomiyama et al. 2011). Nucleus accumbens CRF/dopamine interactions underlie diverse forms of appetitive behaviors (Chen et al. 2012).

**Addictions:** Context matters and the same chemical stimulation can elicit diverse forms of behavior (Berridge 2004). CRF can elicit aggressive behaviors in some situations or fear-related behaviors in others. For instance, housing conditions differentially affect the behavioral effects of CRF in rats (Ruscio et al. 2007).

Addictions reveal obsessions and having one's focus consumed by the object(s) of desire. CRF and the neurotransmitter norepinephrine work together to underlie attention. This is an instance of a peptide and neurotransmitter complementing each other in the organization of action such that attention is embedded in action. This narrowing of focus to one object resulting in the exclusion of others takes precedence and signals the formation of obsessive behaviors.

One system underlying drug addiction is extrahypothalamic CRF and HPA regulation. Feed-forward regulation, particularly at the level of the bed nucleus of the stria terminalis, may underlie some of the increased vulnerability for drug consumption. In other words, vulnerability to relapse, in animal studies, is linked to the overexpression of CRF in critical regions of the brain (such as the bed nucleus of the stria terminalis; Shaham et al. 2000; Koob and Le Moal 2001; McReynolds et al. 2014). An addicted individual is trying to maintain internal stability and adapt to an adverse environment, and may eventually slip into biological and psychological vulnerability (resulting in compromised immunological, neural, and bone tissue).

One hypothesis is that the overactivation of central CRF increases the vulnerability for drug use, relapse, and withdrawal (Contarino and Papaleo 2005). Chronic overactivation of this system can compromise negative inhibition at the level of the PVN, in addition to compromising a number of end organ systems. The cycle of addiction is essentially obsessive focus on a particular substance or object, linked to withdrawal and binge-related behaviors.

Increased activation of CRF exaggerates states of angst reflected in activity in diverse regions of the brain including the lateral amygdala and medial prefrontal cortex (Bijlsma et al. 2011). Information molecules like CRF that are associated with a particular central motive state prepare the individual to perceive relevant environmental stimuli necessary for obtaining the desired outcome (Koob and Bloom 1985). Once an individual is caught in the cycle of addiction, this central motive state can become all-consuming. Craving motivates the appetitive phase, the search for the object (i.e., the drug) supersedes all else in the addicted person's life. The consummation of this motive state is taking the drug, and the addict is then satisfied for a period of time. The appetitive phase in addiction can become more important than social interaction, family, friends, personal needs, and obligations. These addictions can eventually affect diverse information molecules and certainly CRF.

Many neurotransmitter systems and brain areas are engaged during drug use and abuse (Higley and Bennett 1999; Volkow et al. 2016; Koob and Volkow 2010). Each drug involves a different constellation of systems. The reward circuit involves dopamine and opioid peptide neurons that intersect at the ventral tegmental area and the nucleus accumbens and other brain regions. These brain regions are first activated during the initial drug abuse and early binge stages of the addiction cycle, and in the neural circuit, that includes CRF and noradrenergic neurons converge on GABA interneurons in the CEA during development that contribute to the addiction cycle (Koob and Zorrilla 2010; Figure 8.3).

Additionally, a history of chronic adversity plays a role in the dopaminergic systems in the brain. An increase in dopamine release occurs in the frontal cortex of rats exposed to chronic stress when they are continuously put in the same stressful situation (Cuadra et al. 1999). Repeated stimulation of CRF in regions of the brain contributes to the angst that pervades an addict's life and the panic and anxiety of everyday social pathology (Lee et al. 2008).

Social adversity is tied to the diverse information molecules such as norepinephrine and CRF that underlie alcohol and opioid dependence and heightened attention (Chaijale et al. 2013). Again the CRF type 1 receptor is critically important (Ghitza et al. 2006), particularly because beta-2 adrenergic receptor pathways regulate ventral BNST CRF, and CRF affects the ventral tegmental area (VTA). Photomicrographs show that immunostained CRF neurons extend from within the ventral bed nucleus of the stria terminalis to the VTA (Vranjkovic et al. 2014).

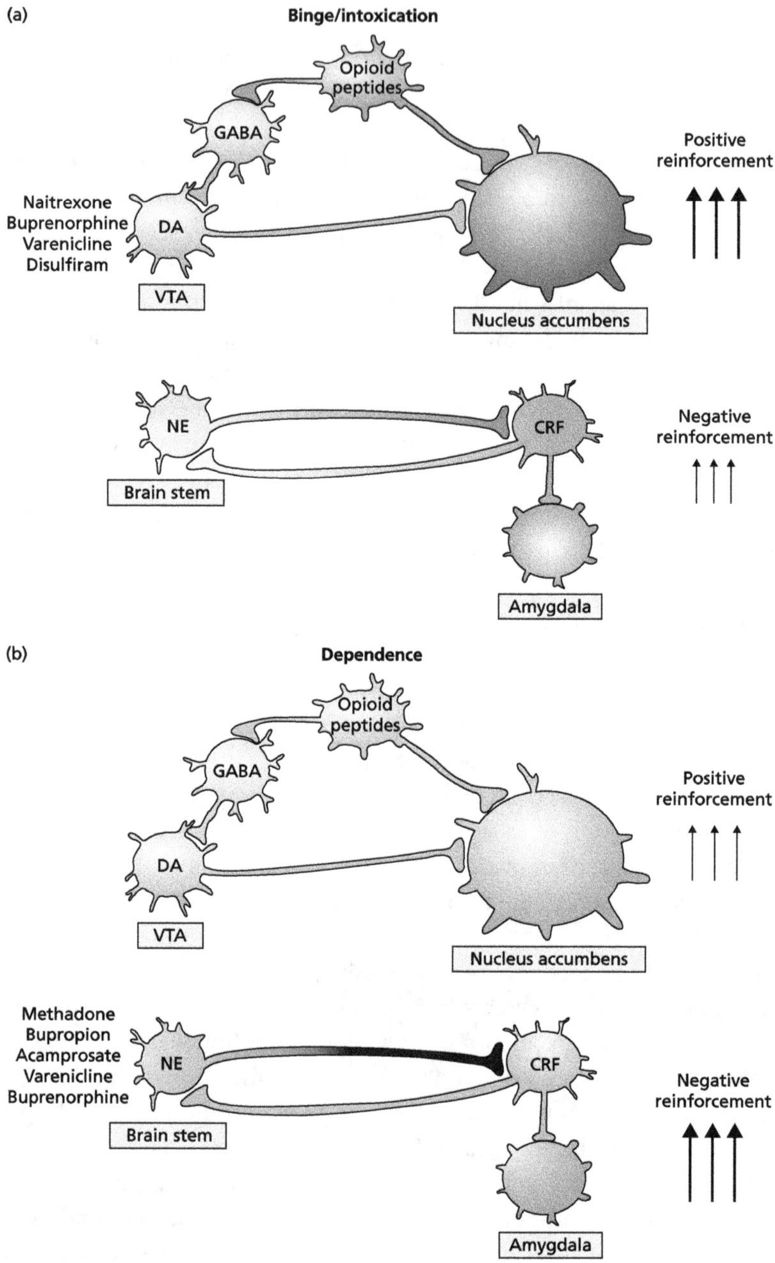


**Figure 8.3** (See color plate section) Neurocircuitry associated with the acute positive reinforcing effects of drugs of abuse and the negative reinforcement of dependence, and changes in the transition from nondependent to dependent drug taking. (a) Key elements of the reward circuit include dopamine and opioid peptide neurons that

*(Continued)*

Adversity of diverse kinds increases the vulnerability to relapse via CRF type 1 receptor activation in several regions of the brain, including brainstem regions such as the dorsal raphe and serotonin sites, or forebrain sites such the as the nucleus accumbens and amygdala (Knapp et al. 2011b). CRF-R1 and CRF-R2 receptors can be visualized by immunofluorescence in the anterior medial dorsal raphe nucleus (Pringle et al. 2008; Day et al. 2004). Interestingly, CRF type 1 receptors targeted in the ventral tegmental region reduced alcohol intake, whereas in the raphe, it reduced both ethanol and water intake (Hwa et al. 2014; Zorrilla and Koop 2010).

Dopamine is also critical to the uptake of substance use, the salience of substance (Berridge 2004), and a person's willingness to obtain it. The fact that CRF is tightly linked to dopamine activation and is critical to the vulnerability of self-administration has been a consistent finding in the literature. Interestingly, social adversity increases the co-activation between CRF and dopamine in the expanded use of cocaine (Boyson et al. 2014).

Chronic duress can lead to a decrease in dopamine in the nucleus accumbens (Gambarana et al. 1999), and a number of studies have shown that prenatal exposure to, for example, cocaine can have long-term consequences on dopaminergic pathways and the response to duress (e.g., Elsworth et al. 2001; Lee and Rivier 1996). Many studies link CRF to adrenergic receptor function in substance use (Park et al. 2013). For instance, CRF and a subset of adrenergic receptors facilitate the withdrawal-related startle responses in heroin users (Park et al. 2013).

The CRF type 1 receptor, which underlies morphine withdrawal, is tied to the brain noradrenergic activation, and catecholaminergic activation in the nucleus accumbens (Funada et al. 2001; Almela et al. 2012). The hypersensitivity that occurs during dependence (e.g., alcohol and heroin) is mediated in

---

**Figure 8.3** *(Continued)*
intersect at both the VTA (ventral tegmental area) and the nucleus accumbens that are activated during initial drug use and the early binge/intoxication stage of the addiction cycle. Drugs approved for the treatment of this component of the addiction cycle include naltrexone, buprenorphine, varenicline, and disulfiram. (b) Key elements of the stress circuit include CRF (corticotrophin-releasing factor) and noradrenergic neurons that converge on GABA interneurons in the central nucleus of the amygdala that are activated during the development. Drugs approved for the treatment of this component of the addiction cycle include methadone, bupropion, acamprosate, varenicline, and buprenorphine. DA, dopamine; NE, norepinephrine.

part via CRF type 1 receptors (Edwards et al. 2011, 2012). Alcohol deprivation induced motivation, for instance, is dependent upon CRF type 1 receptor signaling systems (Schmidt et al. 2010; Sparta et al. 2009). Cytokine or inflammatory signaling systems that sensitize ethanol withdrawal do so via CRF (Knapp et al. 2011b; Whitman et al. 2013). Diverse compounds such as naloxone affect CRF expression (Hamlin et al. 2004), and all forms of withdrawal affect CRF (Negus and Rice 2009), along with overall well-being (Koob 2013; Hamilton et al. 2013; Koob and LeMoal 2001).

Both the ingestion of and withdrawal from nicotine releases CRF (Koob 2010; Grieder et al. 2014). Nicotine withdrawal contributes to social anxiety in addition to elevated CRF (Aydin et al. 2011), and CRF type 1 receptors are tied to a dysphoric state associated with withdrawal (Bruijnzeel et al. 2010; 2012) and hyperalgesia (Baiamonte et al. 2014). Nicotine-dependent rats showed increased sensitivity to thermal stimulus during nicotine withdrawal. In nicotine-dependent rats, levels of CRF and CRF-R1 mRNA are increased in the central amygdala. Systemic administration of the CRF-R1 receptor antagonist, R121919, or a control vehicle resulted in a significant difference between treatment groups of nicotine-dependent rats and their paw withdrawal latency at various time points (6 hours and 72 hours; Baiamonte et al. 2014; Erhardt et al. 2009).

The drugs that produce a pattern of elevated glucocorticoids and CRF in the PVN of the hypothalamus, the central nucleus of the amygdala, and the medial BNST during withdrawal conditions include ethanol (Pich et al. 1993 Adinoff et al. 1996; Merlo-Pich et al. 1995), cocaine (Sarnyai et al. 1998; Richter and Weiss 1999), morphine (Richter et al. 1999), cannabis (de Fonseca et al. 1997), and even nicotine (Rasmussen 1998; Lutfy et al. 2012; George et al. 2007). Rats provided long access to various drugs (cocaine, nicotine, heroin, methamphetamine, and ethanol) has shown an overall increased intake through self-administration over time (Koob 2010; Figure 8.4).

CRF type 1–receptor activation reduces alcohol ingestion in the fawn hooded rat strain (Lodge and Lawrence 2003) and the fawn rat appears to have genetically higher levels of CRF in the central nucleus of the amygdala, which led to vulnerability to fear or anxious states (Altemus et al. 1994; Lodge and Lawrence 2003). Importantly, tonic GABAergic neurons interact with the CRF type 1 receptors (Herman et al. 2013; Partridge et al. 2014, 2016).

One model of GABA CRF interactions is depicted in the next figure (Herman et al. 2013). Phasic $GABA_A$ receptor transmissions occur on both CRF + and CRF- neurons, but more often on CRF + neurons. CRF+ neurons, near the synaptic cleft, project from the CEA to the BNST. CRF- neurons reflect activation of ambient GABA outside of the synaptic cleft. When the CRF + neurons'

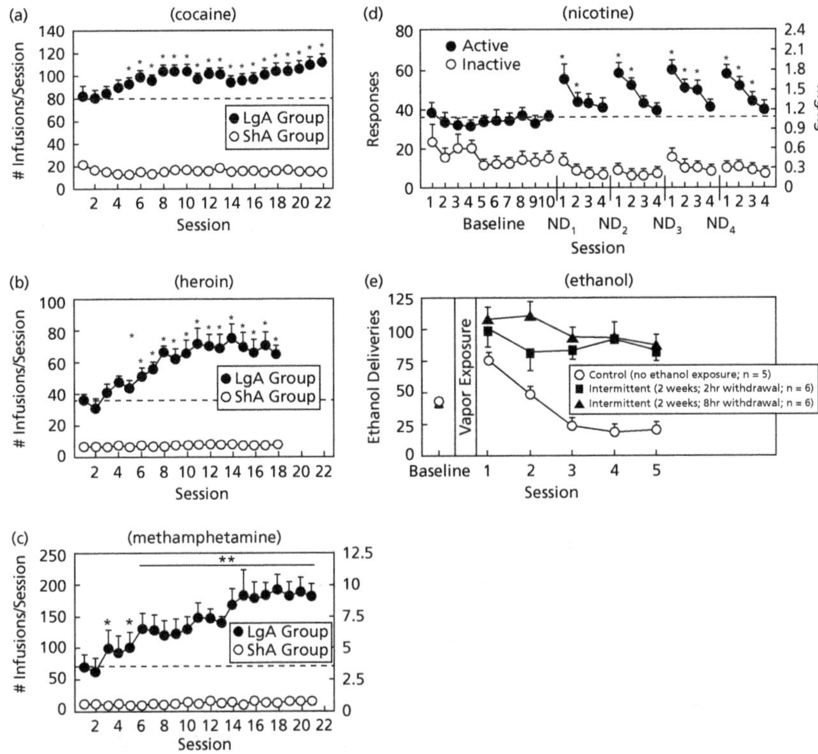


**Figure 8.4** (a) Effect of drug availability on cocaine intake (mean SEM). In 6-hour-long access (LgA) rats (n = 12) but not in 1-hour short-access (ShA) rats (n = 12), mean total cocaine intake started to increase significantly from session 5 ($p < 0.05$; sessions 5–22 compared with session 1) and continued to increase thereafter ($p < 0.05$; session 5 compared with sessions 8–10, 12, 13, 17–22). (b) Effect of drug availability on total intravenous heroin self-infusions (mean SEM). During the escalation phase, rats had access to heroin (40 mg per infusion) for 1 hour (ShA rats, n = 5–6) or 11 hours per session (LgA rats, n = 5–6). Regular 1-hour (ShA rats) or 11-hour (LgA rats) sessions of heroin self-administration were performed 6 days per week. The dotted line indicates the mean (SEM) number of heroin self-infusions of LgA rats during the first 11-hour session. *$p < 0.05$, compared with the first session (paired t-test). (c) Effect of extended access to intravenous methamphetamine self-administration as a function of daily sessions in rats trained to self-administer 0.05 mg/kg/infusion of intravenous methamphetamine during a 6-hour session. Short-access group (ShA), 1-hour session (n = 6). Long-access group (LgA), 6-hour session (n = 4). All data were analyzed using two-way ANOVA (dose escalation session within ShA or LgA group). *$p < 0.05$, **$p < 0.01$, ***$p < 0.001$ vs. day 1. (d) Total 23-hour active and inactive responses after repeated cycles of 72 hours of nicotine deprivation (ND) followed by 4 days of self-administration (*$p < 0.05$ vs. baseline). (e) Ethanol deliveries (mean SEM) in rats trained to respond for 10% ethanol and then either not exposed to ethanol vapor (control, n = 5) or exposed to intermittent ethanol vapor (14 h on/10 h off) for

*(Continued)*

firing rate is increased, GABA release is increased in certain brain regions, e.g., in the BNST. Ethanol can increase this firing rate by decreasing the firing rate of CRF- neurons, through an increase in their tonic conductance (Herman et al. 2013) CRF + neurons have been shown to have significantly higher tonic current than that of CRF- neurons during super fusion of zolpidem (Herman et al. 2013; Figure 8.5).

In rats, decreases in CRF-R1 reduce the behavioral and endocrine response to morphine, cocaine, and alcohol (Skelton et al. 2000; Goeders and Guerin 2000; Goeders 2002; Sillaber et al. 2002; Gilpin et al. 2008; Eisenhardt et al. 2015). Of course, and consistent with a larger body of literature, CRF is enhanced during alcohol consumption (Funk et al. 2006, 2007) and increases the sense of dysphoria (Bruljnzeel et al. 2010). Genetically altered msP (Marchigian Sardinian alcohol-preferring) rats showed significantly higher rates of ethanol self-administration when treated with the CRF-R1 antagonist, antalarmin. Treatment with antalarmin, followed by intermittent stress-inducing foot shocks, resulted in the suppression of alcohol seeking and self-administration in msP rats, but not in controls (Hansson et al. 2006).

---

**Figure 8.4** *(Continued)*

2 weeks and then tested either 2 hours (n = 6) or 8 hours (n = 6) after removal from ethanol vapor. *$p < 0.05$, significant increase in operant self-administration of ethanol in rats receiving intermittent vapor exposure compared with control. No difference was observed between rats exposed to intermittent vapor and tested either 2 or 8 hours after ethanol withdrawal.

(a) From S. H. Ahmed and G. F. Koob, Transition from Moderate to Excessive Drug Intake: Change in Hedonic Set Point, *Science*, 282 (5387), pp. 298–300, DOI:10.1126/science.282.5387.298 Copyright © 1998, American Association for the Advancement of Science. Reprinted with permission from AAAS.

(b) Reprinted by permission from Macmillan Publishers Ltd: *Neuropsychopharmacology*, 22 (4), Serge H Ahmed, John R Walker, and George F Koob, Persistent Increase in the Motivation to Take Heroin in Rats with a History of Drug Escalation, pp. 413–421, Figure 1, doi:10.1016/S0893-133X(99)00133-5, Copyright © 2000, Nature Publishing Group.

(c) Reproduced from Osamu Kitamura, Sunmee Wee, Sheila E. Specio, George F. Koob, and Luigi Pulvirenti, Escalation of methamphetamine self-administration in rats: a dose–effect function, *Psychopharmacology*, 186, (1), pp. 48–53, Figure 4a, Copyright © Springer-Verlag 2006.

(d) Reproduced from Olivier George, Sandy Ghozland, Marc R. Azar, Pietro Cottone, Eric P. Zolrrilla, Loren H. Parsons, Laura E. O'Dell, Heather N. Richardson, and George F. Koob, CRF–CRF1 system activation mediates withdrawal-induced increases in nicotine self-administration in nicotine-dependent rats. *Proceedings of the National Academy of Sciences of the United States of America*, 104 (43), pp. 17198–17203, Figure 2a, DOI: 10.1073/pnas.0707585104, Copyright © 2007, The National Academy of Sciences.

(e) Reproduced from Laura E. O'Dell, Amanda J. Roberts, Ron T. Smith, and George F. Koob, Enhanced Alcohol Self-Administration after Intermittent Versus Continuous Alcohol Vapor Exposure, *Alcoholism: Clinical and Experimental Research*, 28 (11), pp. 1676–1682, DOI: 10.1097/01.ALC.0000145781.11923.4E, Copyright © 2006, John Wiley and Sons.

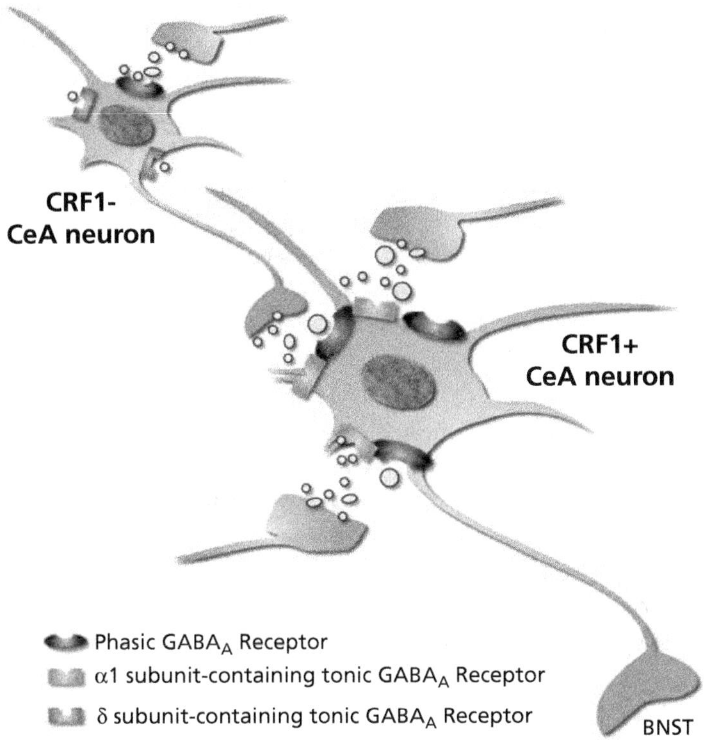


**Figure 8.5** (See color plate section) Proposed microcircuitry of the CEA illustrating local phasic and tonic GABA$_A$ receptor transmission in CRF1 + and CRF1− neurons. Phasic GABA$_A$ receptor transmission is depicted on both CRF1 + and CRF1− neurons, although CRF1 + neurons are depicted with more baseline GABA activity. The CRF1 + neuron possesses α1 subunit-containing tonic GABA$_A$ receptors located in or near the synaptic cleft to reflect the dependence on action potential-dependent GABA release and projects out of the CEA into the BNST. The CRF1− neuron possesses δ subunit-containing GABA$_A$ receptors located outside the synaptic cleft to reflect activation by ambient GABA. We hypothesize that ethanol acts on this CEA circuitry by enhancing tonic conductance in CRF1− neurons through actions at δ-containing GABA$_A$ receptors, resulting in decreased firing of this subpopulation. The ongoing tonic inhibition of CRF1 + neurons is subsequently diminished, resulting in disinhibition and increased firing in this subpopulation. This increased firing then results in increased GABA release in target brain regions such as the BNST.

Increased reactivity to stress characterizes drug withdrawal. Cerebrospinal fluid concentrations of CRF are elevated during the first day of abstinence from alcohol and decrease over the next 3 weeks in humans (Adinoff et al. 1996). While CRF is elevated during withdrawal, several other neurotransmitters or neuropeptides show decreased expression. Moreover, CRF expression in the brain affects a number of neurotransmitters and neuropeptides (e.g., Swanson 1991; Watts 1996; Price and Lucki 2001). For instance, adrenal steroids facilitate CRF-related behavioral sensitization to drugs (e.g., alcohol; Pastor et al. 2012).

I have never believed that there is one definition of *addiction*. There are diverse drugs and objects to which we are vulnerable, and the list is long. One key feature is dependence. In our era, the excessiveness and accessibility of the diverse machines that permeate our life are all-consuming and addictive. CRF is activated by addictive behaviors regardless of the substance or object to which a person is addicted; this is in part due to the role that CRF plays in hypersensitivity (Edwards et al. 2012).

In addition, CRF is tied to the dysregulation of the reward system associated with the chronic use of drugs (Koob and Le Moal 2001). There are indications (e.g., in animal models) that drug consumption can have long-term potentiation in specific neurons from cocaine ingestion (e.g., Ungless et al. 2001) and perhaps relapse from association of the drug with environmental events that are linked to this hyperexcitable state of the brain (Vorel et al. 2001). For instance, social adversity accelerates CRF-dopamine interactions and cocaine self-administration (Boyson et al. 2014) or NPY reduces CRF in the bed nucleus and inhibits alcohol binge drinking (Pleil et al. 2015). This implies that a feedback loop related to CRF is important in developing and sustaining addictive behaviors.

Dopamine neurons that project to forebrain sites underlie normal regulation in terms of tonic and phasic postsynaptic behavioral and physiological responses (Zweifel et al. 2009) that influence anxiety (Zweifel et al. 2011). CRF is co-localized in these neuronal ensembles (Beckerman et al. 2013). In a cocaine-seeking model, episodic social stress elucidated the phasic and tonic nature of CRF in the anterior and posterior VTA (Holly et al. 2016). Perhaps the phasic and tonic regulation of CRF release can be targeted to different brain regions, (i.e., the CEA and BNST), respectively, dependent on the nature of the environment or stimulation. Phasic release from the CEA might be an immediate response to a stimulus, whereas the BNST might release on a more baseline, tonic level: this difference in expression patterns emphasizes the differential regulation of information molecules, specifically CRF within the brain.

## Vulnerability to Drug Abuse: Developmental Adversity

An individual's inborn characteristics can have mediating effects on both response to stress and reaction to drugs. Animals can be selectively bred to have higher basal levels of corticosterone, and this can condition for higher levels of CRF gene expression. Similarly, strains of animals that prefer specific drugs and are more likely to respond to specific drugs can be bred in laboratories. Behavioral and physiological characteristics of selectively bred animals can give insights into characteristics of human vulnerability.

Many studies demonstrate that rats with higher glucocorticoid levels are more likely to self-administer a number of psychotropic agents (Piazza et al. 1993). Rats with higher levels of corticosterone and CRF are more likely to self-administer cocaine, heroin, and amphetamines: in other words, the greater the level of corticosterone that activates the central nervous system, the greater the probability of rats self-administering these agents (Piazza et al. 1989, 1993; Erb et al. 1996). Several studies have shown that stress effects in alcohol dependence models are mediated by glucocorticoid expression. Levels of CRF mRNA are decreased in the PVN of models exposed to corticosterone, but increased in the CEA. The same scenario is found in alcohol-dependent models when exposed to alcohol. This differentiation in response emphasizes the differential brain regulation and response to external stimuli and adversity, as well as the potential role of glucocorticoids in addiction (Figure 8.6).

Prenatal stress can also impair future motor development in primates (Schneider et al. 1992, 1998) and increase vulnerability to high anxiety and drug use (Vallee et al. 1997). Prenatal exposure to diverse drug abuses during various periods in development is also known to influence offspring behaviors and CRF expression in the brain (Sithisarn et al. 2011).

Further, parental deprivation is an important variable for predicting later vulnerability to stress among nonhuman primates. Plasma cortisol levels quickly increase in rhesus monkey infants when they are separated from their mothers and peers (Higley et al. 1992). There is a tendency to have higher systemic levels of cortisol in maternally deprived rhesus monkeys during development than in mother-reared monkeys. Rhesus monkey infants subjected to short-term maternal separation and isolation exhibit higher cortisol levels and perhaps higher levels of central CRF (Kalin et al. 2000, 2016).

These events create a vulnerability to chronic arousal and a vulnerability to, for instance, stress-related alcohol consumption; the alcohol itself is functionally related to CRF gene expression (Barr et al. 2008 2009). In adolescent and young adult rhesus monkeys, it has been determined that there is a CRF

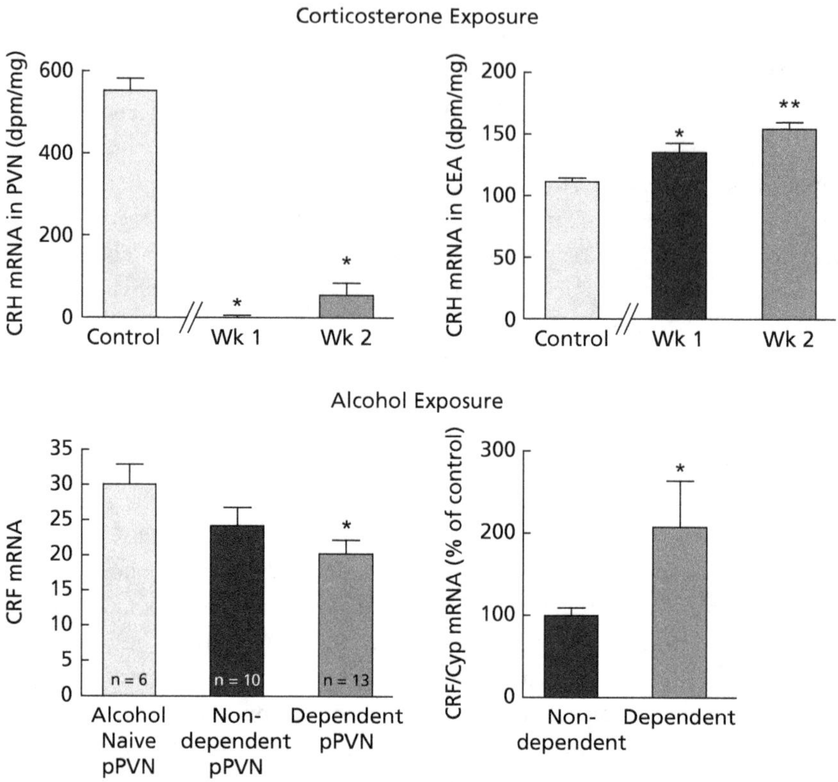


**Figure 8.6** Stress effects in alcohol dependence are mediated by neuroplasticity in glucocorticoid expression in hypothalamic and extrahypothalamic stress systems.

Top left and right: Reprinted from *Brain Research*, 640 (1–2), Shinya Makino, Phillip W. Gold, and Jay Schulkin, Corticosterone effects on corticotropin-releasing hormone mRNA in the central nucleus of the amygdala and the parvocellular region of the paraventricular nucleus of the hypothalamus, pp. 105–112, doi:10.1016/0006-8993(94)91862-7, Copyright © 1994 Elsevier B.V., with permission from Elsevier.

polymorphic genotype that results in statistically lower levels of CRF in their cerebrospinal fluid, as well as statistically higher levels of ACTH in their plasma (Barr et al. 2008 2009).

Aversive events, which lead to higher glucocorticoid levels, also induce an increased likelihood of self-administration of drugs at higher rates. When rats are exposed to noncontingent shock, they self-administer cocaine at much higher rates than do rats who are exposed to contingent shock or no shock (Goeders 1998). The "noncontingent shock rats" were more responsive to lower cocaine doses compared with the rats in the other shock groups, who required two times the dosage before engaging in significant amounts of cocaine self-administration. Levels of plasma corticosterone in the noncontingent shock rats were much higher.

Maternally deprived macaque monkeys have both higher plasma cortisol concentrations and higher central CRF levels in the cerebrospinal fluid as young mature adults (Higley et al. 1992, 1996; Habib et al. 1999 2000). Behaviorally, macaques raised by their peers show higher baseline rates of alcohol consumption as compared with macaques raised by their mothers. However, when subjected to social separation, mother-reared monkeys increase their alcohol consumption during social separation until the consumption is equal to that of peer-reared macaques (Kraemer and McKinney 1985; Higley et al. 1991).

CRF levels are elevated during both positive and negative experiences (Richter et al. 1995; Merali et al. 1998). Expression of CRF in the brain increases the salience of environmental stimuli (Peciña et al. 2006) and has implications for situational learning both in addiction and during a child's development. In the case of addiction, this is relevant in terms of the search for and acquisition of the desired object. The actions of glucocorticoids might also be adaptive mechanisms for reducing overactivity in the physiological stress systems, allowing the animal to cope more effectively with a stressful situation (Piazza and Le Moal 1997).

Both glucocorticoids and CRF have psychomotor effects, as does dopamine (Wise and Bozarth 1987). Cocaine is considered a psychomotor stimulant. The psychomotor and euphoric effects resulting from cocaine consumption may reflect activation of the neuropeptide CRF and neurotransmitter dopamine, as well as their regulation by glucocorticoids (Piazza et al. 1993; Koob 1998). Cocaine activates the pituitary adrenal axis in both rats and nonhuman primates, and CRF is active in both hypothalamic and extrahypothalamic areas during cocaine consumption (Koob 1998). After 3 weeks of binge cocaine administration, rats show higher basal corticosterone levels as compared with rats in a no cocaine placebo group (Kreek 1997; Sarnyai et al. 1998; Schulger et al. 2003).

**CRF and cocaine seizure:** Increases in CRF activity during cocaine consumption, along with mesolimbic dopaminergic activity, increase the salience of cocaine and its associated stimuli (Berridge 2004). Diverse peptides including the interaction between CRF and brain derived trophic factors interact with reward pathways (Walsh et al. 2014) and basic regulatory glucocorticoid/CRF interactions in the hypothalamus (Jeanneteau et al. 2012).

Individual stress responses are also important to the effects of drug consumption, and may be relevant to both the initiation and maintenance of drug abuse. Individual differences in biological responses to stress are measurable in the laboratory. Rats can be categorized as low responders or high responders to stress by measuring corticosterone release and locomotor reactivity in response to a stressful situation. Most animals avoid stressful situations because those situations are unpleasant; however, some animals, and some humans, can be described as stress seeking (Piazza et al. 1993; Goeders 2002). Animal research shows that these individuals are more likely to self-administer addictive substances and are more sensitive to the effects of psychostimulants. Interestingly, an elevation of cocaine self-administration is also tied to dopamine. CRF/dopamine interactions, perhaps via CRF type 1 receptors in the ventral tegmental region and the accumbens underlie cocaine self-administration (Boyson et al. 2014). CRF receptor antagonists were injected into the intra-VTA in stressed and nonstressed rats who self-administered cocaine during a 24-hour binge. It was shown that stress and the addition of either receptor antagonist resulted in significantly more cocaine infusions during the 24-hour binge (Boyson et al. 2014).

Repeated cocaine self-administration in rats can result in the development of seizures linked to amygdala activation (Weiss et al. 1986), and both CRF and glucocorticoids facilitate cocaine-induced seizures (Weiss et al. 1992; Kling et al. 1993). In one experiment, repeated cocaine administration resulted in the development of seizures that were facilitated by both dexamethasone and, to a lesser extent, corticosterone (Kling et al. 1993; Lee et al. 1989). Stimulation of CRF receptors in the bed nucleus of the stria terminalis tends to reduce anxiety (Lee et al. 2008).

Once an individual becomes caught in the cycle of addiction, individual biology is not as relevant to the animal's interpretation of the environment. One result of cocaine-binge administration in rats is the diminution of individual differences in corticosterone hormone responses to stress (Sarnyai et al. 1998). Rats that showed high levels of corticosterone release in response to stress and those that showed low levels of release were put in either cocaine-binge or saline/control conditions. The individual differences were maintained in animals injected with saline over 3- and 6-week periods. However, the differences in response to stress between low responders and high responders disappeared in the rats treated with cocaine during the same periods (Sarnyai et al. 1998).

Importantly, there is some evidence that alcohol consumption, which activates both GABA and CRF in co-localized neurons, in regions like the prefrontal cortex region (Radley and Sawchenko 2015), increases the vulnerability to relapse (George et al. 2012). This indicates that the interactions between substances that an individual uses may also have important implications for the addictive potentials and the outcomes of that individual.

**Vulnerability to relapse and the bed nucleus of the stria terminalis:** Vulnerability to relapse is common for formerly addicted individuals.

Their best defense is to remove themselves from the environment in which they might relapse. But this can be very difficult to accomplish. Avoidance behavior is a kind of anticipatory adaptation, linked perhaps to ensuring internal stability to acquire what is needed and avoid what is not. With relapse so prominent among those who have been addicted to a substance, what is known about some of the mechanisms of relapse? High levels of corticosterone may increase the salience of objects associated with drug administration and the sensation-seeking aspects of drug reward (Piazza et al. 1993) through increased dopamine synthesis and activation (Beutler et al. 2011). This is perhaps facilitated by the effect of corticosterone on CRF and CRF-related cells in regions of the nucleus accumbens.

Rats trained to self-administer cocaine for 12 days, for example, and then withdrawn from the drug, quickly reinstated their self-administration when they were given a foot shock. This stress-inducing event was more potent than simply being exposed to the cocaine (Erb et al. 1998; Erb 2010). This paradigm has been used to assess the relapse for a number of psychotropic drugs of abuse (Erb et al. 2000; Shaham et al. 2000), and CRF's link with other neurotransmitters (e.g., noradrenaline and dopamine; Brown et al. 2009 2012). Rats given long access to intravenous cocaine self-administration displayed significantly higher levels of CRF in the central nucleus of the amygdala and the dorsal raphe, 3 hours post session. The long-access rats showed significantly less CRF in the basolateral nucleus of the amygdala 3 hours post session, but more presession, compared with rats only given short access to cocaine (Zorrilla et al. 2012).

Importantly, while peripheral blockade of corticosterone does not interfere with this effect, interference with central CRF does (Erb et al. 1998). When provided with CRF, animals given a higher dosage pressed a bar to received cocaine more frequently over a 3-hour period. Indeed, CRF is linked to vulnerability to stress-related relapse for former users of heroin (Shaham et al. 1997) and cocaine (Brady et al. 2009; Erb et al. 1998; Buffalari et al. 2012). This reaction was unaffected by a dopamine receptor agonist or opioid receptor antagonist (Shaham et al. 2000).

CRF type 1–receptor activation is critically involved in, for instance, heroin and alcohol dependence (Kwako et al. 2015; Edwards et al. 2012; Kaur et al. 2012). This holds for dependent and nondependent consumption (Cippitelli et al. 2012). Antalarmin, a CRF-R1 antagonist, significantly decreased alcohol consumption in rats who had intermittent exposure to 10% alcohol (Cippitelli et al. 2012). Indeed a genetic association has linked the CRF type 1 receptor to alcohol intake (Treutlein et al. 2006). Its activation is particularly linked to binge-eating, alcohol consumption (Giardino et al. 2013; Karanikas et al. 2013), and stress-related alcohol intake (Molander et al. 2012). Caloric deprivation can result in long-term methylation-related change in CRF and CRF type 1 receptors that may underlie later binge-eating (Pankevich et al. 2010).

One critical brain region for this vulnerability appears to be the bed nucleus of the stria terminalis, and the vulnerability is mediated in part by central CRF (Shaham et al. 2000). Recall that this region of the brain may be linked to the amygdala (Swanson and Petrovich 1998). Neurons within the lateral region, both dorsal and ventral, contain CRF (Swanson and Simmons 1989; Makino et al. 1994; Watts and Sanchez-Watts 1995), and CRF changes within the bed nucleus reflect vagal activation under duress (Nijsen et al. 2001; Dallman and Bhatnagar 2000).

The bed nucleus of the stria terminalis is linked to CRF-related effects on anxiety-like behavioral responses (Davis et al. 1997). CRF infused into the bed nucleus elicits anxious behavior and an increased tendency to self-administer medication. CRF antagonists in the bed nucleus interfere with foot-shock-related relapse, but this does not occur when the CRF antagonist is infused into the amygdala. Conversely, when CRF is infused directly in the bed nucleus, cocaine self-administration increases, and this does not occur when CRF is directly infused into the amygdala (Erb and Stewart 1999).

The bed nucleus of the stria terminalis also appears to be linked to some of the withdrawal symptoms experienced when individuals with a history of drug abuse who are stressed for some reason start to crave the drug again. Thus CRF (in addition to other systems within the bed nucleus of the stria terminalis, e.g., Walker et al. 2000) may contribute to withdrawal symptoms rooted in the bed nucleus of the stria terminalis; an area that has been linked to a number of biological factors underlying drug abuse (Koob 1998; 2015).

Several neurotransmitters (shown in Table 8.2) serve as neurobiological substrates that reinforce the effects of drugs and drug abuse. For example,

**Table 8.2** Neurobiological Substrates for the Acute Reinforcing Effects of Drugs and Abuse

| Drug of Abuse | Neurotransmitter | Sites |
|---|---|---|
| Cocaine and amphetamines | Dopamine serotonin | Nucleus accumbens amygdala |
| Opiates | Dopamine opioid peptides | Ventral tegmental area nucleus accumbens |
| Nicotine | Dopamine opioid peptides | Ventral tegmental area nucleus accumbens amygdala |
| THC | Dopamine opioid peptides | Ventral tegmental area |
| Alcohol | Dopamine opioid peptides serotonin GABA glutamate | Ventral tegmental area nucleus accumbens amygdala |

dopamine seems to play a role in potentiating the effects of cocaine and other amphetamines, opiates, nicotine, THC, and alcohol, from several regions of the brain (Koob and LeMoal 2001). According to Koob and Le Moal, during withdrawal from chronic administration of all major types of abused drugs, reward thresholds are increased. Some of these changes can last for up to a week post drug use (2001). Withdrawal from drugs of abuse can result in neurotransmitter-driven effects that may motivate an abuser to seek out their drug of choice. A decrease in dopamine, opioid peptides, serotonin, and GABA can lead to dysphoria, pain, and anxiety. An increase in CRF will lead to stress. Ultimately, an increase in CRF is seen in both ingestion and withdrawal for every drug of abuse (Table 8.3).

The induction of CRF gene expression in the amygdala and the bed nucleus of the stria terminalis by glucocorticoid hormones may also underlie addictive behaviors. These effects appear to be manifested in the anxiety associated with withdrawal (Marcinkiewcz et al. 2009). Both high levels of corticosterone and central CRF have been linked to addictive behavior. For example, CRF infusions into the lateral ventricle facilitate amphetamine-induced self-administration (Sarnyai et al. 1993). Corticosterone levels are known to influence the expression of amphetamine self-administration (Piazza et al. 1991; Cador et al. 1993). Systemic injections of corticosterone increase the likelihood of amphetamine self-administration, as do stressful events (Maccari et al. 1991) via activation of glucocorticoid receptor sites (Steckler and Holsboer 2001), which increase both corticosterone and central CRF (Heinrichs et al. 1995). In fact, amphetamine increases CRF in regions of the brain such as the nucleus accumbens (Cadet et al. 2014).

Corticosterone and CRF levels have also been shown to influence the dopamine-dependent psychomotor effects of both morphine and cocaine self-administration (Deroche et al. 1994, 1995). Corticosterone is known to

**Table 8.3** CRF Response to Drug Use Conditions

| Drugs of Abuse | CRF Response to Ingestion | CRF Response to Withdrawal |
|---|---|---|
| Cocaine | ↑ | ↑ |
| Amphetamines | ↑ | ↑ |
| Opiates | ↑ | ↑ |
| Nicotine | ↑ | ↑ |
| Marijuana | ↑ | ↑ |
| Alcohol | ↑ | ↑ |

influence self-administration of cocaine such that the more corticosterone that circulates, the greater the probability of self-administration (Goeders and Guerin 1996). Moreover, changes in dopamine levels in the nucleus accumbens are dependent upon adrenal steroid function (Barrot et al. 2000). The further activation of both CRF and dopaminergic neurons by glucocorticoids from feed-forward mechanisms creates (when the glucocorticoids remain elevated) a vulnerability for drug abuse.

As indicated previously, CRF in the brain has been linked to the anxiety associated with heroin and morphine withdrawal (Sarnyai et al. 1995), alcohol withdrawal (Pich et al. 1993), and cocaine withdrawal (Sarnyai et al. 1992a, b, 1993, 1995; Zhou et al. 1996). In addition, one study has demonstrated that cannabinoid withdrawal also elevates CRF in the central nucleus of the amygdala (Rodriguez de Fonseca et al. 1997).

Reducing CRF expression in the amygdala reduces morphine withdrawal systems (Heinrichs et al. 1995). Inhibiting corticosterone synthesis through metyrapone decreases the vulnerability to relapse in cocaine self-administration (Piazza et al. 1994). Corticotrophin-releasing factor is linked to stress-facilitated vulnerability to relapse (Shaham et al. 2000). What this suggests is that the steroid may normally act to facilitate the level of neuropeptide in these extrahypothalamic sites that underlie the behaviors. This is similar to the neuroendocrine mechanisms that facilitate and sustain other central motive states.

Addicts typically spend a great deal of time thinking of how to procure the substance they are addicted to. The actual time of pleasure decreases as the addiction takes hold, and it requires more drug to produce the desired effect. This can lead to the sensation that the individual is constantly seeking a sensation that they are no longer able to achieve.

A person responds to the demands of the external environment by changing his or her internal set points. This is evident in blood pressure changes during the course of an individual day, such as rises in response to stressful, arousing situations that are psychologically experienced as pain, fear, or rage, or during the craving for drugs.

Addiction results in chronic arousal as the individual constantly seeks out the drug resulting in an endless hypervigilant state (Koob and Le Moal 2001). Brainstem sites such as the locus coeruleus, ventral tegmental region, and raphe are all involved in the interaction of both molecules and interact with forebrain sites such as the amygdala, bed nucleus of the stria terminalis, and the PVN of the hypothalamus (Valentino et al. 1995; Ma et al. 2013; Walker et al. 2016; Ryan et al. 2011). These sites underlie, perhaps, the chronic anxiety of the addict. The positive induction of CRF or dopamine by corticosterone in regions of the brain that underlie addiction may contribute to the addiction and physiological overload over time, along with the input from brainstem feed-forward mechanisms of arousal from norepinephrine (Koob 1998).

Physiological systems are designed in part to minimize the impact of drastic changes; this holds for situations ranging from the regulation of food intake to that of drug abuse. Anticipatory responses that serve regulatory physiology can reduce the effect of overingesting as much as underingesting a desired substance. Anticipatory responses that serve to reduce the physiological impact of drug use may also hinder the individual: the addict is stuck in a massive web of salient cues reinforcing the chronic want of the drug (e.g., reward; Koob and LeMoal 2001; Berridge et al. 2004; Heilig 2015).

Maintaining an addiction, in some ways, is like maintaining glucose levels or calcium levels. The body requires the drug in order to maintain equilibrium and to avoid becoming ill, much as the body requires glucose or calcium. The negative reinforcement effects of drugs take hold, central motive states are elicited, and both appetitive and consummatory behaviors are expressed. Further, as psychological and physical tolerance develops, the consummatory phase is less and less satisfying. When an individual becomes drug dependent, there are changes in the internal drug set points; there is an allostatic reaction to the constant presence of a drug such as cocaine, heroin, or alcohol. In rats, a limited period of daily access to a drug leads to maintaining the same levels of intake from one session to the next. An increased period of daily access (6 hours) leads to increased self-administration of the drug (Ahmed and Koob 1998). A higher level of intake is required to gain the same hedonic effect from the drug.

The widening range and availability of consumable objects turns the volume up in terms of arousal. The consumption of and withdrawal from known addictive substances (e.g., heroin, cocaine, alcohol, barbiturates, amphetamine, and marijuana) result in elevated levels of CRF in a number of brain regions: the CEA, PVN, and BNST (e.g., Pich et al. 1993, Koob and Le Moal 2005). Glucocorticoids, at least with regard to the CEA, BNST, and accumbens function, may enhance CRF expression and thereby increase the vulnerability to increased salience and attention to those objects associated with drugs of abuse (Koob and Le Moal 2005; Peciña et al. 2006). One way to understand these events is that a function of glucocorticoids is to magnify the effects of CRF in increased attention to objects and their potential value, for example.

Both consumption of and withdrawal from diverse psychotropic drugs elevate glucocorticoids and CRF in various brain regions (Koob and Le Moal 2003

2005); again, glucocorticoids, like dopamine, are essential for organization of action. In fact, that is one of the ways in which the steroid, by regulating the neurotransmitter, influences behavior; in this case, it is an all-too-real occurrence, namely the vulnerability to self-medication with drugs that are harmful. Feed-forward systems include adrenal steroids and their regulation of the CRF and adrenergic action required that underlies the organization of action under adaptive and pathological conditions.

Drugs of abuse begin with reward, escalate to dependence and addiction, and finally to vulnerability to relapse (Koob and Le Moal 2002 2004; Sommer et al. 2008). Reward declines and dependence expands; salience of cues grows as the importance of the drug accrues and the broad-based learning associated with drug use kicks in. In other words, glucocorticoids regulate a number of neuropeptides and neurotransmitters that are knotted to the organization of action. Importantly, glucocorticoids regulate a number of neurotransmitters (e.g., dopamine) that are tied to some aspect of reward or instrumental response (Dore et al. 2013; DiChiara 1999; Koob and Le Moal 2002 2004; Ikemoto and Panksepp 1999). The regulation of dopamine, for instance, underlies the prediction of reward and the cognitive motor capacity underlying action. The regulation of norepinephrine underlies attention and the regulation of serotonin underlies tonic mood and rhythm. These are all vital in the organization of action (e.g., Torres et al. 1998; Zorrilla et al. 2013).

## Addiction Experiences

Experiences throughout the life cycle impact CRF expression; for instance, juvenile experiences impact corticosterone and CRF release (Jacobson-Pick and Richter-Levin 2010). Neural sensitization means the increased excitability of neurons. The ability of glucocorticoids, CRF, and pre-exposure to psychosocial stressors in animals to precipitate greater stress responses indicates that sensitization processes play a major role in turning normal responses into exaggerated abnormal behavior (Dunn and Swiergiel 1999, 2008; Marks and Tobena 1990; Rosen and Schulkin 1998).

Perhaps consistent and reliable positive touch and contact can mediate the long-term effects of trauma (Liu et al. 1997). It is known that rats who are handled, for example, have reduced levels of stress hormone activation at the level of the PVN, resulting in decreased production of CRF, pituitary ACTH, and adrenal corticosterone (Levine 1975; Levine et al. 1967 Meaney et al. 1988, 1989, 1996). They show diminished responsiveness when challenged; it is as if the experience of being cared for or nurtured somewhat enhances the ability to cope with the impact of adversity. This is an important area for increased study, particularly as it may relate to those recovering from addiction.

Successful drug rehabilitation programs often employ social contact as a vehicle toward mental health (Ouimette et al. 1998). The idea of reaching out to others to maintain abstinence is certainly not new. The effectiveness of this approach might be related to the reduced stress levels and, consequently, reduced CRF levels within the recovering individual. We know that social contact is an important adaptive behavior, and it is known to reduce stress hormones and elevate hormones that can enhance well-being (Carter et al. 1999). An important strategy is to make contact and sustain meaningful relationships when coping with duress (Taylor et al. 2000).

The diverse models of addiction include a constellation of features, including impulse control, tolerance and conditioning, consumption, and incentive features of drug-related events (memory and the degree of liking and wanting) (Koob and Le Moal 2005; Berridge and Robinson 1998; Wise 2005). We know that not all addictive substances are alike, and the diverse array have only grown over time. Most cultures, old and new, have a long list of natural and artificial mind-altering drugs. Thus, CRF blockers by themselves will also not solve substance abuse problems. The human connectedness factor must be addressed (Jaspers 1913).

## Conclusion

Diverse information molecules are activated or deactivated during both acquisition and withdrawal (Zorrilla et al. 2012; Martin-Fardon et al. 2010). The existing literature reminds that CRF in particular is linked to both the rewarding and aversive aspects of psychotropic drugs. Binges are tied to diverse forms of compulsive behaviors reliant on CRF at many regions of the brain (Zhou et al. 1996; Zho et al. 2003, as is the anxiety that pervades these behaviors (Sahuque et al. 2006). CRF activation is linked to fear or anxious states (Lehner et al. 2008; Sanford et al. 2015), which contribute to individuals' in-built tendencies toward anxiety or self-medication (Landgraf and Wigger 2003) via the CRF-R1 receptor (Liebsch et al. 1995; Theoringer et al. 2012), and that includes disinhibition of behavioral vulnerabilities (Isogawa et al. 2013; Douma et al. 2014), panic attacks (Shekhar et al. 2011), and cocaine addiction (Corominas et al. 2010). CRF and CRF type 1 receptors are tied to pain-related angst in the amygdala (Ji et al. 2007; Myers et al. 2005, 2011; Zorrilla et al. 2013). CRF activates both vagally related peripheral and central sites (Nijsen et al. 2000).

Animal research suggests that the amygdala, nucleus accumbens, and ventral striatum are important for associating, acquiring, and responding to the conditioned and reinforcing effects of stimuli associated with addiction (Robbins et al. 1989; Altman et al. 1996; Robbins and Everitt 1999; Ito et al. 2000; Carboni et al. 2000; Thomas and Everitt 2001; Koob 2015). The chronic arousal, fear, and uncertainty that pervade the life of an addicted individual activate many systems in the brain, including the dorsal raphe and amygdala CRF neurons (Zorrilla et al. 2012). Not only are the normal feed-forward systems active, but the circuits that underlie a reward, and probably fear, are vulnerable to breakdown (see Koob and Le Moal 2001; Koob 2015). CRF may potentially engage with hypothalamic and midbrain emotional systems through GABAergic interneurons, as well as the brainstem through norepinephrine systems. These systems activate CRF and help promote motivational withdrawal by recruiting brain stress systems, leading to addiction (Koob 2009 2010).

Cocaine or nicotine-seeking behaviors are locked into both the ventral tegmental pathway and to the bed nucleus of the stria terminalis, which regulates CRF action within the tegmentum (Vranjkovic et al. 2014; Grieder et al. 2014; Boyson et al. 2014). Ventral tegmental CRF neurons are also important for nicotine withdrawal (Grieder et al. 2014). Brain samples from several human subjects have shown that TH and CRF are most often co-localized in the VTA (Grieder et al. 2014). Importantly CRF pathways from the ventral tegmental region to the bed nucleus underlie, for instance, alcohol consumption (Silberman et al. 2013).

CRF is also involved in diverse ways with the ingestion of food resources. Infusion of CRF across the neural axis decreases food intake (Moore et al. 1992; Grill et al. 2000), and appears to do so via a type 1 receptor (Iemolo et al. 2013). Indeed the CRF type 1 receptor and CRF in general are tied to diverse behaviors including fear and conditioned increased heart rate (Gao et al. 2016). CRF is long known to have differential effects on various strains of animals (Glowa and Gold 1991; Wiersma et al. 1998) that reflect genetic differences, or perhaps genetic editing or epigenetic experiences (McCarthy et al. 2009; Bale 2009, 2014), that create a more vulnerable condition for excessive pathological states and a diminution of regulation or avoidance of what is potentially harmful.

By definition, *sensitization* implies that the threshold for activation of the system is lower following the presentation of a stimulus. In other words, the system becomes hyperexcitable. The role of sensitization in the development of addiction is not well-understood (Koob 1989, 1998; Marks and Tobena 1990; Robinson and Berridge 2000). Theoretically, sensitization is an important factor in the etiology of hyperexcitability of the circuits underlying addicts' behavior and to CRF regulation (Richter et al. 1995). Chronic uncertainty (which can engender hyperexcitability) can exacerbate the sensitization process.

Experimental paradigms in which an inescapable foot shock is delivered to rats increase generalized anxiety and can increase the vulnerability of relapse to drug abuse (Erb and Stewart 1999).

Induction of hyperexcitability from repeated exposure to activation may induce a cascade of biological events that include the expression of immediate early genes, such as peptides and structural proteins that render the tissue more sensitive to subsequent stimulation (Rosen and Schulkin 1998), including CRF (Thoeringer et al. 2012). Glutamate receptors, particularly the NMDA type, are important for the development of long-term potentiation and hyperexcitability in the hippocampus and the amygdala (LeDoux 2000). CRF in addition to catecholamines facilitate glutamatergic expression (Silberman and Winder 2013). Foot shock–induced sensitization and facilitation of a fear-conditioned response require NMDA activation; CRF within NMDA-NR1 projects to neurons within the bed nucleus of the stria terminalis (Beckerman et al. 2013).

This hyperexcitability from repeated exposure or of the amygdala during seizure activity can be translated as a physical metaphor for drug addiction. The states of craving and withdrawal are often characterized by psychomotor overactivity (Wise and Bozarth 1987). CRF facilitates movement in every species studied just as the amygdala becomes overactive when glucocorticoids and CRF levels are elevated (Cook 2001; 2002; 2004).

Addiction is not a pleasurable state and is characterized by overactivity and obsession with the search for and consumption of psychotropic drugs. CRF concentrations are increased during aversive and uncertain events, and the state of addiction is riddled with uncertainty.

Chapter 9

# Conclusion: The Study of CRF—An Epistemic and Medical Endeavor

## Introduction

CRF remains as it started: a very reasonable molecule for study that underlies diverse forms of behavioral biological adaptation. CRF is endlessly tied to many other information molecules (Kaminski and Watts 2012; Uban et al. 2013; McEwen et al. 2009). And while this book has emphasized CRF (as I have said throughout the book), it is just one information molecule among others. The activation of CRF can easily promote fear-related states, whether they are innate or learned (Donatti and Leite-Panissi 2011; Rosen et al. 2004; Grillon et al. 2015). Many forms of social duress activate CRF gene expression (e.g., bed nucleus of the stria terminalis; Choi et al. 2006), and diverse forms of social attention and learning are mediated by CRF (Heinrichs et al. 2003). CRF density varies with sociality in many regions of the brain including the nucleus accumbens (Beery et al. 2014 2016). Early social life experience influences CRF expression in older animals (e.g., the nucleus accumbens and ventral tegmental area; Lukkes et al. 2009), as well as vulnerability for depressive and cardiovascular pathology (Wood et al. 2012; Nemeroff and Owens 2004; Gold et al. 1988).

Each of the preceding chapters has highlighted fundamental roles in which CRF underlies biological events. We began with a depiction of just how ancient this one information molecule is: perhaps a half billion years old (Lovejoy and Balmert 1999; Tan et al. 2012; Chand et al. 2013). Its roles are diverse, but historically, CRF is best known within the HPA axis as a releasing factor (Vale et al. 1981; Bale and Chen 2012). At the time of Vale's research on CRF, its role in the HPA axis was its only known function. Like other information molecules, since is original discovery, scientists have uncovered the actual great diversity of function of CRF, such as its role in behavior (Koob and Bloom 1985; Bale and Chen 2012; Chrousos 1995; McNally and Akil 2002).

## CRF and Biological Regulation across Species

CRF is linked to most forms of biological regulation in vertebrates and perhaps in invertebrates; after all CRF has been linked to circadian rhythmicity (a very ancient regulatory system). Clocks mirroring the light–dark rhythms of our planet instantiated in the brain and other end organ systems and perhaps, mediated in part by diverse information molecules, including CRF (Watts and Sanchez-Watts 1995; Swanson 1991).

Neuropeptides were understood as a window into one form of information processing: synaptic transmission in the brain (Swanson 1996; Hokfelt et al. 1980). But the issue, although understood within the HPA axis and then later within extrahypothalamic regions that contain CRF-related cells or receptors (Swanson 1991), is we know now just how broadly represented this information molecule is across bodily tissue.

In vertebrates, CRF and urocortin interact with binding proteins to help ease binding with the CRF receptors. The same is seen in insects, with the ligand DH and its specific receptor and binding protein (Lovejoy et al. 2014; Figure 9.1)

Neuroendocrine cells are well represented in insects (e.g., Sellami et al. 2011), and CRF is found in insects (Husing and Flik 2005; De Loof et al. 2012). Like other peptides, they promote diuretic-related events and feeding behaviors (Van Wielendaele et al. 2012; Hope et al. 2000). CRF in insects is tied to fundamental forms of activity and rest (e.g., movement to approach and avoid events; Matsuda 2013; Huising and Flik 2005; DeLoof et al. 2012).

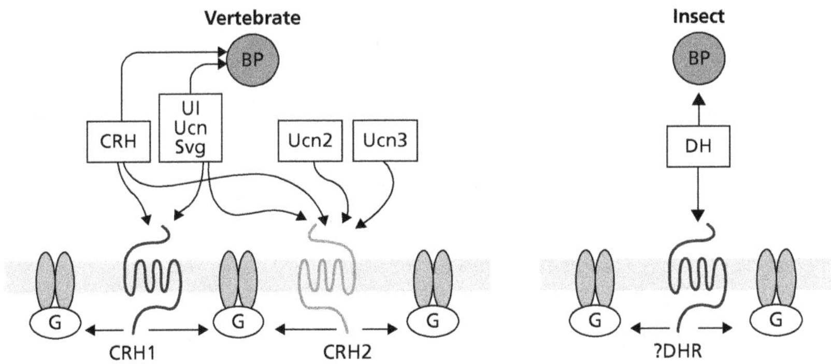


**Figure 9.1** (See color plate section) Scheme of the molecular interaction among CRF ligands, receptors, and the binding protein in chordates and insects.
Reproduced from David A. Lovejoy, Belinda SW Chang, Nathan R. Lovejoy, and Jon del Castillo, Molecular Evolution of GPCRS: CRH/CRH Receptors, *Journal of Molecular Endocrinology*, 52 (3), pp. T43–T60, Figure 2, doi: 10.1530/JME-13-0238, Copyright © 2014, Society for Endocrinology.

Gene duplication and diversification is a major theme in the consideration of CRF evolution and its long history as an information molecule (e.g., as seen in teleost fishes; Grone and Maruska 2015); one key feature is its link to inflammation and to inflammatory signaling, for instance, in mast cells (Theoharides and Cochrane 2004). The finding about skin's production of CRF and the role of CRF in the HPA axis and other signaling systems is quite amazing (Slominski 2009).

Activity is prompted by the activation of CRF, activity that underlies the development of end organ systems, for instance, fetal programing (Sandman et al. 2013), lung development (Class et al. 2008), or, for that matter, whole transformation such as metamorphosis (Denver 1999).

The adaptive system is in the promotion of development or change; CRF is a protean signal for change across diverse end organ systems. Although CRF is most often tied to stress, and indeed diverse conditions of adversity elicit responses from CRF systems, so to do conditions of change and phenotypic plasticity (Denver 1999). A major goal of this book is to emphasize that the existence of CRF is not just a sign of pathology or devolution of function but also an important information molecule that plays a key role in many normal physiological/behavioral functions, alongside other information molecules.

CRF is regulated by both traditional genomic and nongenomic membrane systems; in other words, both a fast-acting and a slower induction of CRF play a role in the regulation of behavior and physiology (Dallman et al. 1987; Jolës and Baram 2009). Rapid-fire activation and regulation are tied to the membrane regulation.

Fundamental to CRF is its tie to inflammation and diverse information molecules involved in the immune response that range from pregnancy (Torricelli et al. 2011a, b) to obesity and the expansion and then deterioration of diverse tissue, including that of brain (Sapolsky 1992; Goebel et al. 2010). Maternal characteristics during pregnancy are tied to CRF levels (Chen et al. 2010). One important role for CRF is when the pregnancy and fetal outcome are in danger.

CRF is a primary signal of danger across diverse events such as cancer and infectious diseases (Wadhwa et al. 2001; Sato et al. 2002). Each chapter has shown both the role of CRF under normal conditions and conditions in which adversity and danger are anticipated or realized.

Glucocorticoids regulate CRF in multiple ways; they can stimulate, restrain, or have no effect at all on CRF neurons in the brain (Swanson and Simmons 1989; Cook 2002; Makino et al. 1994; Watts 2005; Gray et al. 2013; Muglia et al. 1995; Wang et al. 2013). This holds for brain tissue, in addition to diverse end organ systems (e.g., the placenta; Majzoub et al. 1999; Challis et al. 1995; Robinson et al. 1989). Recent findings demonstrate that glucocorticoids upregulate CRF in regions of the prefrontal cortex (Merali et al. 2008; Gray et al. 2013 2015) and that they affect endocannabinoids via fast-acting mechanisms (Hill and McEwen 2009).

Diverse neuropeptides, peptides, and their receptors are involved in many of the biological events described in this book. The degree of peptide expression (e.g., Denver 2009; Campbell et al. 2004) underlies the biology of change, through maintenance and enhanced or degraded expression. The massive peptide, steroid, and transmitter factory in the placenta (Power and Schulkin 2006, Peraglia et al. 2005) influences end organ development (e.g., brain; Bonnin et al. 2011).

Social contact is essential for mammalian development and it attenuates the excessive rise in CRF and glucocorticoids that results from many conditions (Ruscio et al. 2007; Meaney 2001). Indeed, diverse steroids interact with CRF, underlying most forms of behavioral and physiological adaptations, as well as the breakdown and pathology or devolution of function. The birth process across species (Smith 1998; Sandman and Glynn 2009; Majzoub et al. 1999) and metamorphosis (Denver 1998) are events that stand out. After all, what is as dramatic as these events? Well it turns out many events are dramatic, startling, and unbelievably bold and beautiful.

Indeed the very reason for the existence of the mammalian placenta is to act as a storehouse of information molecules, of which CRF is just one among many (Perkins and Linton 1995). Most forms of behavioral and physiological conflict elicit CRF activation in the placenta during pregnancy (Florio et al. 2000, 2003; Gangestad et al. 2012). Indeed maternal/fetal interactions under normal conditions contribute to parturition (Kramer et al. 2009, 2013; Class et al. 2008).

## Role of CRF Receptors

Repeatedly the CRF type 1 receptor (CRF-R1) has been highlighted and is dominant in the literature (Gammie and Stevenson 2006; Labermaier et al. 2014), and information has been revealed through genetic targeting or transgenic manipulations (Weninger et al. 1999; Muglia et al. 2001). This is particularly pronounced in the literature on fear and addiction (Koob 2009). The CRF type 1 receptor is consistently linked to diverse behavioral and physiological events that represent signaling of diverse forms of danger. Adapting to change or facilitating change within the life cycle, CRF also contributes to the devolution of function (e.g., Koob and Bloom 1985; Gammie and Stevenson 2006; Majzoub et al. 1999; Muglia 2000).

One demonstration of the importance of the CRF type 1 receptor is depicted in the next figure (Figure 9.2). CRF-R1 exerts anxiety-inducing effects on glutamate neurons but anxiety-inhibiting effects via dopaminergic neurons. These effects were exhibited by time spent by knockout mouse models in lit compartments and total distance traveled, indicative of anxiety levels in novel environments. These effects were studied in GABA, CNS, and serotonin (5-HT) knockout models, as well, with no changes found in either anxiety or locomotion (Refojo et al. 2011; Figure 9.2).

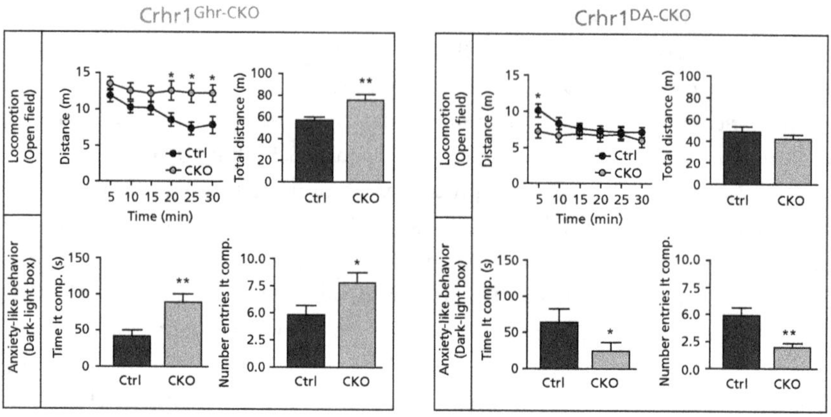


**Figure 9.2** CRFR1 exerts anxiogenic effects acting on Glu neurons but also anxiolytic effects via DA neurons. (Left) CRFr1Glu-CKO mice spent more time in (U = 25.00; **P < 0.01) and showed more entries into the lit compartment (U = 33.5; *P < 0.05; n = 10 to 12 mice per group). CRFr1Glu-CKO mice showed an increase in total distance traveled as compared with controls [analysis of variance (ANOVA) repeated-measures + t-test; *P < 0.05; n = 10 to 12 mice per group]. (Right) CRFr1DA-CKO mice showed increased anxiety-like behavior reflected by a decrease in the time spent in (U = 48.50; *P < 0.05; n = 10 to 12 mice per group) and number of entries into the lit compartment (U = 37; **P < 0.01, n = 10 to 12 mice per group). A difference in locomotion during the first 5 minutes is due to an increased anxiety-related response to a novel environment (ANOVA repeated-measures + t-test; *P < 0.05; n = 10 to 12 mice per group).

From Damian Refojo, Martin Schweizer, Claudia Kuehne, Stefanie Ehrenberg, Christoph Thoeringer, Annette M. Vogl, Nina Dedic, Marion Schumacher, Gregor von Wolff, Charilaos Avrabos, Chadi Touma, David Engblom, Günther Schütz, Klaus-Armin Nave, Matthias Eder, Carsten T. Wotjak, Inge Sillaber, Florian Holsboer, Wolfgang Wurst, and Jan M. Deussing, Glutamatergic and Dopaminergic Neurons Mediate Anxiogenic and Anxiolytic Effects of CRHR1, *Science*, 333 (6051), pp. 1903–1907 DOI: 10.1126/science.1202107, Copyright © 2011, American Association for the Advancement of Science. Reprinted with permission from AAAS.

And while a dominant focus has been on the CRF type 1 receptor, throughout the book I have mentioned that the CRF type 2 receptor and urocortin have been linked, although much less so, to the diverse phenomena described in this book (e.g., Risbrough et al. 2009; Takahashi 2001; Klampfl et al. 2014; Weiser et al. 2008; Wolf et al. 2013; Issler et al. 2014; Skorzewska et al. 2011 Henry et al. 2006; Ohmura et al. 2010; Voltolini et al. 2014; Martin-Fardon et al. 2010; Funk and Koob 2007; Chen et al. 2006; Tabarin et al. 2007; Hope et al. 2000; Bale et al. 2000, 2003; Morisot and Contarino 2016; Holly et al. 2015; Slater et al. 2016; Uchoa et al. 2015; Bosch et al. 2016; Komuro et al. 2016; Rinker et al. 2016).

## CRF and Addiction

Figure 9.3 shows the many information molecules in addition to CRF that are tied to diverse events in addiction. Addiction is a multidimensional phenomenon. What stands out in the end are two events: the focus on one thing as a consummatory object and the physical withdrawal from its consumption.

In drug addiction, the central nucleus of the amygdala is most likely affected at GABAergic synapses. In the case of ethanol, GABA release is enhanced through both direct and indirect action of CRF-R1 acted on by GABAergic afferents and interneurons. CRF-receptor antagonists and some drugs (e.g., gabapentin and κ-Opioid antagonists) decrease presynaptic GABA release in dependent animal models (Koob 2015). Within the ventral tegmental area (VTA), research has shown that CRF acts to reinstate stress-induced cocaine seeking in rats through a GABA-dependent pathway. In a cocaine-seeking model with an extinguished-conditioned response, the expected effect of CRF injection (i.e., reinstatement of cocaine seeking) was blocked by a GABA-B antagonist (Blacktop et al. 2016).

As described previously, other brain regions, information molecules, and systems are involved in drug addiction. CRF antagonism within the VTA lessened anxiogenic effects in cocaine self-administering rats (Ettenberg et al. 2015). In an experiment with stressed mice given intermittent access to alcohol, CRF-R1 antagonists infused into the VTA reduced the intake of alcohol, and increased the output of dopamine to the nucleus accumbens (Hwa et al. 2015). Another study considered the effect of CRF-receptor-dependent mechanisms on neurons of the central nucleus of the amygdala exposed to alcohol. Alcohol resulted in an increase in the spontaneous excitatory postsynaptic currents (measuring glutamatergic transmission); the effect of alcohol on transmission was nullified by CRF-receptor type 1 and 2 antagonists (Silberman et al. 2015).

Withdrawal, not just ingestion, of addictive substances results in increased levels of CRF and other information molecules. Methamphetamine withdrawal in one study resulted in increased levels of CRF neurons in the amygdala, striatal μ–opioid receptor binding, and oxytocin receptor upregulation in the lateral septum (Georgiou et al. 2016). In another study, both methamphetamine self-administration and withdrawal treatment groups resulted in no change in CRF-immunoreactive fiber density, but did show oxytocin-immunoreactivity fiber density changes in the nucleus accumbens and subthalamic nucleus (Baracz et al. 2015). Clearly, many information molecules are tied to and interact during addiction: CRF being just one important molecule.

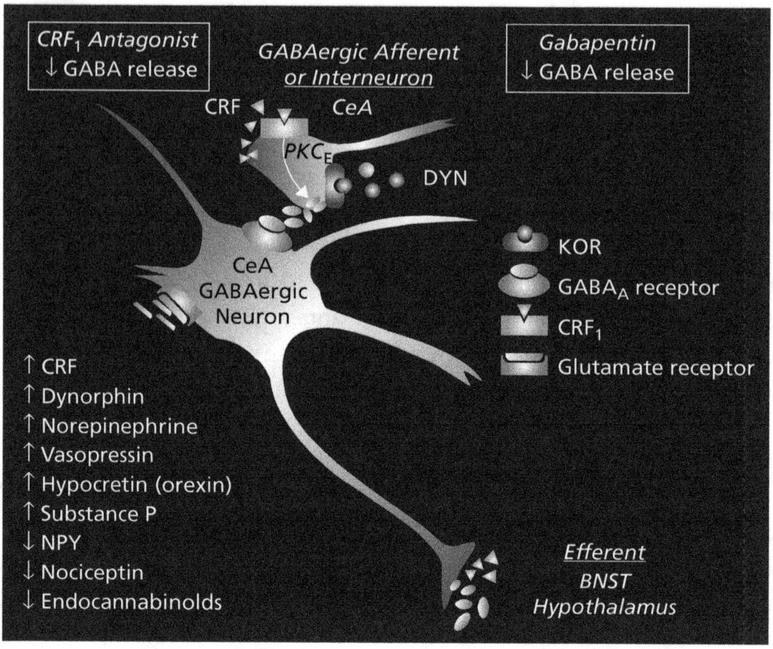


**Figure 9.3** (See color plate section) Cellular neuroadaptive mechanisms in the central nucleus of the amygdala in drug addiction. Simplified schematic of rodent central nucleus of the amygdala circuitry and hypothetical sites of ethanol and CRF action on GABAergic synapses. Most neurons in the CEA are GABAergic inhibitory projection neurons or interneurons that contain CRF or other neuropeptides as co-transmitters. (Upper synapse) Ethanol may enhance the release of GABA (filled ellipsoids) from GABAergic afferents or interneurons either via release from the same terminal as CRF (triangles), which then acts on CRF1 receptors on the terminal to elicit (arrow) release of more GABA via a PKCε-mediated mechanism, or direct activation of CRF1 receptors to elicit the release of more GABA (Bajo et al. 2008). CRF1 antagonists and the drug gabapentin decrease presynaptic GABA release in dependent animals (Roberto et al. 2008, 2010). κ-Opioid antagonists have similar effects as CRF1 antagonists in rats that present an escalation in cocaine intake (Kallupi et al. 2013). Thus, CRF, dynorphin, and ethanol augment the inhibition of CEA projection interneurons (co-containing CRF, opioids, or NPY), leading to the excitation of downstream (e.g., BNST) neurons through disinhibition. The activation of presynaptic cannabinoid CB1 or NPY receptors (not shown) may reduce GABA release onto CEA inhibitory projection neurons, increasing their excitability and release of GABA onto downstream targets, such as in the BNST. CRF, corticotrophin-releasing factor; GABA, γ-aminobutyric acid; CEA, central nucleus of the amygdala; DYN, dynorphi; KOR, κ opioid receptor; NPY, neuropeptide Y; BNST, bed nucleus of the stria terminalis.

## CRF and Change

One primary function of CRF is production and mediation of change. Life changes occur across the life cycle; promoting change in all animal studies to date (e.g., insects, fish, amphibians, reptiles, and mammals), and CRF elicits a diverse range of behaviors from aggression to anxiety (Backstrom et al. 2011). A fruitful understanding of CRF with regard to regulatory behaviors is the context in which CRF is elicited and the pathways that underlie diverse forms of behavior can be differentiated, and the co-location of neurotransmitter systems (Berridge 2004). Indeed, the penetration into core neurotransmitter systems is one of the outstanding stories of CRF diversification of function (e.g., GABA, dopamine, norepinephrine, and serotonin).

Throughout the life cycle there are sex specific changes in methylation-mediated CRF expression (Sterrenburg et al. 2011) that underlie changes to pain and anxiety (Tran et al. 2014) and vulnerabilities that are afflictive (Buffalari et al. 2012; Sanford et al. 2016). Moreover, marked sex differences are apparent in the CRF literature. In fetal development (Sandman et al. 2013; Bale 2009) and in psychotropic drug abuse (Buffalari et al. 2012), the core anatomy of the CRF system and sexual dimorphism signaling systems (Valentino et al. 2013) influence behavior (Bale 2011; McEwen 2009; Lenglos et al. 2013). CRF-related sex differences are noted in fear (Brunton et al. 2011; Gilman et al. 2015) and in glucocorticoid/CRF relationships (Bingham et al. 2013).

Early life experience with food availability is another event of unexpected change and adversity with long-lasting consequences (i.e., responses to adversity and lung development; Brunson et al. 2005; Muglia et al. 1995). As noted earlier, studies of macaques show that variable food availability during development resulted in greater CRF levels in the cerebrospinal fluid during adulthood and greater fear-related behavioral responses (Coplan et al. 2001). The changes in diverse regulatory systems, as a result of early trauma, stem from a lack of predictability and extreme demands put upon the CRF system.

Injections of the CRF-receptor antagonist (D-Phe-CRF(12-41)) decreased frustration and stress-induced binge-eating in animals with a history of intermittent food restriction. In animals with a history of food restriction, given the highest concentration of CRF injection, there was significantly less stress-induced binge-eating than in those given only a vehicle injection. And those animals with a history of food restriction given a vehicle injection showed more stress as compared with animals with no history of food restriction at both time points (15 minutes into test session and after 2 hour test session) (Di Bonaventura et al. 2014; Figure 9.4).

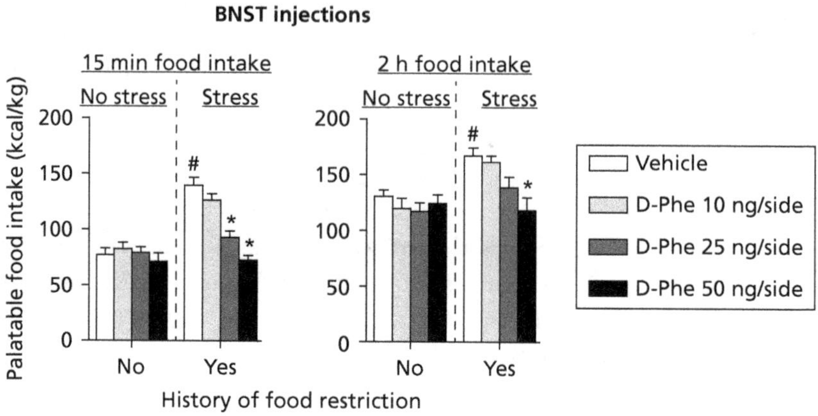


**Figure 9.4** BNST and ventricular injections of the CRF-receptor antagonist D-Phe-CRF(12–41) decreased frustration stress-induced binge eating in rats with a history of intermittent food restriction. BNST injections. Mean + /– SEM palatable food intake (kilocalories per kilogram) during the first 15 minutes (left) and the total 2-hour (right) test session. *$p < 0.05$, different from the vehicle condition; n 6–8 per group. # $p < 0.05$, different from the vehicle no-restriction plus no-stress group.
Reproduced from Maria Vittoria Micioni Di Bonaventura, Roberto Ciccocioppo, Adele Romano, Jennifer M. Bossert, Kenner C. Rice, Massimo Ubaldi, Robyn St. Laurent, Silvana Gaetani, Maurizio Massi, Yavin Shaham, and Carlo Cifani, Role of Bed Nucleus of the Stria Terminalis Corticotrophin-Releasing Factor Receptors in Frustration Stress-Induced Binge-Like Palatable Food Consumption in Female Rats with a History of Food Restriction, *The Journal of Neuroscience*, 34 (34), pp. 11316–11324, Figure 4a, doi: 10.1523/JNEUROSCI.1854-14.2014, Copyright © 2014, Society for Neuroscience.

## Co-localization of CRF and Other Neurotransmitters

The co-localization of CRF and interaction with major neurotransmitters is now well known; this holds for serotonin, dopamine, GABA, and NE (e.g., Beckerman et al. 2013). A longitudinally sectioned dendritic profile shows GFP and CRF expression, as does that of an axon terminal (Figure 9.5).

This is also site specific (BNST) and functional, and diversification of function and the ubiquitous way in which this peptide or neuropeptide is involved explains its diverse and numerous functions. A role for CRF in this region is exemplified by cocaine relapse (Erb et al. 2001), anxiety (Davis et al. 1997; Lee et al. 2008), and social contact (e.g., parental behavior and contact with parents; Gos et al. 2014; Carter et al. 1999). CRF involvement across a wide spectrum of behaviors is apparent.

Some neurotransmitters associated with CRF within the BNST are depicted alongside their effects in the next table. CRF, norepinephrine (NE), and dopamine (DA) all enhance glutamatergic input in the dorsal BNST (dBNST). NE

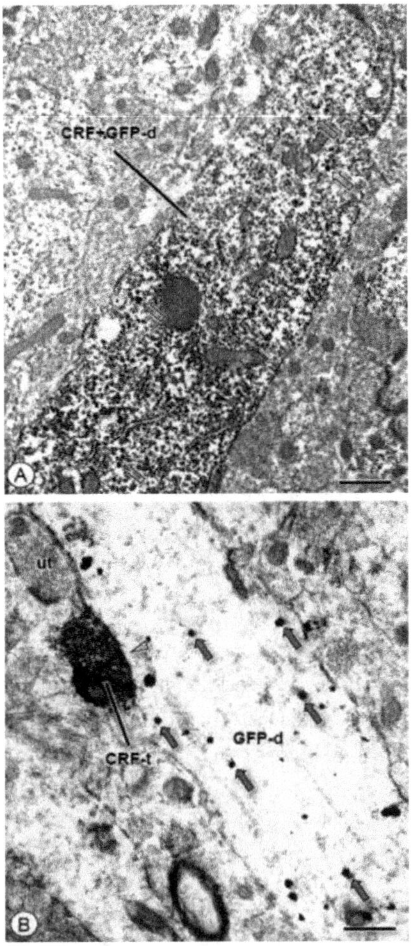


**Figure 9.5** (a) A longitudinally sectioned dendritic profile (CRF + GFP-d) expresses the diffuse immunoperoxidase immunoreaction product for GFP. Immunogold particles for CRF (unfilled arrows) are also present in this profile. (b) An axon terminal (CRF-t) labeled for CRF forms a symmetric inhibitory-type synapse (open arrow head) with a dendritic profile (GFP-d) containing IGS labeling for GFP (gray arrows). Scale bars: 0.5 μm.
Reprinted from *Experimental Neurology*, 239, Marc A. Beckerman, Tracey A. Van Kempen, Nicholas J. Justice, Teresa A. Milner, and Michael J. Glass, *Corticotropin-releasing factor in the mouse central nucleus of the amygdala: Ultrastructural distribution in NMDA-NR1 receptor subunit expressing neurons as well as projection neurons to the bed nucleus of the stria terminalis*, pp. 120–132, doi:10.1016/j.expneurol.2012.10.009, Copyright © 2012 Elsevier Inc., with permission from Elsevier.

also depolarizes CRF neurons in the dBNST (as does DA) and attenuates glutamatergic transmission in the ventral BNST. Serotonin (5-HT) inhibits and depolarizes portions of the dBNST, and it attenuates glutamergic transmission in the dBNST (Daniel and Rainnie 2016, Table 9.1).

**Table 9.1** A summary of the behavioral and physiological action of corticotrophin releasing factor (CRF), norepinephrine (NE), dopamine (DA), and serotonin (5-HT) in the BNST.

| Modulator | Physiological Action |
|---|---|
| CRF | – Enhances glutamatergic input in dBNST |
| NE | – Enhances glutamatergic input in dBNST<br>– Depolarizes CRF neurons in dBNST<br>– Attenuates glutamatergic transmission in vBNST<br>– Chronic stress results in LTD of glutamatergic input |
| DA | – Enhances glutamatergic input dBNST<br>– Depolarizes CRF neurons in dBNST |
| 5-HT | – Attenuates glutamatergic input in dBNST<br>– Inhibits a large portion of the dBNST and depolarizes a small portion of the dBNST<br>– The inhibitory/excitatory balance may shift after chronic stress |

Adapted with permission from Macmillan Publishers Ltd: *Neuropsychopharmacology*, 41 (1), Sarah E Daniel and Donald G Rainnie, Drug Addiction, pp. 103–125, doi:10.1038/npp.2015.178, Copyright © 2016, Nature Publishing Group.

Co-localization with diverse neurotransmitters and peptide systems has been emphasized throughout the book. And drug targeting is devoted toward understanding CRF signaling systems (Refojo and Holsboer 2009). CRF-R1 is tied to the regulation of most of the biology in this book; fear and anxiety are just one context. But we have come to see that CRF is not the same in terms of function and physiology, depending upon where it is expressed in the brain (Rainnie et al. 2004; Sanford et al. 2016; Partridge et al. 2014, 2016).

## Transgenic Animal Models and CRF

Altering CRF in animals has provided a beautiful array of many genetically engineered CRF models and shown the impact of CRF on diverse forms of behavior and physiology (e.g., mouse, Chen et al. 2015; rat, Pomrenze et al. 2015). Using viral vectors, many animal models have been created with manipulated brain regions like the BNST, CEA, basolateral amygdala, and globus pallidus (GPe). Effects on phenotype are varied in these cases and include such changes as decreased binding density to receptors, both decreased and increased anxiety, and enhanced fear memory expression (Gaffrod et al. 2014). Some models resulted in changes to the HPA axis, locomotion, and fear conditioning (Giardino et al. 2001). CRF deletion and overexpression models, as well as receptor mutant models, have been constructed, resulting in various phenotypes with increased and decreased levels of pre-existing information molecules. Other phenotypes of note resulting from these transgenic models include Cushing syndrome, adrenal hypertrophy, and increased REM sleep (Laryea et al. 2012).

## Therapeutic Implications of CRF

The potential therapeutic implications are at the heart of CRF studies. The wide range of biology in which CRF participates, including that of inflammation (Benou et al. 2005), angst, social contact, food intake, and brain deterioration are but a subset of an inviting list of possibilities. The therapeutic endeavors have continued an orientation toward the type 1 receptor. And there are a wide variety of CRF type 1–receptor antagonists, such as NOVA, antalarmin, NBI 34031 (Ising et al. 2007) and NBI 27914 (Grigoriadis 2003; Grigoriadis et al. 2009). In animal models, therapeutic attempts using CRF antagonists have proven promising and successful in reversing the effects of genetic alterations. A recent study explored the use of the CRF-R1 antagonist, R121919, in an Alzheimer's disease transgenic mouse model, with positive results (i.e., reduction of cellular and synaptic deficits, beta amyloid, and C-terminal fragment-β levels in males and females, and it prevented cognitive impairment in females; Zhang et al. 2015, Kunzel et al. 2003, 2005). But most of the therapeutic results when introduced to human patients through clinical trials to date have been mild, underwhelming, and disappointing (e.g., Dunlop et al. 2014; Held et al. 2004; Bailey et al. 2011; Zobel et al. 2000; Coric et al. 2010). The potential for therapeutic interventions in human patients are depicted in Table 9.2.

Thus far, very few therapeutic interventions have been successful. One reason the therapeutics have not worked is targeting the peptide alone. It is likely that it will be necessary to couple the peptide to neurotransmitters and then targeting these to specific regions of the brain. After all, the CRF event occurs in a context with other neurotransmitters (e.g., dopamine, GABA, and NP), and each seems to play a site-specific role.

Many ongoing studies using optogenetics have targeted specific regions of the brain; the review of prior research suggested that these regions were relevant to study (e.g., bed nucleus); thus, many studies target what may require short-term manipulations (e.g., optogenetics). Recent research using these methods has explored the BNST, and experiments have shown that inhibition of CRF neurons in the CEA results in reduced retention of fear memories and freezing behavior in response to contextual fear conditioning. A green laser light was used effectively to turn off the CEA CRF neurons involved in a fear learning response (Sanford et al. 2016; Asok et al. 2016a; Partridge et al. 2016).

**Table 9.2** Summary of potential therapeutic indications for CRF-receptor antagonists.

| Potential Therapeutic Indication | Key Evidence for CRF Antagonist Involvement | Results |
|---|---|---|
| Depression | CRF is hypersecreted in patients with major depression and has been implicated in melancholic and atypical depression, as well as in post-traumatic stress disorder. | Experimental results have been varied, with some studies presenting significant results, and others insignificant. Sleep and weight gain have also been studied. |
| Anxiety | CRF is hypersecreted in patients with anxiety disorders. Antagonism of both CRF-R1 and CRF-R2 receptors has been shown to reduce anxiogenic behavior. | Experimental results have been varied, with one study presenting significant results in experimental models of human anxiety, and insignificant in patients with generalized anxiety disorder. One study showed an overall increase in fear but no change in overall anxiety. |
| Infantile seizures | Selective CRF-R1 receptor antagonists reverse limbic seizures in neonatal rats. | Treatment was not able to penetrate the blood-brain barrier. The effect on peripheral stress hormones was minimal. |
| Substance abuse | Patients suffering from alcoholism can have elevated CRF levels in their cerebral spinal fluid. | There is no effect on alcohol craving, emotional responses, or anxiety. |
| Irritable bowel syndrome | Irritable bowel syndrome has been linked to psychiatric disorders. CRF-R1 receptor antagonists block stress-induced hyperalgesia and colonic motor function. | Results were insignificant in colonic GC24, gastric emptying; orocecal transit; ascending colon emptying; half time; and stool frequency, consistency, and ease of passage in one experiment. In another, antagonists improved gastrointestinal motility, visceral perception, and anxiety/mood in response to gut stimulation. |

Adapted from Dimitri E. Grigoriadis, Corticotropin-Releasing Factor Receptor Antagonists: Potential Novel Therapies for Human Disease, *Celltransmissions* 19 (4), pp. 3–10, Table 1, Copyright © 2003, Sigma-Aldrich Co. LLC.

Current clinical studies involving CRF antagonists include stress-induced craving in alcoholism, post-traumatic stress disorder, and social anxiety disorder, for example (clinicaltrials.gov as of 2015).

**Final comments**: Importantly, CRF is a diverse scientific object; the CRF known to one area of study and end organ system is not necessarily known by others; hence, a major reason why I put this book together. CRF is a moving object; even the chemical signature varies with species and function in various end organ systems.

It has been a remarkable journey, still ongoing, since many basic questions have still not been answered.

The CRF systems are a scientific object and a concept. This text gives some sense of CRF as a scientific biography (see Daston 1998). As an object of study, CRF has features common to other biographical depictions. But CRF stands out; it is tied to theory and indeed many forms of scientific inquiry. The scientific community had a concept of CRF before it had a specific molecule called CRF. As noted previously, the original work on CRF in the HPA axis provided a fixed reference for the concept as an actual object, whose chemical structure and diverse functions we now know vary only slightly with species.

Subsequent research discovered that it was a primary neuropeptide and one widely distributed in the body and across species. Considering CRF as a peptide hormone is a rich concept that has been fruitful to study, as it is embedded in diverse scientific communities. A search for articles containing the keyword *corticotrophin releasing factor* produces unsurprising results from the last few decades (Figure 9.6). According to the U.S. National Library

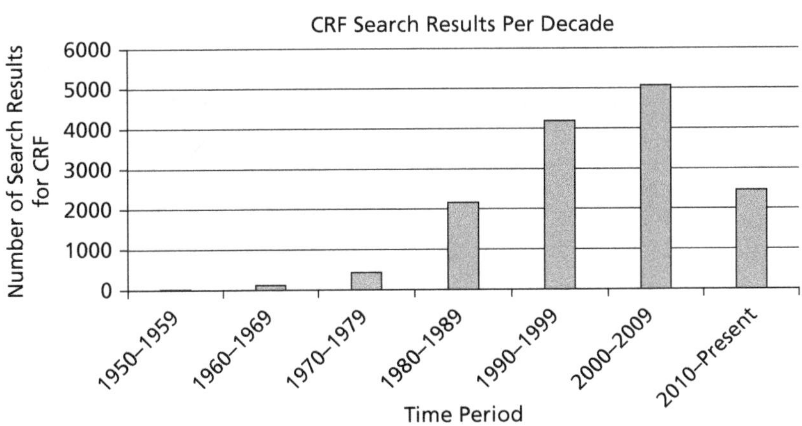


**Figure 9.6** Results from a search using the U.S. National Library of Medicine's PubMed.gov database, for the term *corticotrophin-releasing factor* (CRF) since 1950.

of Medicine's PubMed.gov database, articles containing this keyword were first published in the 1950s, and have been increasingly published ever since. It must be noted that these searches only retrieve articles to which the information molecule was attributed. The period of research on the concept of this information molecule, before it was named (e.g., Harris 1937), was not recognized by the terms of this search. Additionally, the distribution of search results between *corticotropin releasing factor and corticotropin releasing hormone* did not differ greatly.

With the advent of new biotechnology and lab techniques, it is easier than ever to fully explore peptides and neurotransmitters in various in vitro and in vivo models. The expansive and diverse nature of CRF leaves it perfectly poised to be studied widely in the lab. While this ease of experimentation opens up many opportunities to researchers, it also makes the field of scientific inquiry and obtaining grants all the more competitive. A subject that may have once only been studied by a few researchers around the world, CRF is now studied by many, leaving some excellent scientists unconsidered, unfunded, and unappreciated by the scientific community at large. Not everyone can be published, funded, and recognized in the competitive world of CRF research, so sadly I have not covered everything and everyone involved in this exciting field. And I apologize for those left out.

In the years ahead, the scientific community will learn much more about the biological basis of this fundamental peptide, as we learn about other chemical properties in the body. It is an exciting prospect to understand the neurobiological basis of resilience, along with the study of epigenetics and the long-term regulation of gene products by environment and context. These are rich ongoing lines of research. Human social contact is a fundamental form of adaptation (Jaspers 1913), but mastering being alone, something Henry David Thoreau (1971) knew and wrote about, is one important goal and self-mastery is a normative goal within a less egocentric sense of oneself. It is so great to be a part of the process of inquiry in which there is a rich international community.

# References


1. **Abbott, D. H., Keverne, E. B., Bercovitch, F. B., Shively, C. A., Mendoza, S. P., Saltzman, W., Snowdon, C. T., Ziegler, T. E., Banjevic, M., Garland, T. Jr.**, and **Sapolsky, R. M.** (2003). Are subordinates always stressed? A comparative analysis of rank differences in cortisol levels among primates. *Hormones and Behavior* **43**, pp. 67–82.
2. **Abou-Seif, C., Shipman, K., Allars, M., Norris, M., Chen, Y., Smith, R.**, and **Nicholson, R.** (2012). Tissue specific epigenetic differences in CRH gene expression. *Frontiers in Bioscience* 17, pp. 713–725.
3. **Adams Waldorf, K. M., Gravett, M. G., McAdams, R. M., Paolella, L. J., Gough, G. M., Carl, D. J., Bansal, A., Liggitt, H. D., Kapur, R.P., Reitz, F. B.**, and **Rubens, C. E.** (2011). Choriodecidual group B streptococcal inoculation induces fetal lung injury without intra-amniotic infection and preterm labor in Macaca nemestrina. *PloS One* **6**, p. e28972.
4. **Adinoff, B., Anton, R., Linnoila, M., Guidotti, A., Nemeroff, C. B.**, and **Bissette, G.** (1996). Cerebrospinal fluid concentrations of corticotropin-releasing hormone (CRH) and diazepam-binding inhibitor (DBI) during alcohol withdrawal and abstinence. *Neuropsychopharmacology* **15**, pp. 288–295.
5. **Aggleton, J.** (2000). *The amygdala*. 2nd ed. New York: Oxford University Press.
6. **Akana, S. F., Chu, A., Soriano, L.**, and **Dallman, M. F.** (2001). Corticosterone exerts site-specific and state-dependent effects in prefrontal cortex and amygdala on regulation of adrenocorticotropic hormone, insulin and fat depots. *Journal of Neuroendocrinology* **13**, pp. 625–637.
7. **AlAhmed, S.**, and **Herbert, J.** (2010). Effect of agomelatine and its interaction with the daily corticosterone rhythm on progenitor cell proliferation in the dentate gyrus of the adult rat. *Neuropharmacology* **59**, pp. 375–379.
8. **Albeck, D. S., McKittrick, C. R., Blanchard, D. C., Blanchard, R. J., Nikulina, J., McEwen, B. S.**, and **Sakai, R. R.** (1997). Chronic social stress alters levels of corticotropin-releasing factor and arginine vasopressin mRNA in rat brain. *Journal of Neuroscience* **17**, pp. 4895–4903.
9. **Alderman, S. L.**, and **Bernier, N. J.** (2007). Localization of corticotropin-releasing factor, urotensin I, and CRF-binding protein gene expression in the brain of the zebrafish, Danio rerio. *Journal of Comparative Neurology* **502**, pp. 783–793.
10. **Alderman, S. L.**, and **Bernier, N. J.** (2009). Ontogeny of the corticotropin-releasing factor system in zebrafish. *General and Comparative Endocrinology* **164**, pp. 61–69.
11. **Alderman, S. L., McGuire, A., Bernier, N. J.**, and **Vijayan, M. M.** (2012). Central and peripheral glucocorticoid receptors are involved in the plasma cortisol response to an acute stressor in rainbow trout. *General and Comparative Endocrinology* **176**, pp. 79–85.
12. **Alderman, S. L., Raine, J. C.**, and **Bernier, N. J.** (2008). Distribution and regional stressor-induced regulation of corticotrophin-releasing factor binding protein in rainbow trout (Oncorhynchus mykiss). *Journal of Neuroendocrinology* **20**, pp. 347–358.

13. **Almela, P., Navarro-Zaragoza, J., García-Carmona, J. A., Mora, L., Hidalgo, J., Milanés, M. V.,** and **Laorden, M. L.** (2012). Role of corticotropin-releasing factor (CRF) receptor-1 on the catecholaminergic response to morphine withdrawal in the nucleus accumbens (NAc). *PLoS One* **8**, p. e47089.
14. **Alon, T., Zhou, L., Perez, C. A., Garfield, A. S., Friedman, J. M.,** and **Heisler, L. K.** (2009). Transgenic mice expressing green fluorescent protein under the control of the corticotropin-releasing hormone promoter. *Endocrinology* **150**, pp. 5626–5632.
15. **Alsat, E., Guibourdenche, J., Luton, D., Frankenne, F.,** and **Evain-Brion, D.** (1997). Human placental growth hormone. *American Journal of Obstetrics and Gynecology* **177**, pp. 1526–1534.
16. **Alsop, D.,** and **Vijayan, M.** (2009). The zebrafish stress axis: molecular fallout from the teleost-specific genome duplication event. *General and Comparative Endocrinology* **161**, pp. 62–66.
17. **Altemus, M., Smith, M., Diep, V., Aulakh, C.,** and **Murphy, D.** (1994). Increased mRNA for corticotropin releasing hormone in the amygdala of Fawn-Hooded rats: a potential animal model of anxiety. *Anxiety* **1**, pp. 251–257.
18. **Altman, J., Everitt, B. J., Robbins, T. W., Glautier, S., Markou, A., Nutt, D., Oretti, R., Phillips, G. D.,** and **Robbins, T.W.** (1996). The biological, social and clinical bases of drug addiction: commentary and debate. *Psychopharmacology* **125**, pp. 285–345.
19. **Andres, A. L., Regev, L., Phi, L., Seese, R. R., Chen, Y., Gall, C. M.,** and **Baram, T. Z.** (2013). NMDA receptor activation and calpain contribute to disruption of dendritic spines by the stress neuropeptide CRH. *Journal of Neuroscience* **33**, pp. 16945–16960.
20. **Androulidaki, A., Dermitzai, E., Venihaki, M., Karagianni, E., Rassouli, O., Andreakou, E., Stournaras, C. Margioris, A.,** and **Tsatsanis, C.** (2009). Corticotropin-releasing factor promotes breast cancerl cell motility and invasiveness. *BioMed Central* **8**, p. **30**.
21. **Anisman, H., Prakash, P., Merali, Z.,** and **Poulter, M.** (2007). Corticotrpin releasing hormone receptor alterations elicited by acute and chronic unpredictable stressor challenges in stressor-susceptible and resilient strains of mice. *Behavioral Brain Research* **181**, pp. 180–190.
22. **Arai, M., Assil, I. Q.,** and **Abou-Samra, A. B.** (2001). Characterization of three corticotropin-releasing factor receptors in catfish: a novel third receptor is predominantly expressed in pituitary and urophysis. *Endocrinology* **142**, pp. 446–451.
23. **Armini, R., Bernabe, C., Rosa, C., Siller, C., Schimitel, F., Tufik, S., Klein, D.,** and **Schenberg, L.** (2015). In a rat model of panic, corticotropin responses to dorsal periaqueductal gray stimulation depend on physical exertion. *Psychoneuroendocrinology* **53**, pp. 136–147.
24. **Arnold, A. P.,** and **Breedlove, S. M.** (1985). Organizational and activational effects of sex steroids on brain and behavior: a reanalysis. *Hormones and Behavior* **19**, pp. 469–498.
25. **Asok, A., Ayers, L., Awoyemi, B., Schulkin, J.,** and **Rosen, J.** (2013). Immediate early gene and neuropeptide expression following exposure to the predator odor 2,5-dihydro-2,4,5-trimethylthiazoline (TMT). *Behavioural Brain Research* **248**, pp. 85–93.
26. **Asok, A., Draper, A. I., Hoffman, A., Schulkin, J., Lupica, C. R.,** and **Rosen, J. B.** (2016a, Forthcoming). A peptidergic pathway that gates sustained fear.

27. **Asok, A., Schulkin, J.,** and **Rosen, J.** (2016b). Corticotropin releasing factor type-1 receptor antagonism in the dorsolateral bed nucleus of the stria terminalis disrupts contextually conditioned fear; but not unconditioned fear to a predator odor. *Psychoneuroendocrinology* **70**, pp. 17–24.

28. **Athanassakis, I., Farmakiotis, V., Aifanatis, I., Gravanis, A.,** and **Vassiliadis, S.** (1999). Expression of corticotrophin-releasing hormone in the mouse uterus: participation in embryo implantation. *Journal of Endocrinology* **163**, pp. 221–227.

29. **Avishai-Eliner, S., Brunson, K. L., Sandman, C. A.,** and **Baram, T. Z.** (2002). Stressed-out, or in (utero)? *Trends in Neurosciences* **25**, pp. 518–524.

30. **Aydin, C., Oztan, O.,** and **Isgor, C.** (2011a). Effects of a selective Y2R antagonist, JNJ-31020028, on nicotine abstinence-related social anxiety-like behavior, neuropeptide Y and corticotropin releasing factor mRNA levels in the novelty-seeking phenotype. *Behavioural Brain Research* **222**, pp. 332–341.

31. **Aydin, C., Oztan, O.,** and **Isgor, C.** 2011b Vulnerability to nicotine abstinence-related social anxiety-like behavior: molecular correlates in neuropeptide Y, Y2 receptor and corticotropin releasing factor. *Neurosci Lett* **490**(3), pp. 220–225.

32. **Babb, J., Masini, C., Day, H.,** and **Campeau, S.** (2013). Sex differences in activated CRF neurons within stress-related neurocircutiry and HPA axis hormones follow restraint in rats. *Neuroscience* **234**, pp. 40–52.

33. **Backström, T.,** and **Winberg, S.** (2013). Central corticotropin releasing factor and social stress. *Frontiers in Neuroscience* 7, pp. 1–10.

34. **Backström, T., Pettersson, A., Johansson, V.,** and **Winberg, S.** (2011). CRF and urotensin I effects on aggression and anxiety-like behavior in rainbow trout. *Journal of Experimental Biology* **214**, pp. 907–914.

35. **Baerwald, C. G. O., Mok, C. C., Tickly, M., Lau, C. S., Wordsworth, B. P., Ollier, B., Panayi, G. S.,** and **Lanchbury, J. S.** (2000). Corticotropin releasing hormone (CRH) promoter polymorphisms in various ethnic groups of patients with rheumatoid arthritis. *Zeitschrift für Rheumatologie* **59**, pp. 29–34.

36. **Baerwald, C., Mok, C., Fife, M., Tikly, M., Lau, C., Wordsworth, B., Ollier, B., Panayi, G.,** and **Lanchbury, J.** (1999). Distribution of corticotropin-releasing hormone promoter polymorphism in different ethnic groups: evidence for natural selection in human populations. *Immunogenetics* **49**, pp. 894–899.

37. **Baiamonte, B., Valenza, M., Roltsch, E., Whitaker, A., Bayness, B., Sabino, V.,** and **Gilpin, N.** (2014). Nicotine dependence produces hyperalgesia: role of corticotropin-releasing factor-1 receptors (CRF1Rs) in the central amygdala (CeA). *Neuropharmocology* 77, pp. 217–223.

38. **Bailey, J. E., Papadopoulos, A., Diaper, A., Phillips, S., Schmidt, M. E., van der Ark, P., Dourish, C. T., Dawson, G. R.,** and **Nutt, D. J.** (2011). Preliminary evidence of anxiolytic effects of the CRF1 receptor antagonist R317573 in the 7.5% CO2 proof-of-concept experimental model of human anxiety. *Journal of Psychopharmacology* **25**, pp. 1199–1206.

39. **Baker, M., Funder, J.,** and **Kattoula, S.** (2013). Evolution of hormone selectivity in glucocorticoid and mineralcorticoid receptors. *Journal of Steroid Biochemistry and Molecular Biology* **137**, pp. 57–70.

40. **Bakshi, V. P., Smith-Roe, S., Newman, S. M., Grigoriadis, D. E.,** and **Kalin, N. H.** (2002). Reduction of stress-induced behavior by antagonism of corticotropin-releasing

hormone 2 (CRH2) receptors in lateral septum or CRH1 receptors in amygdala. *Journal of. Neuroscience* **22**, pp. 2926–2935.

41. **Bale, T.** (2009). Neuroendocrine and immune influences on the CNS: It's a matter of sex. *Neuron* **64**, pp. 13–16.
42. **Bale, T.** (2011). Sex differences in prenatal epigenetic programming of stress pathways. *Stress* **14**, pp. 348–356.
43. **Bale, T.** (2014). Lifetime stress experience: transgenerational epigenetics and germ cell programming. *Dialogues in Clinical Neuroscience* **16**, pp. 297–305.
44. **Bale, T. L.** (2015). Epigenetic and transgenerational reprogramming of brain development. *Nature Reviews Neuroscience* **16**, pp. 332–344.
45. **Bale, T. L.**, and **Vale, W. W.** (2004). CRF and CRF receptors: role in stress responsivity and other behaviors. *Annual Review of Pharmacology and Toxicology* **44**, pp. 525–557.
46. **Bale, T. L.**, **Anderson, K. R.**, **Roberts, A. J.**, **Lee, K. F.**, **Nagy, T. R.**, and **Vale, W. W.** (2003). Corticotropin-releasing factor receptor-2-deficient mice display abnormal homeostatic responses to challenges of increased dietary fat and cold. *Endocrinology* **144**, pp. 2580–2587.
47. **Bale, T.**, and **Chen, A.** (2012). Minireview: CRF and Wylie Vale: A story of 41 amino acids and a texan with grit. *Endocrinology* **153**, pp. 2556–2561.
48. **Bale, T.**, **Contarino, A.**, **Smith, G.**, **Chan, R.**, **Gold, L.**, **Sawchenko, P.**, **Koob, G.**, **Vale, W.**, and **Lee, K.** (2000). Mice deficient for corticotropin-releasing hormone receptor-2 display anxiety-like behavior and are hypersensitive to stress. *Nature Genetics* **24**, pp. 410–414.
49. **Bale, T.**, and **Epperson, C.** (2015). Sex differences and stress across the lifespan. *Nature Neuroscience* **18**, pp. 1413–1420.
50. **Bale, T.**, **Giordano, F.**, and **Vale, W. W.** (2003). A new role for corticotropin-releasing factor receptor 2: suppression of vascularization. *Trends in Cardiovascular Medicine* **13**, pp. 68–71.
51. **Balthazart, J.**, and **Ball, G. F.** (1998). New insights into the regulation and function of brain estrogen synthase (aromatase). *Trends in Neurosciences* **21**, pp. 243–249.
52. **Bangasser, D.** (2013). Sex differences in stress-related receptors: "micro" differences with "macro" implications for mood and anxiety disorders. *Biology of Sex Differences* **4**, pp. 1–15.
53. **Bangasser, D. A.**, and **Valentino, R. J.** (2012). Sex differences in molecular and cellular substrates of stress. *Cellular and Molecular Neurobiology* **32**, pp. 709–723.
54. **Bangasser, D. A.**, **Reyes, B. A.**, **Piel, D.**, **Garachh, V.**, **Zhang, X. Y.**, **Plona, Z. M.**, **Van Bockstaele, J.**, **Beck, S. G.**, and **Valentino, R. J.** (2013). Increased vulnerability of the brain norepinephrine system of females to corticotropin-releasing factor overexpression. *Molecular Psychiatry* **18**, pp. 166–173.
55. **Bangasser, D. A.**, **Wiersielis, K. R.**, and **Khantsis, S.** (2015). Sex differences in the locus coeruleus-norepinephrine system and its regulation by stress. *Brain Research* **1641**, pp. 177–188.
56. **Bangasser, D. A.**, **Zhang, X.**, **Garachh, V.**, **Hanhauser, E.**, and **Valentino, R. J.** (2011). Sexual dimorphism in locus coeruleus dendritic morphology: a structural basis for sex differences in emotional arousal. *Physiology & Behavior* **103**, pp. 342–351.
57. **Banks, W.** (2015). Peptides and the blood-brain barrier. *Peptides* **72**, pp. 1–4.

58. **Banks, W.** (2015). The blood-brain barrier in neuroimmunology: Tales of separation and assimilation. *Brain, Behavior, and Immunity* **44**, pp. 1–8.
59. **Banks, W.**, and **Erickson, M.** (2010). The blood-brain barrier and immune function and dysfunction. *Neurobiology of Disease* **37**, pp. 26–32.
60. **Bao, A. M.**, and **Swaab, D. F.** (2007). Gender difference in age-related number of corticotropin-releasing hormone-expressing neurons in the human hypothalamic paraventricular nucleus and the role of sex hormones. *Neuroendocrinology* **85**, pp. 27–36.
61. **Bao, A.**, **Meynen, G.**, and **Swaab, D.** (2008). The stress system in depression and neurodegeneration: focus on the human hypothalamus. *Brain Research Reviews* **57**, pp. 531–553.
62. **Bao, A.**, and **Swaab, D.** (2010). Corticotropin-releasing hormone and arginine vasopressin in depression: focus on the human postmortem hypothalamus. *Vitamins and Hormones* **82**, pp. 339–365.
63. **Baracz, S. J.**, **Parker, L. M.**, **Suraev, A. S.**, **Everett, N. A.**, **Goodchild, A. K.**, **McGregor, I. S.**, and **Cornish, J. L.** (2015). Chronic methamphetamine self-administration dysregulates oxytocin plasma levels and oxytocin receptor fibre density in the nucleus accumbens core and subthalamic nucleus of the rat. *Journal of Neuroendocrinology* **28(4)**.
64. **Baram, T. Z.**, and **Schultz, L.** (1991). Corticotropin-releasing hormone is a rapid and potent convulsant in the infant rat. *Developmental Brain Research* **61**, pp. 97–101.
65. **Baram, T.**, **Mitchell, W.**, and **Haden, E.** (1996). Inhibition of pituitary-adrenal secretion by a corticotropin releasing hormone antagonist in humans. *Molecular Psychiatry* **1**, pp. 320–324.
66. **Baram, T. Z.**, **Mitchell, W. G.**, **Brunson, K.**, and **Haden, E.** (1999). Infantile spasms: hypothesis-driven therapy and pilot human infant experiments using corticotropin-releasing hormone receptor antagonists. *Developmental Neuroscience* **21**, pp. 281–289.
67. **Bardou, I.**, **Dipatrizio, N.**, **Brothers, H.**, **Kaercher, R.**, **Branger, K.**, **Mitchem, M.**, **Hopp, S.**, **Wenk, G.**, and **Marchalant, Y.** (2012). Pharmacological manipulation of cannabinoid neurotransmission reduces neuroinflammation associated with normal aging. *Health* **4**, pp. 679–684.
68. **Barker, D. J.** (1997). The fetal origins of coronary heart disease. *Acta Paediatr Supplement* **422**, pp. 78–82.
69. **Barker, D. J.** (2004). The developmental origins of well-being. *Philosophical Transactions of the Royal Society of London—Series b: Biological Sciences* **359**, pp. 1359–1366.
70. **Barna, I.**, **Balint, E.**, **Baranyi, J.**, **Bakos, N.**, **Makara, G.**, and **Haller, J.** (2003). Gender-specific effect of maternal deprivation on anxiety and corticotropin-releasing hormone mRNA expression in rats. *Brain Research Bulletin* **62**, pp. 85–91.
71. **Barr, C.**, **Dvoskin, R.**, **Yuan, Q.**, **Lipsky, R.**, **Gupte, M.**, **Hu, X.**, **Zhou, Z.**, **Schwandt, M.**, **Lindell, S.**, **McKee, M.**, **Becker, M.**, **Kling, M.**, **Gold, P.**, **Higley, D.**, **Heilig, M.**, **Suomi, S.**, and **Goldman, D.** (2008). CRH haplotype as a factor influencing cerebrospinal fluid levels of corticotropin-releasing hormone, hypothalamic-pituitary-adrenal axis activity, temperament, and alcohol consumption in rhesus macaques. *Archives of General Psychiatry* **65**, pp. 934–944

72. **Barr, C., Dvoskin, R. L., Gupte, M., Sommer, W., Sun, H., Schwandt, M. L., Lindell, S. G., Kasckow, J. W., Suomi, S. J., Goldman, D., Higley, J. O.,** and **Heilig, M.** (2009). Functional CRH variation increase stress-induced alcohol consumption in primates. *Proceedings of the National Academy of Sciences of the United States of America* **106**, pp. 14593–14598.

73. **Barrios, A., Ghosh, R., Fang, C., Emmons, S.,** and **Barr, M.** (2012). PDF-1 neuropeptide signaling modulates a neural circuit for mate-searcing behavior in C. elegans. *Nature Neuroscience* **15**, pp. 1675–7682.

74. **Barrot, M., Rettori, M. C., Guardiola-Lemaître, B., Jarry, C., Le Moal, M.,** and **Piazza, P. V.** (2000). Interactions between imidazoline binding sites and dopamine levels in the rat nucleus accumbens. *European Journal of Neuroscience* **12**, pp. 4547–4551.

75. **Bastek, J.,** and **Elovitz, M.** (2013). The role and challenges of biomarkers in spontaneous preterm birth and preeclampsia. *Fertility and Sterility* **99**, pp. 1117–1123.

76. **Bastek, J., Hirshberg, A., Chandraskaran, S., Owen, C., Heiser, L., Araujo, B., McShea, M., Ryan, M.,** and **Elovitz, M.** (2013). Biomarkers and cervical length to predict spontaneous preterm bith in asymptomatic high-risk women. *Obstetrics and Gynecology* **122**, pp. 283–289.

77. **Bastek, J. A., Gomez, L. M.,** and **Elovitz, M. A.** (2011). The role of inflammation and infection in preterm birth. *Clinics in Perinatology* **38**, pp. 385–406.

78. **Beaton, E. A., Schmidt, L. A., Schulkin, J., Antony, M. M., Swinson, R. P.,** and **Hall, G. B.** (2008). Different neural responses to stranger and personally familiar faces in shy and bold adults. *Behavioral Neuroscience* **122**, p. **704**.

79. **Beaton, E., Schmidt, L., Schulkin, J., Antony, M., Swinson, P.,** and **Hall, G.** (2015). Different fusiform activity to stranger and personally familiar faces in shy and social adults. *Social Neuroscience* **4**, pp. 308–316.

80. **Becker, J. B., Berkley, K. J., Geary, N., Hampson, E., Herman, J. P.,** and **Young, E.** (2007). *Sex differences in the brain: from genes to behavior*. Oxford: Oxford University Press.

81. **Becker, K., Abraham, A., Kindler, J., Helmeke, C.,** and **Braun, K.** (2007). Exposure to neonatal separation stress alters exploratory behavior and corticotropin releasing factor expression in neurons in the amygdala and hippocampus. *Developmental Neurobiology* **67**, pp. 617–629.

82. **Beckerman, M. A., Van Kempen, T. A., Justice, N. J., Milner, T. A.,** and **Glass, M. J.** (2013). Corticotropin-releasing factor in the mouse central nucleus of the amygdala: ultrastructural distribution in NMDA-NR1 receptor subunit expressing neurons as well as projection neurons to the bed nucleus of the stria terminalis. *Experimental Neurology* **239**, pp. 120–132.

83. **Beery, A. K., Bicks, L., Mooney, S. J., Goodwin, N. L.,** and **Holmes, M. M.** (2016). Sex, social status, and CRF receptor densities in naked mole-rats. *Journal of Comparative Neurology* **524**, pp. 228–243.

84. **Beery, A., Vahaba, D.,** and **Grunberg, D.** (2014). Corticotropin releasing factor receptor densities vary with photoperiod and sociality. *Hormones and Behavior* **66**, pp. 779–786.

85. **Behan, D., De Souza, E., Lowry, P., Potter, E., Sawchenko, P.,** and **Vale, W.** (1995). Corticotropin releasing factor (CRF) binding protein: a novel regulator of CRF and related peptides. *Frontiers in Neuroendocrinology* **16**, pp. 362–382.

86. **Behan, D., Heinrichs, S., Troncoso, J., Liu, X., Kawas, C., Ling, N.,** and **De Souza, E.** (1995). Displacement of corticotropin releasing factor from its binding protein as a possible treatment for Alzheimer's disease. *Nature* **378**, pp. 284–287.
87. **Bell, M., Bhatnagar, S., Liang, J., Soriano, L., Nagy, T., Dallman, M.** (2000). Voluntary sucrose ingestion, like corticosterone replacement, prevents the metabolic deficits of adrenalectomy. *Journal of Neuroendocrinology* **12**, pp. 461–470.
88. **Ben-Efraim, Y. J., Wasserman, D., Wasserman, J.,** and **Sokolowski, M.** (2011). Gene–environment interactions between CRHR1 variants and physical assault in suicide attempts. *Genes, Brain and Behavior* **10**, pp. 663–672.
89. **Benou, C., Wang, Y., Imitola, J., Vlerken, L., Chandras, C., Karalis, K.,** and **Khoury, S.** (2005). Corticotropin-releasing hormone contributes to the peripheral inflammatory response in experimental autoimmune encephalomyelitis. *Journal of Immunology* **174**, pp. 5407–5413.
90. **Bentley, P. J.** (1982). *Comparative vertebrate endocrinology*. 2nd ed. Cambridge: Cambridge University Press.
91. **Berghorn, K. A., Albrecht, E. D.,** and **Pepe, G. J.** (1991). Responsivity of the baboon fetal pituitary to corticotropin-releasing hormone in utero at midgestation. *Endocrinology* **129**, pp. 1424–1428.
92. **Berkowitz, G. S., Lapinski, R. H., Lockwood, C. J., Florio, P., Blackmore-Prince, C.,** and **Petraglia, F.** (1996). Corticotropin-releasing factor and its binding protein: maternal serum levels in term and preterm deliveries. *American Journal of Obstetrics and Gynecology* **174**, pp. 1477–1483.
93. **Bernier, N., Alderman, S.,** and **Bristow, E.** (2008). Heads or tails? Stressor-specific expression of corticotropin releasing factor and urotensin I in the preoptic area and caudal neurosecretory system of rainbow trout. *Journal of Endocrinology* **196**, pp. 637–648.
94. **Berridge, K. C.** (2000). Measuring hedonic impact in animals and infants: microstructure of affective taste reactivity patterns. *Neuroscience & Biobehavioral Reviews* **24**, pp. 173–198.
95. **Berridge, K. C.** (2004). Motivation concepts in behavioral neuroscience. *Physiology and Behavior* **81**, pp. 179–209.
96. **Berridge, K., Pecina, S.,** and **Schulkin, J.** (2004). Appetitive effect of CRF in accumens shell on conditioned approach for sucrose reward. Program No 437.12. In 2004 Abstract Viewer and Itinerary Planner, Society for Neuroscience online.
97. **Berridge, K. C.,** and **Robinson, T. E.** (1998). What is the role of dopamine in reward: hedonic impact, reward learning, or incentive salience. *Brain Research Review* **28**, pp. 306–369.
98. **Beutler, L. R., Wanat, M. J., Quintana, A., Sanz, E., Bamford, N. S., Zweifel, L. S.,** and **Palmiter, R. D.** (2011). Balanced NMDA receptor activity in dopamine D1 receptor (D1R) and D2R-expressing medium spiny neurons is required for amphetamine sensitization. *Proceedings of the National Academy of Sciences of the United States of America* **108**, pp. 4206–4211.
99. **Bhatnagar, S., Bell, M., Liang, J., Soirano, L., Nagy, T.,** and **Dallman, M.** (2000). Corticosterone facilitates saccharin intake in adranlectomized rats: does corticosterone increase stimulus salience? *Journal of Neuroendocrinology* **12**, pp. 453–460.

100. **Bijlsma, E. Y., Van Leeuwen, M. L. F., Westphal, K. G. C., Olivier, B.,** and **Groenink, L.** (2011). Local repeated corticotropin-releasing factor infusion exacerbates anxiety- and fear-related behavior: differential involvement of the basolateral amygdala and medial prefrontal cortex. *Neuroscience* **173**, pp. 82–92.

101. **Binder, E. B.,** and **Nemeroff, C. B.** (2010). The CRF system, stress, depression and anxiety—insights from human genetic studies. *Molecular Psychiatry* **15**, pp. 574–588.

102. **Bingham, B. C., Rani, C. S., Frazer, A., Strong, R.,** and **Morilak, D. A.** (2013). Exogenous prenatal corticosterone exposure mimics the effects of prenatal stress on adult brain stress response systems and fear extinction behavior. *Psychoneuroendocrinology* **38**, pp. 2746–2757.

103. **Binneman, B., Feltner, D., Kolluri, S., Shi, Y., Qiu, R.,** and **Stiger, T.** (2008). A 6-week randomized, placebo-controlled trial of CP-316,311 (a selective CRH1 antagonist) in the treatment of major depression. *American Journal of Psychiatry* **165**, pp. 617–620.

104. **Bisits, A., Madsen, G., McLean, M., O'Callaghan, S., Smith, R.,** and **Giles, W.** (1998). Corticotropin-releasing hormone: a biochemical predictor of preterm delivery in a pilot randomized trial of the treatment of preterm labor. *American Journal of Obstetrics and Gynecology* **178**, pp. 862–866.

105. **Blacktop, J. M., Vranjkovic, O., Mayer, M., Van Hoof, M., Baker, D. A.,** and **Mantsch, J. R.** (2016). Antagonism of GABA-B but not GABA-A receptors in the VTA prevents stress-and intra-VTA CRF-induced reinstatement of extinguished cocaine seeking in rats. *Neuropharmacology* **102**, pp. 197–206.

106. **Blanchard, D., Spencer, R., Weiss, S., Blanchard, R., McEwen, B.,** and **Sakai, R.** (1994). Visible burrow system as a model of chronic social stress: behavioral and neuroendocrine correlates. *Psychoneuroendocrinology* **20**, pp. 117–134.

107. **Blanchard, R., Nikulina, J., Sakai, R., McKittrick, C., McEwen, B.,** and **Blanchard, C.** (1998). Behavioral and endocrine change following chronic predatory stress. *Physiology and Behavior* **63**, pp. 561–569.

108. **Blank, T., Nijholt, I., Eckart, K.,** and **Spiess, J.** (2002). Priming of long-term potentiation in mouse hippocampus by corticotropin-releasing factor and acute stress: Implications by hippocampus-dependent learning. *Journal of Neuroscience* **22**, pp. 3788–3794.

109. **Blank, T., Nijholt, I., Vollstaedt, S.,** and **Spiess, J.** (2003). The corticotropin releasing factor receptor 1 antagonist CP-154, 526 reverses stress-induced learning deficits in mice. *Behavioral Brain Research* **138**, pp. 207–213.

110. **Blass, E. M.** (1996). Mothers and their infants: Peptide-mediated physiological, behavioral and affective changes during suckling. *Regulatory Peptides* **66**, pp. 109–112.

111. **Bloch, G.,** and **Meshi, A.** (2007). Influences of octopamine and juvenile hormone on locomotor behavior and period gene expression in the honeybee, Apis mellifera. *Journal of Comparative Physiology* A **193**, pp. 181–199.

112. **Bluthgen, N., Sumpter, J., Odermatt, A.,** and **Fent, K.** (2013). Effects of low concentrations of the antiprogestin minepristone (RU486) in adults and embryos of zebrafish (danio rerio): 2. Gene expression analysis and in vitro analysis. *Aquatic Toxciology* **144**, pp. 96–104.

113. **Bohler, H. C., Zoeller, R. T., King, J. C., Rubin, B. S., Weber, R.,** and **Merriam, G. R.** (1990). Corticotropin releasing hormone mRNA is elevated on the afternoon of

proestrus in the parvocellular paraventricular nuclei of the female rat. *Molecular Brain Research* **8**, pp. 259–262.

114. **Bonnin, A., Goeden, N., Chen, K., Wilson, M. L., King, J., Shih, J. C., Blakely, R. D., Deneris, E. S.,** and **Levitt, P.** (2011). A transient placental source of serotonin for the fetal forebrain. *Nature* **472**, pp. 347–350.
115. **Boonstra, R., Hik, D., Singleton, G. R.,** and **Tinnikov, A.** (1998). The impact of predator-induced stress on the snowshoe hare cycle. *Ecological Monographs* **68**, pp. 371–394.
116. **Boorse, G. C.,** and **Denver, R. J.** (2006). Widespread tissue distribution and diverse functions of corticotropin-releasing factor and related peptides. *General and Comparative Endocrinology* **146**, pp. 9–18.
117. **Boorse, G., Crespi, E., Dautzenberg, F.,** and **Denver, R.** (2005). Urocortins of the South African clawed frog, Xenopus laevis: conservation of structure and function in tetrapod evolution. *Endocrinology* **146**, pp. 4851–4860.
118. **Boorse, G. C.,** and **Denver, R. J.**, (2003). Endocrine mechanisms underlying plasticity in metamorphic timing in spadefoot toads. *Integrative and Comparative Biology* **43**, pp. 646–657.
119. **Borelli, K., Albrecht-Souza, L., Fedoce, A., Fabri, D., Resstel, L.,** and **Brandao, M.** (2013). Conditioned fear is modulated by CRH mechanisms in the periaqueductal gray columns. *Hormones and Behavior* **63**, pp. 791–799.
120. **Borelli, K., Souza, L., Fedoce, A., Fabri, D., Besstel, L.,** and **Brandao, M.** (2013). Conditioned fear is modulated by CRH mechanisms in the periaqueductal columns. *Hormones and Behavior* **63**, pp. 791–799.
121. **Bosch, O., Nair, H., Ahern, T., Neumann, I.,** and **Young, L.** (2009). The CRF system mediates increased passive stress-coping behavior following the loss of a bonded partner in a monogamous rodent. *Neuropsychopharmacology* **34**, pp. 1406–1415.
122. **Bowlby, J.** (1977). The making and breaking of affectional bonds. II. Some principles of psychotherapy. The fiftieth Maudsley Lecture. *British Journal of Psychiatry* **130**, pp. 421–431.
123. **Bowman, M. E., Lopata, A., Jaffe, R. B., Golos, T. G., Wickings, J.,** and **Smith, R.** (2001). Corticotropin-releasing hormone-binding protein in primates. *American Journal of Primatology* **53**, pp. 123–130.
124. **Boyson, C. O., Holly, E. N., Shimamoto, A., Albrecht-Souza, L., Weiner, L. A., DeBold, J. F.,** and **Miczek, K. A.** (2014). Social stress and CRF-dopamine interactions in the VTA: role in long-term escalation of cocaine self-administration. *Journal of Neuroscience* **34**, pp. 6659–6667.
125. **Brady, K., McRae, A., Maria, M., DeSantis, S, Simpson, A., Waldrop, A., Back, S.,** and **Kreek, M.** (2009). Response to CRH infusion in cocaine-dependent individuals. *Archives of General Psychiatry* **66**, pp. 422–430.
126. **Braun, K.,** and **Poeggel, G.** (2001). Recognition of Mother's voice evokes metabolic activation in the medial prefrontal cortex and lateral thalamus of Octodon degus pups. *Neuroscience* **103**, pp. 861–864.
127. **Braun, K., Seidel, K., Weigel, S., Roski, C.,** and **Poeggel, G.** (2011). Paternal deprivation alters region-and age-specific interneuron expression patterns in the biparental rodent, Octodon degus. *Cerebral Cortex* **21**, pp. 1532–1546.

128. **Braun, S.**, and **Scheich, H.** (1997). Influence of experience on the representation of the "mothering call" in frontoparietal and auditory cortex of pups of the rodent Octodon degus: FDG mapping. *Journal of Comparative Physiology* A **181**, pp. 697–709.

129. **Brawand, D., Wahli, W.**, and **Kaessmann, H.** (2008). Loss of egg yolk genes in mammals and the origin of lactation and placentation. *PLOS Biology* **6**, pp. 507–515.

130. **Bremner, J. D., Licinio, J., Darnell, A., Krystal, J. H., Owens, M. J., Southwick, S. M., Nemeroff, C. B.**, and **Charney, D. S.** (1997). Elevated CSF corticotropin-releasing factor concentrations in posttraumatic stress disorder. *Am J Psychiatry* **154(5)**, pp. 624–629.

131. **Bridges, R. S.**, and **Freemark, M. S.** (1995). Human placental lactogen infusions into the medial preoptic area stimulate maternal behavior in steroid-primed, nulliparous female rats. *Hormones and Behavior* **29**, pp. 216–226.

132. **Bridges, R. S., Rigero, B. A., Byrnes, E. M., Yang, L.**, and **Walker, A. M.** (2001). Central infusions of the recombinant human prolactin receptor antagonist, S179D-PRL, delay the onset of maternal behavior in steroid-primed, nulliparous female rats 1. *Endocrinology* **142**, pp. 730–739.

133. **Bridgham, J., Carroll, S. M.**, and **Thornton, J. W.** (2006). Evolution of hormone-receptor complexity bymolecular exploitation. *Science* **312**, pp. 97–101.

134. **Brinks, V., Berger, S., Gass, P., de Kloet, E. R.**, and **Oitzl, M. S.** (2009). Mineralocorticoid receptors in control of emotional arousal and fear memory. *Hormones and Behavior* **5**, pp. 232–238.

135. **Broad, K. D., Curley, J. P.**, and **Keverne, E. B.** (2006). Mother–infant bonding and the evolution of mammalian social relationships. *Philosophical Transactions of the Royal Society B: Biological Sciences* **36**, pp. 2199–2214.

136. **Broad, K. D., Keverne, E. B.**, and **Kendrick, K. M.** (1995). Corticotrophin releasing factor mRNA expression in the sheep brain during pregnancy, parturition and lactation and following exogenous progesterone and oestrogen treatment. *Molecular Brain Research* **29**, pp. 310–316.

137. **Brown, M. R.**, and **Fisher, L. A.** (1983). Central nervous system effects of corticotropin releasing factor in the dog. *Brain Research* **280**, pp. 75–79.

138. **Brown, M. R., Fisher, L. A., Rivier, J., Spiess, J., Rivier, C.**, and **Vale, W.** (1982). Corticotropin-releasing factor: effects on the sympathetic nervous system and oxygen consumption. *Life Sciences* **30**, pp. 207–210.

139. **Brown, Z. J., Tribe, E., D'souza, N. A.**, and **Erb, S.** (2009). Interaction between noradrenaline and corticotrophin-releasing factor in the reinstatement of cocaine seeking in the rat. *Psychopharmacology* **203**, pp. 121–130.

140. **Brown, Z., Kupferschmidt, D.**, and **Erb, S.** (2012). Reinstatement of cocaine seeking in rats by the pharmacological stressors, corticotropin-releasing factor and yohimbine: role for D1/5 dopamine receptors. *Psychopharmacology* **224**, pp. 431–440.

141. **Bruijnzeel, A. W., Ford, J., Rogers, J. A., Scheick, S., Ji, Y., Bishnoi, M.**, and **Alexander, J. C.** (2012). Blockade of CRF1 receptors in the central nucleus of the amygdala attenuates the dysphoria associated with nicotine withdrawal in rats. *Pharmacology Biochemistry and Behavior* **101**, pp. 62–68.

142. **Bruijnzeel, A. W., Small, E., Pasek, T. M.**, and **Yamada, H.** (2010). Corticotropin-releasing factor mediates the dysphoria-like state associated with alcohol withdrawal in rats. *Behavioural Brain Research* **210**, pp. 288–291.

143. **Brunson, K. L., Eghbal-Ahmadi, M., Bender, R., Chen, Y.,** and **Baram, T. Z.** (2001). Long-term, progressive hippocampal cell loss and dysfunction induced by early-life administration of corticotropin-releasing hormone reproduce the effects of early-life stress. *Proceedings of the National Academy of Sciences* **98**, pp. 8856–8861.

144. **Brunson, K. L., Kramár, E., Lin, B., Chen, Y., Colgin, L. L., Yanagihara, T. K., Lynch, G.,** and **Baram, T. Z.** (2005). Mechanisms of late-onset cognitive decline after early-life stress. *Journal of Neuroscience* **25**, pp. 9328–9338.

145. **Brunson, K. L., Grigoriadis, D. E., Lorang, M. T.,** and **Baram, T. Z.** (2002). Corticotropin releasing hormone (CRH) downregulates the function of its receptor (CRF1) and induces CRF1 expression in hippocampal and cortical regions of the immature rat brain. *Experimental Neurology* **176**, pp. 75–86.

146. **Brunton, P., Donadio, M.,** and **Russell, J.** (2011). Sex differences in prenatally programmed anxiety behavior in rats: differential corticotropin releasing hormone receptor mRNA expression in the amygdaloid complex. *Stress* **14**, pp. 634–643.

147. **Bryce, C. A.,** and **Floresco, S. B.** (2016). Perturbations in effort-related decision-making driven by acute stress and corticotropin-releasing factor. *Neuropsychopharmacology* **41(8)**, pp. 2147–2159.

148. **Buckley, T. M.,** and **Schatzberg, A. F.** (2005). On the interactions of the hypothalamic-pituitary-adrenal (HPA) axis and sleep: normal HPA axis activity and circadian rhythm, exemplary sleep disorders. *Journal of Clinical Endocrinology & Metabolism* **90**, pp. 3106–3114.

149. **Buffalari, D. M., Baldwin, C. K., Feltenstein, M. W.,** and **See, R. E.** (2012). Corticotrophin releasing factor (CRF) induced reinstatement of cocaine seeking in male and female rats. *Physiology & Behavior* **105**, pp. 209–214.

150. **Butler, R. K., Oliver, E. M., Sharko, A. C., Parilla-Carrero, J., Kaigler, K. F., Fadel, J. R.,** and **Wilson, M. A.** (2016). Activation of corticotropin releasing factor-containing neurons in the rat central amygdala and bed nucleus of the stria terminalis following exposure to two different anxiogenic stressors. *Behavioural Brain Research* **304**, pp. 92–101.

151. **Byrne, J., Wing, D., Fraser, M., Fassett, M. Godowin, T.,** and **Challis, J.** (2004). Mifepristone: Effect on plasma corticotropin-releasing hormone, adrenocorticotropic hormone, and cortisol in term pregnancy. *Journal of Perinatology* **24**, pp. 416–420.

152. **Byrnes, E. M., Casey, K.,** and **Bridges, R. S.** (2012). Reproductive experience modifies the effects of estrogen receptor alpha activity on anxiety-like behavior and corticotropin releasing hormone mRNA expression. *Hormones and Behavior* **61**, pp. 44–49.

153. **Byrnes, E. M., Casey, K., Carini, L. M.,** and **Bridges, R. S.** (2013). Reproductive experience alters neural and behavioural responses to acute oestrogen receptor α activation. *Journal of Neuroendocrinology* **25**, pp. 1280–1289.

154. **Caberlotto, L., Rimondini, R., Hansson, A., Eriksson, S.,** and **Heilig, M.** (2004). Corticotropin-releasing hormone (CRH) mRNA expression in rat central amygdala in cannabinoid tolerance and withdrawal: evidence for an allostatic shift? *Neuropsychopharmacology* **29**, pp. 15–22.

155. **Cabral, A., Portiansky, E., Sánchez-Jaramillo, E., Zigman, J. M.,** and **Perello, M.** (2016). Ghrelin activates hypophysiotropic corticotropin-releasing factor neurons independently of the arcuate nucleus. *Psychoneuroendocrinology* **67**, pp. 27–39.

156. **Cadet, J. L., Brannock, C., Ladenheim, B., McCoy, M. T., Krasnova, I. N., Lehrmann, E. Becker, K.G.,** and **Jayanthi, S.** (2014). Enhanced upregulation of CRH mRNA expression in the nucleus accumbens of male rats after a second injection of methamphetamine given thirty days later. *PloS One* **9**, p. e84665.

157. **Cador, M., Dulluc, J.,** and **Mormede, P.** (1993). Modulation of the locomotor response to amphetamine by corticosterone. *Neuroscience* **56**, pp. 981–988.

158. **Cameron, N. M., Shahrokh, D., Del Corpo, A., Dhir, S. K., Szyf, M., Champagne, F. A.,** and **Meaney, M. J.** (2008). Epigenetic programming of phenotypic variations in reproductive strategies in the rat through maternal care. *Journal of Neuroendocrinology* **20**, pp. 795–801.

159. **Campbell, B., Morrison, J., Walker, E.,** and **Merchant, K.** (2004). Differential regulation of behavioral, genomic and neuroendocrine responses by CRF infusions in rats. *Pharmacology, Biochemistry and Behavior* **77**, pp. 447–455.

160. **Campbell, E. A., Linton, E. A., Wolfe, C. D., Scraggs, P. R., Jones, M. T.,** and **Lowry, P. J.** (1987). Plasma corticotropin-releasing hormone concentrations during pregnancy and parturition. *Journal of Clinical Endocrinology and Metabolism* **64**, pp. 1054–1059.

161. **Campbell, R., Satoh, N.,** and **Degnan, B.** (2004). Piecing together evolution of the vertebrate endocrine system. *Trends in Genetics* **20**, pp. 359–366.

162. **Campeau, S., Akil, H.,** and **Watson, S. J.** (1997a). Lesions of the medial geniculate nuclei specifically block corticosterone release and induction of c-fos mRNA in the forebrain associated with audiogenic stress in rats. *Journal of Neuroscience* **17**, pp. 5979–5992.

163. **Campeau, S., Falls, W. A., Cullinan, W. E., Helmreich, D. L., Davis, M.,** and **Watson, S. J.** (1997b). Elicitation and reduction of fear: behavioural and neuroendocrine indices and brain induction of the immediate-early gene c-fos. *Neuroscience* **78**, pp. 1087–1104.

164. **Cannon, W. B.** (1929). *Bodily changes in pain, hunger, fear and rage, an account of recent researches into the function of emotional excitement.* New York: Appleton and Co.

165. **Cannon, W. B.** (1932). *The wisdom of the body.* New York: W. W. Norton.

166. **Cannon, W. B.** (1927). The James–Lange theory of emotions. A critical examination and an alternative theory. *American Journal of Psychology* **39**, pp. 106–124.

167. **Capellini, I., Venditti, C.,** and **Barton, R. A.** (2011). Placentation and maternal investment in mammals. *The American Naturalist* **177**, pp. 86–98.

168. **Carboni, E., Silvagni, A., Rolando, M. T.,** and **Di Chiara, G.** (2000). Stimulation of in vivo dopamine transmission in the bed nucleus of stria terminalis by reinforcing drugs. *Journal of Neuroscience* **20**, p. RC102.

169. **Cardinal, R. N., Pennicott, D. R., Lakmali, C., Robbins, T. W.,** and **Everitt, B. J.** (2001). Impulsive choice induced in rats by lesions of the nucleus accumbens core. *Science* **292**, pp. 2499–2501.

170. **Cardoso, J., Felix, R., Bergqvist, C., Larhammar, D.** (2014). New insights into the evolution of vertebrate CRH (corticotropin-releasing hormone) and invertebrate DH44 (diuretic hormone 44) receptors in metazoans. *General and Comparative Endocrinology* **209**, pp. 162–170.

171. **Carlson, D. E., Chiu, W. C., Fiedler, S. M.,** and **Hoffman, G. E.** (2007). Central neural distribution of immunoreactive Fos and CRH in relation to plasma ACTH and corticosterone during sepsis in the rat. *Experimental Neurology* **205**, pp. 485–500.

172. **Carter, C. S.** (2007). Sex differences in oxytocin and vasopressin: implications for autism spectrum disorders? *Behavioural Brain Research* **176**, pp. 170–186.

173. **Carter, C. S.**, **Lederhendler, I. I.**, and **Kirkpatrick, B.** (1999). *The integrative neurobiology of affiliation*. Cambridge: MIT Press.

174. **Carter, C.**, and **Altemus, M.** (1997). Integrative functions of lactational hormones in social behavior and stress managementa. *Annals of the New York Academy of Sciences* **807**, pp. 164–174.

175. **Carter, M.**, **Borg, J.**, and **de Lecea, L.** (2009). The brain hypocretins and their receptors: mediators of allostatic arousal. *Current Opinion in Pharmacology* **9**, pp. 39–45.

176. **Cavanaugh, D.**, **Ggeratowski, J.**, **Wooltorton, J.**, **Spaethling, J.**, **Hector, C.**, **Zheng, X.**, **Johnson, E.**, **Eberwine, J.**, and **Sehgal, A.** (2014). Identification of a circadian output circuit for rest: activity rhythms in drosophila. *Cell* **157**, pp. 689–701.

177. **Cavigelli, S. A.**, **Dubovick, T.**, **Levash, W.**, **Jolly, A.**, and **Pitts, A.** (2003). Female dominance status and fecal corticoids in a cooperative breeder with low reproductive skew: ring-tailed lemurs. *Hormones and Behavior* **43**, pp. 166–179.

178. **Ceccaldi, P.**, **Saada, J.**, **Nicolas, M.**, **Ducarme, G.**, **Blot, P.**, **Guibourdenche, J.**, and **Luton, D.** (2012). Modulation of free corticotropin-releasing hormone, adrenal and placental steroid hormone levels induced by mifepristone during pregnancy. *Fetal Dagnosis and Therapy* **32**, pp. 267–270.

179. **Ceccatelli, S.**, **Cintra, A.**, **Hökfelt, T.**, **Fuxe, K.**, **Wikström, A. C.**, and **Gustafsson, J. Å.** (1989). Coexistence of glucocorticoid receptor-like immunoreactivity with neuropeptides in the hypothalamic paraventricular nucleus. *Experimental Brain Research* **78**, pp. 33–42.

180. **Chaijale, N.**, **Curtis, A.**, **Wood, S.**, **Zhang, X.**, **Bhatnagar, S.**, **Reyes, B.**, **Bockstaele, E.**, and **Valentino, R.** (2013). Social tress engages opioid regulation of locus coeruleus norepinephrine neurons and induces a state of cellular and physical opiate dependence *Neuropsychopharmacology* **38**, pp. 1833–1843.

181. **Challis, J.** (2001). Understanding preterm birth. *Clinical Investigative Medicine* **24**, pp. 60–67.

182. **Challis, J.** (2004). Maternal corticotropin-releasing hormone, fetal growth, and preterm birth. *American Journal of Obstetrics and Gynecology* **191**, pp. 1059–1060.

183. **Challis, J.**, **Bloomfield, F. H.**, **Bocking, A. D.**, **Casciani, V.**, **Chisaka, H.**, **Connor, K.**, **Dong, X.**, **Gluckman, P.**, **Harding, J.E.**, **Johnstone, J.**, **Li, W.**, **Lye, S.**, **Okamura, K.**, and **Premyslova, M.** (2005). Fetal signals and parturition. *Journal of Obstetrics and Gynaecology Research* **31**, pp. 492–499.

184. **Challis, J.**, **Matthews, S.**, **VanMeir, C.** and **Ramirez, M.** (1995) Current topic: the placental corticotropin-releasing hormone– adrenocorticotropin axis. *Placenta* **16**, pp. 481–502.

185. **Challis, J. R. G.**, **Matthews, S. G.**, **Gibb, W.**, and **Lye, S. J.** (2000). Endocrine and paracrine regulation of birth at term and preterm. *Endocrine Reviews* **21**, pp. 514–550.

186. **Chan, E. C.**, **Thompson, M.**, **Madsen, G.**, **Falconer, J.**, **Smith, R.** (1998). Differential processing of corticotropin-releasing hormone by the human placenta and hypothalamus. *Biochemical and Biophysical Research Communications* **153**, pp. 1229–1235.

187. **Chand, D.**, **Lannoy, L.**, **Tucker, R.**, and **Lovejoy, D.** (2013). Origin of chordate peptides by horizontal protozoan gene transfer in early metaozoans and protists: evolution

of the teneurin C-terminal associated peptides (TCAP). *General and Comparative Endocrinology* **1898**, pp. 144–150.

188. **Chandrasekar, G.**, **Lauter, G.**, and **Hauptmann, G.** (2007). Distribution of corticotropin-releasing hormone in the developing zebrafish brain. *Journal of Comparative Neurology* **505**, pp. 337–351.
189. **Chang, C. L.**, and **Hsu, S. Y.** (2004). Ancient evolution of stress-regulating peptides in vertebrates. *Peptides* **25**, pp. 1681–1688.
190. **Chang, F. C.**, and **Opp, M. R.** (1998). Blockade of corticotropin-releasing hormone receptors reduces spontaneous waking in the rat. *American Journal of Physiology-Regulatory, Integrative and Comparative Physiology* **275**, pp. R793–R802.
191. **Chattarji, S.**, **Tomar, A.**, **Suvrathan, A.**, **Ghosh, S.**, and **Rahman, M.** (2015). Neighborhood matters: divergent patterns of stress induced plasticity across the brain. *Nature Neuroscience* **18**, pp. 1364–1373.
192. **Chatzaki, E.**, **Anton, P. A.**, **Million, M.**, **Lambropoulou, M.**, **Constantinidis, T.**, **Kolios, G.**, **Taché, Y.**, and **Grigoriadis, D. E.** (2013). Corticotropin-releasing factor receptor subtype 2 in human colonic mucosa: down-regulation in ulcerative colitis. *World Journal of Gastroenterology* **19**, pp. 1416–1423.
193. **Chen, A.**, **Perrin, M.**, **Brar, B.**, **Li, C.**, **Jamieson, P.**, **DiGruccio, M.**, **Lewis, K.**, and **Vale, W.** (2005). Mouse corticotropin-releasing factor receptor type 2α gene: isolation, distribution, pharmacological characterization and regulation by stress and glucocorticoids. *Molecular Endocrinology* **19**, pp. 441–458.
194. **Chen, A.**, **Vaughan, J.**, and **Vale, W. W.** (2003). Glucocorticoids regulate the expression of the mouse urocortin II gene: a putative connection between the corticotropin-releasing factor receptor pathways. *Molecular Endocrinology* **17**, pp. 1622–1639.
195. **Chen, A.**, **Zorrilla, E.**, **Smith, S.**, **Rousso, D.**, **Levy, C.**, **Vaughan, J.**, **Donaldson, C.**, **Roberts, A.**, **Lee, K.**, and **Vale, W.** (2006). Urocortin 2-deficient mice exhibit gender specific alterations in circadian hypothalamus pituitary adrenal axis and depressive like behavior. *Journal of Neuroscience* **26**, pp. 5500–5510.
196. **Chen, C.**, **Opazo, J.**, **Erez, O. Uddin, M.**, **Forgas, J.**, **Goodman, M.**, **Grossman, L.**, **Romero, R.**, and **Wildman, D.** (2008). The human progesterone receptor shows evidence of adaptive evolution associated with its ability ot act as a transcription factor. *Molecular Phylogenettics and Evolution* **47**, pp. 637–649.
197. **Chen, M.**, **Chiu, T.**, and **Lee, E.** (1992). Noradrenergic mediation of the memory-enhancing effect of corticotropin-releasing factor in the locus coeruleus of rats. *Psychoneuroendocrinology* **17**, pp. 113–124.
198. **Chen, N.**, **Jupp, B.**, **Sztainberg, Y.**, **Lebow, M.**, **Brown, R.**, **Kim, J.**, **Chen, A.**, and **Lawrence, A.** (2014). Knockdown of CRF1 receptors in the ventral tegmental area attenuates cue and acute food deprivation stress induced cocaine seeking in mice. *Journal of Neuroscience* **34**, pp. 11560–11570.
199. **Chen, Y.**, **Allars, M.**, **Pan, X.**, **Maiti, K.**, **Angeli, G.**, **Smith, R.**, and **Nicholson, R.** (2013). Effects of corticotrophin releasing hormone (CRH) on cell viability and differentiation in the human BeWo choriocarcinoma cell line: a potential syncytialisation inducer distinct from cyclic adenosine monophosphate (cAMP). *Reproductive Biology and Endocrinology* **11**, p. **30**.

200. **Chen, Y., Andres, A., Frotscher, M.,** and **Baram, T.** (2012). Tuning synaptic transmission in the hippocampus by stress: the CRH system. *Frontiers in Cellular Neuroscience* **6**, pp. 1–4.
201. **Chen, Y., Bender, R. A., Brunson, K. L., Pomper, J. K., Grigoriadis, D. E., Wurst, W.,** and **Baram, T. Z.,** (2004). Modulation of dendritic differentiation by CRF in the developing hippocampus. *Proceedings of National Academy of Sciences of the United States of America* **101**, pp. 15782–15787.
202. **Chen, Y., Dubé, C. M., Rice, C. J.,** and **Baram, T. Z.** (2008). Rapid loss of dendritic spines after stress involves derangement of spine dynamics by corticotropin-releasing hormone. *The Journal of Neuroscience* **28**, pp. 2903–2911.
203. **Chen, Y., Holzman, C., Chung, H., Senagore, P., Talge, N. M.,** and **Siler-Khodr, T.** (2010). Levels of maternal serum corticotropin-releasing hormone (CRH) at midpregnancy in relation to maternal characteristics. *Psychoneuroendocrinology* **35**(6), pp. 820–832.
204. **Chen, Y., Kramar, E., Chen, L. Y., Babayan, A. H., Andres, A. L., Gall, C. M., Lynch, G.,** and **Baram, T. Z,** (2013). Impairment of synaptic plasticity by the stress mediator CRH involves selective destruction of thin dendritic spines via RhoA signaling. *Molecular Psychiatry* **18**, pp. 485–496.
205. **Chen, Y., Molet, J., Gunn, B. G., Ressler, K.,** and **Baram, T. Z.** (2015). Diversity of reporter expression patterns in transgenic mouse lines targeting corticotropin-releasing hormone-expressing neurons. *Endocrinology* **156**, pp. 4769–4780.
206. **Chen, Y., Rada, P., Butzler, B., Leibowitz, S.,** and **Hoebel, B.** (2012). Corticotropin-releasing factor in the nucleus accumbens shell induces swim depression, anxiety, and anhedonia along with changes in local dopamine/acetylcholine balance. *Neuroscience* **206**, pp. 155–166.
207. **Chen, Y., Rex, C. S., Rice, C. J., Dubé, C. M., Gall, C. M., Lynch, G.,** and **Baram, T. Z.** (2010). Correlated memory defects and hippocampal dendritic spine loss after acute stress involve corticotropin-releasing hormone signaling. *Proceedings of the National Academy of Sciences of the United States of America* **107**, pp. 13123–13128.
208. **Chen, Y., Sheng, H., Xu, Y., Zhang, Y.,** and **Ni, X.** (2012). Actviation of CRHR2 exers an inhibitory effect on the expression of collapsing response mediator protein 3 in hippocampal neurons. *Neuropeptides* **46**, pp. 93–98.
209. **Cheney, D. L.,** and **Seyfarth, R. M.** (2007). *Baboon metaphysics: the evolution of a social mind.* University of Chicago Press.
210. **Cheng, Y. H., Nicholson, R. C., King, B., Chan, E. C., Fitter, J. T.,** and **Smith, R.** (2000). Glucocrticoid stimulation of corticotropin-releasing hormone gene expression requires a cyclic adenosine 3', 5' monophosphate regulatory element in human primary placental cytophoblast cells. *Journal of Clinical Endocrinology and Metabolism* **85**, pp. 1937–1945.
211. **Choi, D. C., Furay, A. R., Evanson, N. K., Ostrander, M. M., Ulrich-Lai, Y. M.,** and **Herman, J. P.** (2007). Bed nucleus of the stria terminalis subregions differentially regulate hypothalamic–pituitary–adrenal axis activity: implications for the integration of limbic inputs. *The Journal of Neuroscience* **27**, pp. 2025–2034.
212. **Choi, D. C., Nguyen, M. M., Tamashiro, K. L., Ma, L. Y., Sakai, R. R.,** and **Herman, J. P.** (2006). Chronic social stress in the visible burrow system modulates stress-related

gene expression in the bed nucleus of the stria terminalis. *Physiology & Behavior* **89**, pp. 301–310.

213. **Choleris, E.**, **Devidze, N.**, **Kavaliers, M.**, and **Pfaff, D. W.** (2008). Steroidal/neuropeptide interactions in hypothalamus and amygdala related to social anxiety. *Progress in Brain Research* **170**, pp. 291–303.
214. **Choleris, E.**, **Gustafsson, J. Å.**, **Korach, K. S.**, **Muglia, L. J.**, **Pfaff, D. W.**, and **Ogawa, S.** (2003). An estrogen-dependent four-gene micronet regulating social recognition: a study with oxytocin and estrogen receptor-α and-β knockout mice. *Proceedings of the National Academy of Sciences* **100**, pp. 6192–6197.
215. **Choleris, E.**, **Ogawa, S.**, **Kavaliers, M.**, **Gustafsson, J. Å.**, **Korach, K. S.**, **Muglia, L. J.**, and **Pfaff, D. W.** (2006). Involvement of estrogen receptor α, β and oxytocin in social discrimination: a detailed behavioral analysis with knockout female mice. *Genes Brain and Behavior* **5**, pp. 528–539.
216. **Choy, M. Y.**, **Leung, T. N.**, and **Lau, T. K.** (2004). Corticotropin-releasing hormone peptide and human first-trimester placental growth. *Early Human Development* **79**, pp. 77–80.
217. **Chrousos, G. P.** (1995). The hypothalamic-pituitary-adrenal axis and immune-mediated inflammation. *New England Journal of Medicine* **332**, pp. 1351–1362.
218. **Chrousos, G. P.**, **Loriaux, D. L.**, **Brandon, D.**, **Shull, J.**, **Renquist, D.**, **Hogan, W.**, **Tomita, M.** and **Lipsett, M. B.** (1984). Adaptation of the mineralocorticoid target tissues to the high circulating cortisol and progesterone plasma levels in the squirrel monkey. *Endocrinology* **115**, pp. 25–32.
219. **Chu, S. C.**, **Yu, C. H.**, **Chen, P. N.**, **Hsieh, Y. S.**, and **Kuo, D. Y.** (2016). Role of oxidative stress in disrupting the function of negative glucocorticoid response element in daily amphetamine-treated rats. *Psychoneuroendocrinology* **71**, pp. 1–11.
220. **Cintra, A.**, **Fuxe, K.**, **Härfstrand, A.**, **Agnati, L. F.**, **Wikström, A. C.**, **Okret, S.**, **Vale, W.**, and **Gustafsson, J. Å.** (1987). Presence of glucocorticoid receptor immunoreactivity in corticotrophin releasing factor and in growth hormone releasing factor immunoreactive neurons of the rat di-and telencephalon. *Neuroscience Letters* **77**, pp. 25–30.
221. **Cintra, A.**, **Fuxe, K.**, **Solfrini, V.**, **Agnati, L. F.**, **Tinner, B.**, **Wikström, A. C.**, **Staines, W.**, **Okret, S.**, and **Gustafsson, J. Å.** (1991). Central peptidergic neurons as targets for glucocorticoid action. Evidence for the presence of glucocorticoid receptor immunoreactivity in various types of classes of peptidergic neurons. *The Journal of Steroid Biochemistry and Molecular Biology* **40**, pp. 93–103.
222. **Cippitelli, A.**, **Damadzic, R.**, **Singley, E.**, **Thorsell, A.**, **Ciccocioppo, R.**, **Eskay, R. L.**, and **Heilig, M.** (2012). Pharmacological blockade of corticotropin-releasing hormone receptor 1 (CRH1R) reduces voluntary consumption of high alcohol concentrations in non-dependent Wistar rats. *Pharmacology Biochemistry and Behavior* **100**, pp. 522–529.
223. **Class, Q. A.**, **Buss, C.**, **Davis, E. P.**, **Gierczak, M.**, **Pattillo, C.**, **Chicz-DeMet, A.**, and **Sandman, C. A.** (2008). Low levels of corticotropin-releasing hormone during early pregnancy are associated with precocious maturation of the human fetus. *Developmental Neuroscience* **30**(6), pp. 419–426.
224. **Clifton, V.**, **Smith, C.**, and **Wright, I.** (2012). Microvascular effects of CRH in human skin vary in relation to gender. *Journal of Clinical Endocrinology and Metabolism* **87**, pp. 267–270.

225. **Clifton, V. L.**, **Read, M. A.**, **Leitch, I. M.**, **Giles, W. B.**, **Boura, A. L.**, **Robinson, P. J.**, and **Smith, R.** (1995). Corticotropin-releasing hormone induced vasodilation in the human fetal-placental circulation: involvement of the nitric oxide-cylcic guanosine 3'5' monophosphate-mediated pathway. *Journal of Clinical Endocrinology and Metabolism* **80**, pp. 2888–2893.

226. **Clifton, V. L.**, **Telfer, J. F.**, **Thompson, A. J.**, **Cameron, I. T.**, **Teoh, T. G.**, **Lye, S. J.**, and **Challis, J. R.** (1998). Corticotropin-releasing hormone and proopiomelanocortin-derived peptides are present in human myometrium. *Journal of Clinical Endocrinology and Metabolism* **83**, pp. 3716–3721.

227. **Clinchy, M.**, **Schulkin, J.**, **Zanette, L.**, **Sheriff, M.**, **McGowan, P.**, and **Boonstra, R.** (2011). The neurological ecology of fear: insights neuroscientists and ecologists have to offer one another. *Frontiers in Behavioral Neuroscience* **5**, pp. 1–6.

228. **Coddington, E.**, **Lewis, C.**, **Rose, J. D.**, & **Moore, F. L.** (2007). Endocannabinoids mediate the effects of acute stress and corticosterone on sex behavior. *Endocrinology* **148**, pp. 493–500.

229. **Coe, C. L.**, **Wiener, S. G.**, **Rosenberg, L. T.**, and **Levine, S.** (1985). Endocrine and immune responses to separation and maternal loss in nonhuman primates. In: Reite (ed.), *The psychobiology of attachment and separation*. New York: Acamdeic Press, pp. 22–42.

230. **Cohen, H.**, **Kozlovsky, N.**, **Matar, M. A.**, **Zohar, J.**, and **Kaplan, Z.** (2014). Distinctive hippocampal and amygdalar cytoarchitecture changes underlie specific patterns of behavioral disruption following stress exposure in an animal model of PTSD. *European Neuropsychopharm* **24**, pp. 1925–1944.

231. **Cole, R. D.**, **Kawasumi, Y.**, **Parikh, V.**, and **Bangasser, D. A.** (2016). Corticotropin releasing factor impairs sustained attention in male and female rats. *Behavioural Brain Research* **296**, pp. 30–34.

232. **Colucci, J.**, **Ortland, E.** (2012). Evolution of ligand specificity in steroid hormone nuclear receptors. *Evolution* **2**, p. **6**.

233. **Conlon, J. M.**, and **Larhammar, D.** (2005). The evolution of neuroendocrine peptides. *General and Comparative Endocrinology* **142**, pp. 53–59.

234. **Conrad, C. D.**, **MacMillan, D. D., 2nd**, **Tsekhanov, S.**, **Wright, R. L.**, **Baran, S. E.**, and **Fuchs, R. A.** (2004). Influence of chronic corticosterone and glucocorticoid receptor antagonism in the amygdala on fear conditioning. Neurobiology of Learning and Memory **81**, pp. 185–199.

235. **Contarino, A.**, and **Papaleo, F.** (2005). The corticotropin relasing factor receptor 1 pathway mediates the negative affective states of opiate withdrawal. *Proceedings of the National Academy of Sciences* **102**, pp. 18649–18654.

236. **Contoreggi, C.**, **Herning, R. I.**, **Na, P.**, **Gold, P. W.**, **Chrousos, G.**, **Negro, P. J.**, **Better, W.**, and **Cadet, J. L.** (2003). Stress hormone responses to corticotropin-releasing hormone in substance abusers without severe comorbid psychiatric disease. *Biological Psychiatry* **54**, pp. 873–878.

237. **Cook, C.J.** (2002). Glucocorticoid feedback increases the sensitivity of the limbic system to stress. *Physiology & Behavior* **75**, pp. 455–464.

238. **Cook, C.** (2004). Stress induces CRF release in the paraventricular nucleus, and both CRF and GABA release in the amygdala. *Physiology and Behavior* **82**, pp. 751–762.

239. **Cook, C. J.** 2001. Measuring of extracellular cortisol and corticotropin-releasing hormone in the amygdala using immunosensor coupled microdialysis. *J Neurosci Methods* **110**(1–2), pp. 95–101.

240. **Cook, C.J.** (2002). Glucocorticoid feedback increases the sensitivity of the limbic system to stress. *Physiology & Behavior* **75**, pp. 455–464.

241. **Coplan, J. D.**, and **Lydiard, R. B.** (1998). Brain circuits in panic disorder. *Biological Psychiatry* **44**, pp. 1264–1276.

242. **Coplan, J. D.**, **Andrews, M. W.**, **Rosenblum, L. A.**, **Owens, M. J.**, **Friedman, S.**, **Gorman, J. M.**, and **Nemeroff, C. B.** (1996). Persistent elevations of cerebrospinal fluid concentrations of corticotropin-releasing factor in adult nonhuman primates exposed to early-life stressors: implications for the pathophysiology of mood and anxiety disorders. *Proceedings of the National Academy of Sciences of the United States of America* **93**, pp. 1619–1623.

243. **Coplan, J. D.**, **Smith, E. L. P.**, **Altemus, M.**, **Scharf, B. A.**, **Owens, M. J.**, **Nemeroff, C. B.**, **Gorman, J.M.**, and **Rosenblum, L. A.** (2001). Variable foraging demand rearing: sustained elevations in cisternal cerebrospinal fluid corticotropin-releasing factor concentrations in adult primates. *Biological Psychiatry* **50**, pp. 200–204.

244. **Cordero, M.I.**, and **Sandi, C.** (1998). A role for brain glucocorticoid receptors in contextual fear conditioning: Dependence upon training intensity. *Brain Research* **786**, pp. 11–17.

245. **Coric, V.**, **Feldman, H. H.**, **Oren, D. A.**, **Shekhar, A.**, **Pultz, J.**, **Dockens, R. C.**, **Wu, X.**, **Gentile, K. A.**, **Huang, S. P.**, **Emison, E.**, **Delmonte, T.**, **D'Souza, B. B.**, **Zimbroff, D. L.**, **Grebb, J. A.**, **Goddard, A. W.**, and **Stock, E. G.** (2010). Multicenter, randomized, double-blind, active comparator and placebo-controlled trial of a corticotropin-releasing factor receptor-1 antagonist in generalized anxiety disorder. *Depression and Anxiety* **27**, pp. 417–25.

246. **Corodimas, K. P.**, **LeDoux, J. E.**, **Gold, P. W.**, and **Schulkin, J.** (1994). Corticosterone potentiation of conditioned fear in rats. *Annals of the New York Academy of Sciences* **746**, pp. 392–393.

247. **Corominas, M.**, **Roncero, C.**, and **Casas, M.** (2010). Corticotropin releasing factor and neuroplasticity in cocaine addiction. *Life Sciences* **86**, pp. 1–9.

248. **Costine, B.**, **Oberlander, J.**, **Davis, M.**, **Penatti, C.**, **Porter, D.**, **Leaton, R.**, and **Henderson, L.** (2010). Chronic anabolic androgenic steroid exposure alters corticotropin releasing factor expression and anxiety-like behaviors in the female mouse. *Psychoneuroendocrinology* **35**, pp. 1473–1485.

249. **Craig, W.** (1918). Appetites and aversions as constituents of instincts. *Biological Bulletin* **34**, pp. 91–107.

250. **Cratty, M. S.**, **Ward, H. E.**, **Johnson, E. A.**, **Azzaro, A. J.**, and **Birkle, D. L.** (1995). Prenatal stress increases corticotropin-releasing factor (CRF) content and release in rat amygdala minces. *Brain Research* **675**, pp. 297–302.

251. **Creel, S.** (2001). Social dominance and stress hormones. *Trends in Ecology and Evolution* **16**, pp. 491–497.

252. **Crespi, E. J.**, and **Denver, R. J.** (2005). Ancient origins of human developmental plasticity. *American Journal of Human Biology* **17**, pp. 44–54.

253. **Crews, D.** (2005). Evolution of neuroendocrine mechanisms that regulate sexual behavior. *Trends in Endocrinology and Metabolism* **16**, pp. 354–361.

254. **Crews, D., Robker, R.,** and **Mendonça, M.** (1993). Seasonal fluctuations in brain nuclei in the red-sided garter snake and their hormonal control. *The Journal of Neuroscience* **13**, pp. 5356–5364.

255. **Crompton, R., Clifton, V. L., Bisits, A. T., Read, M. A., Smith, R.,** and **Wright, I. M.** (2003). Corticotropin-releasing hormone causes vasodilation in human skin via mast cell-dependent pathways. *The Journal of Clinical Endocrinology & Metabolism* **88**, pp. 5427–5432.

256. **Croteau, J. D., Schulkin, J.,** and **Shepard, J. D.** (2017). Behavioral effects of chronically elevated corticosterone in subregions of the medial prefrontal cortex.

257. **Cuadra, G., Zurita, A., Lacerra, C.,** and **Molina, V.** (1999). Chronic stress sensitizes frontal cortex dopamine release in response to a subsequent novel stressor: reversal by naloxone. *Brain Research Bulletin* **48**, 303–308.

258. **Cullinan, W. E., Ziegler, D. R., & Herman, J. P.** (2008). Functional role of local GABAergic influences on the HPA axis. Brain Structure and Function **213**: 63–72.

259. **Curtis, A., Bethea, T.,** and **Valentino, R.** (2006). Sexually dimorphic responses of the brain norepinephrine system to stress and corticotropin-releasing factor. *Neuropsychopharmacology* **31**, pp. 544–554.

260. **Cusulin, J. I., Füzesi, T., Watts, A. G., Bains, J. S.** (2013) Characterization of corticotropin-releasing hormone neurons in the paraventricular nucleus of the hypothalamus of CRH-IRES-Cre mutant mice. *PLoS One* **8**, p. e64943.

261. **Dabrowska, J., Hazra, R., Guo, J., DeWitt, S.,** and **Rainnie, D.** (2013). Central CRF neurons are not created equal: phenotypic differences in CRF-containing neurons of the rat paraventricular hypothalamus and the bed nucleus of the stria terminalis. *Frontiers in Neuroscience* 7, pp. 1–14.

262. **Dabrowska, J., Hazra, R., Ahern, T. H., Guo, J. D., McDonald, A. J., Mascagni, F., Muller, J. F., Young, L. J.,** and **Rainnie, D.G.** (2011). Neuroanatomical evidence for reciprocal regulation of the corticotrophin-releasing factor and oxytocin systems in the hypothalamus and the bed nucleus of the stria terminalis of the rat: Implications for balancing stress and affect. *Psychoneuroendocrinology* **36**(9), pp. 1312–1326.

263. **Dallman, M. F., Warne, J. P., Foster, M. T.,** and **Pecoraro, N. C.** (2007). Glucocorticoids and insulin both modulate caloric intake through actions on the brain. *The Journal of Physiology* **583**, pp. 431–436.

264. **Dallman, M. F.** (2003). Stress by any other name? *Hormones and Behavior* **43**, pp. 18–20.

265. **Dallman, M. F.** (2005). Fast glucocorticoid actions on the brain: back to the future. *Frontiers in Neuroendocrinology* **26**, pp. 103–108.

266. **Dallman, M. F., Akana, S. F., Strack, A. M., Hanson, E. S.,** and **Sebastian, R. J.** (1995). The neural network that regulates energy balance is responsive to glucocorticoids and insulin and also regulates HPA axis responsivity at a site proximal to CRF neurons. *Annals of the New York Academy of Sciences* **771**, pp. 730–742.

267. **Dallman, M. F., Akana, S. F., Cascio, C. S., Darlington, D. N., Jacobson, L.,** and **Levin, N.** (1987). Regulation of ACTH secretion: variations on a theme of B. *Recent Progress in Hormone Research* **43**, pp. 113–173.

268. **Dallman, M. F., Akana, S. F., Levin, N., Walker, C. D., Bradbury, M. J., Suemaru, S.,** and **Scribner, K. S.** (1994). Corticosteroids and the control of function in the

hypothalamo-pituitary–adrenal HPA) axis. *Annals of the New York Academy of Sciences* **746**, pp. 22–31.

269. **Dallman, M. F.**, and **Bhatnagar, S.** (2000). *Chronic stress and energy balance: role of the hypothalamo-pituitary–adrenal axis*. New York: Oxford University Press.
270. **Dallman, M. F., Engeland, W. C., Rose, J. C., Wilkinson, C. W., Shinsako, J.**, and **Siedenburg, F.** (1978). Nycthemeral rhythm in adrenal responsiveness to ACTH. *American Journal of Physiology* **235**, pp. R210–R218.
271. **Dallman, M. F., Pecoraro, N., Akana, S. F., La Fleur, S. E., Gomez, F., Houshyar, H., Bell, M. E., Bhatnagar, S., Laugero, K. D.**, and **Manalo, S.** (2003). Chronic stress and obesity: a new view of "comfort food." *Proceedings of the National Academy of Sciences of the United States of America* **100**, pp. 11696–11701.
272. **Dallman, M. F., Pecoraro, N. C.**, and **la Fleur, S.E.** (2005). Chronic stress and comfort foods: self-medication and abdominal obesity. *Brain Behavior and Immunity* **19**, pp. 275–280.
273. **Daniel, S. E.**, and **Rainnie, D. G.** (2016). Stress modulation of opposing circuits in the bed nucleus of the stria terminalis. *Neuropsychopharmacology* **41**, pp. 103–125.
274. **Darwin, C.** (1859, 1958). *The origin of species*. New York: Mentor Book.
275. **Darwin, C.** (1872, 1998). *The expression of the emotions in man and animals*. Oxford: Oxford University Press.
276. **Darwin, C.** (1874). *The descent of man, and selection in relation to sex*. Chicago: Rand, McNally and Co.
277. **Darwin, C.** (1881). *The formation of vegetable mould through the action of worms, with observations on their habits*. London: John Murray.
278. **Daston, L.** (1998). *Biographies of scientific objects*. Chicago: University of Chicago Press.
279. **Davis, E. P., Glynn, L. M., Dunkel Schetter, C., Hobel, C., Chicz-Demet, A.**, and **Sandman, C. A.** (2005). Corticotropin releasing hormone during pregnancy is associated with infant temperament. *Developmental Neuroscience* **27**, pp. 299–305.
280. **Davis, M.** (1998). Are different parts of the extended amygdala involved in fear versus anxiety? *Biological Psychiatry* **44**, pp. 1239–1247.
281. **Davis, M.**, and **Shi, C.** (1999). The extended amygdala: are the central nucleus of the amygdala and the bed nucleus of the stria terminalis differentially involved in fear versus anxiety? *Annals of the New York Academy of Sciences* **877**, pp. 281–291.
282. **Davis, M., Walker, D. L.**, and **Lee, Y.** (1997). Amygdala and bed nucleus of the stria terminalis: differential roles in fear and anxiety measured with the acoustic startle reflex. *Philosophical Transactions of the Royal Society of London Series B Biological Sciences* **352**, pp. 1675–1687.
283. **Davis, M., Walker, D. L., Miles, L.**, and **Grillon, C.** (2010). Phasic vs sustained fear in rats and humans: role of the extended amygdala in fear vs anxiety. *Neuropsychopharmacology* **35**, pp. 105–135.
284. **Day, H. E., Greenwood, B. N., Hammack, S. E., Watkins, L. R., Fleshner, M., Maier, S. F.**, and **Campeau, S.** (2004). Differential expression of 5HT-1A, $\alpha$1b adrenergic, CRF-R1, and CRF-R2 receptor mRNA in serotonergic, $\gamma$-aminobutyric acidergic, and catecholaminergic cells of the rat dorsal raphe nucleus. *Journal of Comparative Neurology* **474**, pp. 364–378.

285. **De Castro e Silva, E., Fregoneze, J.B.,** and **Johnson, A.K.** (2006). Corticotropin-releasing hormone in the lateral parabrachial nucleus inhibits sodium appetitie in rats. *American Journal of Physiology Regulatory Integrative and Comparative Physiology* **290**, pp. R1136–R1141.
286. **de Fonseca, F. R., Carrera, M. R. A., Navarro, M., Koob, G. F.,** and **Weiss, F.** (1997). Activation of corticotropin-releasing factor in the limbic system during cannabinoid withdrawal. *Science* **276**, pp. 2050–2054.
287. **De Groef, B., Goris, N., Arkcens, L., Kuhn, E.,** and **Darras, V.** (2003). Corticotropin-releasing hormone (CRH)-induced thyrotropin release is directly mediated through CRH receptor type 2 on thyrotrypes. *Endocrinology* **144**, pp. 5537–5544.
288. **De Groef, B., Van der Geyten, S., Darras, V. M.,** and **Kühn, E. R.** (2006). Role of corticotropin-releasing hormone as a thyrotropin-releasing factor in non-mammalian vertebrates. *General and Comparative Endocrinology* **146**, pp. 62–68.
289. **De Kloet, E. R., Vreugdenhil, E., Oitzl, M. S.,** and **Joels, M.** (1998). Brain corticosteroid receptor balance in health and disease 1. *Endocrine Reviews* **19**, pp. 269–301.
290. **De Kloet, E. R.** (1991). Brain corticosteroid receptor balance and homeostatic control. *Frontiers in Neuroendocrinology* **12**, pp. 95–164.
291. **De Kloet, E. R., Joels, M.,** and **Holsboer, F.** (2005). Stress and the brain: from adaptation to disease. *Neuroscience* **3**, pp. 463–475.
292. **De Lecea, L., Jones, B., Boutrel, B., Borlgand, S., Nishino, S., Bubser, M.,** and **DiLeone, R.** (2006). Addiction and arousal: alternative roles of hypothalamic peptdies. *Journal of Neuroscience* **26**, pp. 10372–10375.
293. **De Loof, A., Lindemans, M., Liu, F., De Groef, B.,** and **Schoofs, L.** (2012). Endocrine archeology: do insects retain ancestrally inherited counterparts of the vertebrate releasing hormones GnRH, GHRH, TRH, and CRF?. *General and Comparative Endocrinology* **177**(1), pp. 18–27.
294. **De Quervain, D. J., Roozendaal, B.,** and **McGaugh, J. L.** (1998). Stress and glucocorticoids impair retrieval of long-term spatial memory. *Nature* **394**, pp. 787–790.
295. **De Souza, E. B., Perrin, M. H., Insel, T. R., Rivier, J., Vale, W. W.,** and **Kuhar, M. J.** (1984). Corticotropin-releasing factor receptors in rat forebrain: autoradiographic identification. *Science* **224**, pp. 1449–1451.
296. **De Souza, E., Insel, T., Perrin, M., Rivier, J, Vale, W.,** and **Kuhar, M.** (1985). Coritcotropin-releasing factor receptors are widely distributed within the rat central nervous system: an autoradiographic study. *Journal of Neuroscience* **5**, pp. 3189–3203.
297. **De Vries, A., Guptaa, T., Cardillo, S., Cho, M.,** and **Carter, C.** (2002). Corticotropin-releasing factor induces social prefeernces in male prairie voles. *Psychoneuroendocrinology* **27**, pp. 705–714.
298. **De Vries, G. J.,** and **Simerly, R. B.** (2002). Anatomy, development, and function of sexually dimorphic neural circuits in the mammalian brain. *Hormones, Brain and Behavior* **4**, pp. 137–191.
299. **Deak, T., Nguyen, K. T., Ehrlich, A. L., Watkins, L. R., Spencer, R. L., Maier, S. F., Licinio, J.,Wong, M., Chrousos, G. P.,Webster, E.,** and **Gold, P. W.** (1999). The impact of thenonpeptides corticotropin-releasing hormone antagonist antalarmin on behavioral and endocrine responses to stress. *Endocrinology* **140**, pp. 79–86.

300. **DeCasper, A. J., Fifer, W. P.** (1980). Of human bonding: Newborns prefer their mothers' voices. *Science* **208**, pp. 1174–1176.

301. **Deisseroth, K.** (2015). Optogenetics: 10 years of microbial opsins in neuroscience. *Nature Neuroscience* **18**, pp. 1213–1225.

302. **Delaney, A. J.**, and **Sah, P.** (2001). Pathway-specific targeting of GABAA receptor subtypes to somatic and dendritic synapses in the central amygdala. *Journal of Neurophysiology* **86**, pp. 717–723.

303. **Delgado, M. R.**, **Nearing, K. I.**, **LeDoux, J. E.**, and **Phelps, E. A.** (2008). Neural circuitry underlying the regulation of conditioned fear and its relation to extinction. *Neuron* **11**, pp. 829–838.

304. **Denton, D. A.** (1982). *Hunger for salt: an anthropological, physiological, and medical analysis*. New York: Springer-Verlag.

305. **Denver, R.** (1997). Environmental stress as a developmental cue: corticotropin-releasing hormone is a proximate mediator of adaptive phenotypic plasticity in amphibian metamorphosis. *Hormones and Behavior* **31**, pp. 169–179.

306. **Denver, R.** (1999). Evolution of the corticotropin-releasing hormone signaling system and its role in stress-induced phenotypic plasticity. *Annals of the New York Academy of Sciences* **897**, pp. 46–53.

307. **Denver, R.** (2007). Endocannabinoids link rapid membrane-mediated corticosteroid actions to behavior. *Endocrinology* **148**, pp. 490–492.

308. **Denver, R.**, and **Crespi, E. J.** (2006). Stress hormones and human developmental plasticity: less from tadpoles. *NeoReview* 7, pp. 183–187.

309. **Denver, R. J.** (2009). Structural and functional evolution of vertebrate neuroendocrine stress systems. *Annals of the New York Academy of Sciences* **1163**, pp. 1–16.

310. **Denver, R. J.** (1998). Hormonal correlates of environmentally induced metamorphosis in the western spadefoot toad, Scaphiopus hammondii. *General Comparative Endocrinology* **110**, pp. 326–336.

311. **Deroche, V.**, **Marinelli, M.**, **Maccari, S.**, **Le Moal, M.**, **Simon, H.**, and **Piazza, P. V.** (1995). Stress-induced sensitization and glucocorticoids. I. Sensitization of dopamine-dependent locomotor effects of amphetamine and morphine depends on stress-induced corticosterone secretion. *The Journal of Neuroscience* **15**, pp. 7181–7188.

312. **Deroche, V.**, **Piazza, P. V.**, **Le Moal, M.** and **Simon, H.** (1994). Social isolation-induced enhancement of the psychomotor effects of morphine depends on corticosterone secretion. *Brain Research* **640**, pp. 136–139.

313. **Dethier, V.** (1961). Behavioral aspects of protein ingestion by the blowfly Phormia regina Meigen. *The Biological Bulletin* **121**, pp. 456–470.

314. **Dethier, V.** (1964). Microscopic brains. *Science* **143**, pp. 1138–1145.

315. **Dethier, V. J.**, and **Stellar, E.** (1961). *Animal behavior: its evolutionary and neurological basis*. Englewood Cliffs, NJ: Prentice-Hall.

316. **Dewey, J.** (1925, 1989). *Experience and nature*. La Salle, IL: Open Court.

317. **Di Bonaventura, M. V. M.**, **Ciccocioppo, R.**, **Romano, A.**, **Bossert, J. M.**, **Rice, K. C.**, **Ubaldi, M.**, **St. Laurent, R.**, **Gaetani, S.**, **Massi, M.**, **Shaham, Y.**, and **Cifani, C.** (2014). Role of bed nucleus of the stria terminalis corticotrophin-releasing factor receptors in frustration stress-induced binge-like palatable food consumption in female rats with a history of food restriction. *The Journal of Neuroscience* **34**, pp. 11316–11324.

318. **Di Chiara, G.** (1999). Drug addiction as dopamine-dependent associative learning disorder. *European Journal of Pharmacology* **375**, pp. 13–30.

319. **Di Stefano, V.**, **Wang, B.**, **Parobchak, N.**, **Roche, N.**, and **Rosen, T.** (2015). RelB/p52-mediated NF-κB signaling alters histone acetylation to increase the abundance of corticotropin-releasing hormone in human placenta. *Science Signaling* **8**, p. ra85.

320. **Dingle, H.** (2002). Hormonal mediation of insect life histories. *Hormones, Brain and Behavior* **3**, pp. 237–279.

321. **Diorio, D.**, **Viau, V.**, and **Meaney, M. J.** (1993). The role of the medial prefrontal cortex (cingulate gyrus) in the regulation of hypothalamic-pituitary-adrenal responses to stress. *The Journal of Neuroscience* **13**, pp. 3839–3847.

322. **Dixson, A. F.**, and **George, L.** (1982). Prolactin and parental behaviour in a male New World primate. *Nature* **299**, pp. 551–553.

323. **Dobzhansky, T.** (1962). *Mankind evolving: the evolution of the human species.* New Haven, CT: Yale University Press.

324. **Donaldson, Z. R.**, and **Young, L. J.** (2008). Oxytocin, vasopressin, and the neurogenetics of sociality. Science **322**, pp. 900–904.

325. **Donatti, A.**, and **leite-Panissi, C.** (2011). Activation of corticotropin-releasing factor receptors from the basolateral or central amygdala increases the tonic immobility response in guinea pigs: an innate fear behavior. *Behavioral Brain Research* **225**, pp. 23–30.

326. **Donelan, J.**, **Boucher, W.**, **Papadopoulou, N.**, **Lytinas, M.**, **Papaliodis, D.**, **Dobner, P.**, and **Theoharides, T.** (2006). Corticotropin releasing hormone induces skin vascular permeability through a neurotensin dependent process. *Proceedings of the National Academy of Sciences* **103**, pp. 7759–7764.

327. **Dong, H. W.**, and **Swanson, L. W.** (2006). Projections from bed nuclei of the stria terminalis, anteromedial area: cerebral hemisphere integration of neuroendocrine, autonomic, and behavioral aspects of energy balance. *Journal of Comparative Neurology* **494**, pp. 142–178.

328. **Dong, H. W.**, **Petrovich, G. D.**, and **Swanson, L. W.** (2001). Topography of projections from amygdala to bed nuclei of the stria terminalis. *Brain Research Reviews* **38**, pp. 192–246.

329. **Donley, M. P.**, **Schulkin, J.**, and **Rosen, J. B.** (2005). Glucocorticoid receptor antagonism in the basolateral amygdala and ventral hippocampus interferes with long-term memory of contextual fear. *Behavioural Brain Research* **164**, pp. 197–205.

330. **Dore, R.**, **Iemolo, A.**, **Smith, K.**, **Wang, X.**, **Cottone, P.**, and **Sabino, V.** (2013). CRF mediates the anxiogenic and anti-rewarding, but not the anorectic effects of PACAP. *Neuropsychopharmacology* **38**, pp. 2160–2169.

331. **Douma, T.**, **Millan, M.**, **Boulay, D.**, **Griebel, G.**, **Verdouw, P.**, **Westphal, K.**, **Olivier, B.**, and **Groeniuk, L.** (2014). CRF1 receptor antagonists do not reverse pharmacological disruption of prepulse inhibition in rodents. *Psychopharmacology* **231**, pp. 1289–1303.

332. **Doyon, C.**, **Gilmour, K.**, **Trudeau, V.**, and **Moon, T.** (2003). Corticotropin-releasing factor and neuropeptide Y mRNA levels are elevated in the preoptic area of socially subordinate rainbow trout. *General and Comparative Endocrinology* **133**, pp. 260–271.

333. **Doztler, S. A.**, **Digeronimo, R. J.**, **Yoder, B. A.**, and **Siler-Khodr, T.** (2004). Distribution of corticotropin-releasing hormone in the fetus, newborn, juvenile and adult baboon. *Pediatric Research* **55**, pp. 120–125.

334. **Drevets, W. C.**, **Price, J. L.**, **Bardgett, M. E.**, **Reich, T.**, **Todd, R. D.**, and **Raichle, M. E.** (2002). Glucose metabolism in the amygdla in depression: relationship to diagnostic subtype and plasma cortisol levels. *Pharmacology Biochemistry and Behavior* **71**, pp. 431–447.

335. **Ducottet, C.**, **Griebel, G.**, and **Belzung, C.** (2003). Effects of the selective nonpeptide corticotropin-releasing factor receptor 1 antagonist antalarmin in the chronic mild stress model of depression in mice. *Progress in Neuro-Psychopharmacology and Biological Psychiatry* **27**, pp. 625–631.

336. **Dulac, C.**, **O'Connell, L. A.**, and **Wu, Z.** (2014). Neural control of maternal and paternal behaviors. *Science* **345**, pp. 765–770.

337. **Dunlop, B. W.**, **Rothbaum, B. O.**, **Binder, E. B.**, **Duncan, E.**, **Harvey, P. D.**, **Jovanovic, T.**, **Kelley, M. E.**, **Kinkead, B.**, **Kutner, M.**, **Iosifescu, D. V.**, **Matthew, S. J.**, **Neylan, T. C.**, **Kilts, C. D.**, **Nemeroff, C. B.**, and **Mayberg, H. S.** (2014). Evaluation of a corticotropin releasing hormone type 1 receptor antagonist in women with posttraumatic stress disorder: study protocol for a randomized controlled trial. *Trials* **15**, p. **240**.

338. **Dunn, A. J.**, and **Berridge, C. W.** (1990). Corticotropin-releasing factor as the mediator of stress responses. *Psychobiology of Stress* **579**, pp. 183–191.

339. **Dunn, A. J.**, and **Swiergiel, A. H.** (2008). Effects of acute and chronic stressors and CRF in rat and mouse tests for depression. *Annals of the New Academy of Sciences* **1148**, pp. 118–126.

340. **Duvarci, S.**, and **Pare, D.** (2014). Amygdala microcircuits controlling learned fear. *Neuron* **82**, pp. 966–980.

341. **Ebensperger, L. A.**, **Ramírez-Otarola, N.**, **León, C.**, **Ortiz, M. E.**, and **Croxatto, H. B.** (2010). Early fitness consequences and hormonal correlates of parental behaviour in the social rodent, Octodon degus. *Physiology & Behavior* **101**, pp. 509–517.

342. **Eberhart, J. A.**, **Keverne, E. B.**, and **Meller, R. E.** (1983). Social influences on circulating levels of cortisol and prolactin in male talapoin monkeys. *Physiology & Behavior* **30**, pp. 361–369.

343. **Eberhart, J. A.**, **Keverne, E. B.**, and **Meller, R. E.** (1980). Social influences on plasma testosterone levels in male talapoin monkeys. *Hormones and Behavior* **14**, pp. 247–266.

344. **Edwards, S.**, **Guerrero, M.**, **Ghoneim, O.M.**, **Roberts, E.**, and **Koob, G.F.** (2011). Evidence that vasopressin $V_{1b}$ receptors mediate the transition to excessive drinking in ethanol-dependent rats. *Addiction Biology* **17**, pp. 76–85. doi:10.1111/j.1369-1600.2010.00291.x

345. **Edwards, S.**, **Vendruscolo, L. F.**, **Schlosburg, J. E.**, **Misra, K. K.**, **Wee, S.**, **Park, P. E.**, **Schulteis, G.**, and **Koob, G. F.** (2012). Development of mechanical hypersensitivity in rats during heroin and ethanol dependence: alleviation by CRF 1 receptor antagonism. *Neuropharmacology* **62**, pp. 1142–1151.

346. **Eede, F.**, **Venkin, T.**, **Del-Favero, J.**, **Norrback, K.**, **Souery, D.**, **Nilsson, L.**, **Bossche, B. Hulstijn, W.**, **Sabbe, B.**, **Cosyns, P.**, **Mendlewicz, J.**, **Adolfsson, R.**, **Broeckhoven, C.**, and **Claes, S.** (2007). Single nucleotide polymorphisms analysis of corticotropin-releasing factor-binding protein gene in recurrent major depressive disorder. *Psychiatry Research* **153**, pp. 17–25.

347. **Ehlers, C. L.**, **Reed, T. K.**, and **Henriksen, S. J.** (1986). Effects of corticotropin-releasing factor and growth hormone-releasing factor on sleep and activity in rats. *Neuroendocrinology* **42**, pp. 467–474.

348. **Eick, G.**, **Colucci, J.**, **Harms, M.**, **Ortlund, E.**, and **Thornton, J.** (2012). Evolution of minimal specificity and promiscuity in steroid hormone receptors. *PLos Genetics* **8**, p. e1003072.

349. **Eisenhardt, M.**, **Hansson, A. C.**, **Spanagel, R.**, and **Bilbao, A.** (2015). Chronic intermittent ethanol exposure in mice leads to an up-regulation of CRH/CRHR1 signaling. *Alcoholism: Clinical and Experimental Research* **39**, pp. 752–762.

350. **Elharrar, E.**, **Warhaftig, G.**, **Issler, O.**, **Sztainberg, Y.**, **Dikshtein, Y.**, **Zahut, R.**, **Redlus, L.**, **Chen, A.**, and **Yadid, G.** (2013). Overexpression of corticotropin-releasing factor receptor type 2 in the bed nucleus of stria terminalis improves posttraumatic stress disorder-like symptoms in a model of incubation of fear. *Biological Psychiatry* **74**, pp. 827–836.

351. **Elliott, E.**, **Ezra-Nevo, G.**, **Regev, L.**, **Neufeld-Cohen, A.**, and **Chen-A.** (2010). Resilience to social stress coincides with functional DNA methylation of the CRF gene in adult mice. *Nature Neuroscience* **13**, pp. 1351–1353.

352. **Elster, J.**, **Skog, O. J.** (1999). *Getting hooked: rationality and addiction*. Cambridge: Cambridge University Press.

353. **Elsworth, J. D.**, **Morrow, B. A.**, and **Roth, R. H.** (2001). Prenatal cocaine exposure increases mesoprefrontal dopamine neuron responsivity to mild stress. *Synapse* **42**, pp. 80–83.

354. **Emanuel, R. L.**, **Robinson, B. G.**, **Seely, E. W.**, **Graves, S. W.**, **Kohane, I.**, **Saltzman, D.**, **Barbieri, R.**, and **Majzoub, J. A.** (1994). Corticotrophin releasing hormone levels in human plasma and amniotic fluid during gestation. *Clinical Endocrinology* **40**, pp. 257–262.

355. **Emanuel, R.**, **Torday, J.**, **Asokananthan, N.**, and **Sunday, M.** (2000). Direct effects of corticotropin-releasing hormone and thyrotropin-releasing hormone on fetal lung explants. *Peptdies* **21**, pp. 1819–1829.

356. **Engh, A. L.**, **Beehner, J. C.**, **Bergman, T. J.**, **Whitten, P. L.**, **Hoffmeier, R. R.**, **Seyfarth, R. M.**, and **Cheney, D. L.** (2006). Behavioural and hormonal responses to predation in female chacma baboons (Papio hamadryas ursinus). *Proceedings Biological Sciences* **273**, pp. 707–712.

357. **Engineer, N.**, **Darwin, L.**, **Nishigandh, D.**, **Ngianga-Bakwin, K.**, **Smith, S. C.**, and **Grammatopoulos, D. K.** (2013). Association of glucocorticoid and type 1 corticotropin-releasing hormone receptors gene variants and risk for depression during pregnancy and post-partum. *Journal of Psychiatric Research* **47**, pp. 1166–1173.

358. **Erb, S.**, **Shaham, Y.**, and **Stewart, J.** 1996 Stress reinstates cocaine-seeking behavior after prolonged extinction and a drug-free period. *Psychopharmacology* **128**(4), pp. 408–412.

359. **Eraslan, I.**, **Akyazi, E.**, and **Ergul-Ekiz, E.** (2015). Noise stress induced changes in mRNA levels of corticotropin-releasing hormone family molecules and glucocorticoid receptors in the rat brain. *Folia Biologica* **61**, pp. 66–73.

360. **Erb, S.** (2010). Evaluation of the relationship between anxiety during withdrawal and stress-induced reinstatement of cocaine seeking. *Progress in Neuro-Psychopharmacology & Biological Psychiatry* **34**, pp. 798–807.

361. **Erb, S., Hitchcott, P. K., Rajabi, H., Mueller, D., Shaham, Y.,** and **Stewart, J.** (2000). Alpha-2 adrenergic receptor agonists block stress-induced reinstatement of cocaine seeking. *Neuropsychopharmacology* **23**, pp. 138–150.

362. **Erb, S., Salmaso, N., Rodaros, D.,** and **Stewart, J.** (2001). A role for the CRF-containing pathway from central nucleus of the amygdala to bed nucleus of the stria terminalis in the stress-induced reinstatement of cocaine seeking in rats. *Psychopharmacology* **158**, pp. 360–365.

363. **Erb, S., Shaham, Y.,** and **Stewart, J.** (1998). The role of corticotropin-releasing factor and corticosterone in stress- and cocaine-induced relapse to cocaine seeking in rats. *Journal of neuroscience* **18**, pp. 5529–5536.

364. **Erb, S.,** and **Stewart, J.** (1999). A role for the bed nucleus of the stria terminalis, but not the amygdala, in the effects of corticotropin-releasing factor on stress-induced reinstatement of cocaine seeking. *Journal of Neuroscience* **19**, p. RC35.

365. **Ergang, P., Vodička, M., Soták, M., Klusoňová, P., Behuliak, M., Řeháková, L., Zach, P.,** and **Pácha, J.** (2015). Differential impact of stress on hypothalamic–pituitary–adrenal axis: gene expression changes in Lewis and Fisher rats. *Psychoneuroendocrinology* **53**, pp. 49–59.

366. **Erhardt, A., Muller, M., Rodel, A., Welt, T., Ohl, F., Holsboer, F.,** and **Keck, M. E.** (2009). Consequences of chronic social stress on behaviour and vasopressin gene expression in the PVN of DBA/2OlaHsd mice-influence of treatment with the CRHR1-antagonist R121919/NBI 30775. *Journal of Psychopharmacology* **23**, pp. 31–39.

367. **Erickson, K., Drevets, W.,** and **Schulkin, J.** (2003). Glucocorticoid regulation of diverse cognitive functions in normal and pathological emotional states. *Neuroscience & Biobehavioral Reviews* **27**, pp. 233–246.

368. **Erickson, K., Gabry, K. E., Lindell, S., Champoux, M., Schulkin, J., Gold, P., Suomi, S. J.,** and **Higley, J. D.** (2005). Social withdrawal behaviors in nonhuman primates and changes in neuroendocrine and monoamine concentrations during a separation paradigm. *Developmental Psychobiology* **47**, pp. 196–197.

369. **Erickson, K., Thorsen, P., Chrousos, G., Grigoriadis, D.E., Khongsaly, O., McGregor, J.,** and **Schulkin, J.** (2001). Preterm birth: associated neuroendocrine, medical and behavioral risk factors. *Journal of Clinical Endocrinology and Metabolism* **86**, pp. 2544–2552.

370. **Erkut, Z. A., Hofman, M. A., Ravid, R.,** and **Swaab, D. F.** (1995). Increased activity of hypothalamic corticotropin-releasing hormone neurons in multiple sclerosis. *Journal of Neuroimmunology* **62**, pp. 27–33.

371. **Erkut, Z., Pool, C.,** and **Swaab, D.** (1998). Glucocorticoids suppress corticotropin-releasing hormone and vasopressin expression in human hypothalamic neurons. *Journal of Clinical Endocrinology and Metabolism* **83**, pp. 2066–2073.

372. **Eschenbach, D. A., Hillier, S., Critchlow, C., Stevens, C., DeRouen, T.,** and **Holmes, K. K.** (1988). Diagnosis and clinical manifestations of bacterial vaginosis. *American Journal of Obstetrics and Gynecology* **158**, pp. 819–828.

373. **Esposito, P., Chandler, N., Kandere, K. Basu, S., Jacobson, S., Connolly, R., Tutor, D.,** and **Theoharides, T.** (2002). Corticotropin-releasing hormone and brain mast cells regulate blood-brain-barrier permeability induced by acute stress. *Journal of Pharmacology and Experimental Therapeutics* **303**, pp. 1061–1066.

374. **Esposito, P., Gheorghe, D., Kandere, K., Pang, X., Connolly, R., Jacobson, S.,** and **Theoharides, T.** (2001). Acute stress increases permeability of the blood-brain-barrier through activation of brain mast cells. *Brain Research* **888**, pp. 117–127.

375. **Ettenberg, A., Cotten, S. W., Brito, M. A., Klein, A. K., Ohana, T. A., Margolin, B., Wei, A.,** and **Wenzel, J. M.** (2015). CRF antagonism within the ventral tegmental area but not the extended amygdala attenuates the anxiogenic effects of cocaine in rats. *Pharmacology, Biochemistry, and Behavior* **138**, pp. 148–155.

376. **Evans, A. K., Strassmann, P. S., Lee, I. P.,** and **Sapolsky, R. M.** (2014). Patterns of Toxoplasma gondii cyst distribution in the forebrain associate with individual variation in predator odor avoidance and anxiety-related behavior in male Long–Evans rats. *Brain, Behavior, and Immunity* **37**, pp. 122–133.

377. **Even, N., Devaud, J. M.,** and **Barron, A. B.** (2012). General stress responses in the honey bee. *Insects* **3**, pp. 1271–1298.

378. **Ewer, J.,** and **Reynolds, S.** (2002). Neuropeptide control of molting in insects. *Hormones, Brain and Behavior* **3**, pp. 1–92.

379. **Fahrbach, S. E.,** and **Weeks, J. C.** (2002). Hormonal regulation of neural and behavioral plasticity in insects. *Hormones, Brain, and Behavior* **3**, pp. 331–358.

380. **Fanselow, M.S.** (1991). The midbrain periaqueductal gray as a coordinator of action in response to fear and anxiety. In: Depaulis, A., and Bandler, R. (eds.), *The midbrain periaqueductal gray matter: functional, anatomical and neurochemical organization.* New York: Plenum Publishing. 151–173.

381. **Faria, M. P., Miguel, T. T., Gomes, K. S.,** and **Nunes-de-Souza, R. L.** (2016). Anxiety-like responses induced by nitric oxide within the BNST in mice: Role of CRF1 and NMDA receptors. *Hormones and Behavior* **79**, pp. 74–83.

382. **Faure, A., Richard, J. M.,** and **Berridge, K. C.** (2010). Desire and dread from the nucleus accumbens: cortical glutamate and subcortical GABA differentially generate motivation and hedonic impact in the rat. *PloS One* **5**, p. e11223.

383. **Febo, M., Numan, M.,** and **Ferris, C. F.** (2005). Functional magnetic resonance imaging shows oxytocin activates brain regions associated with mother–pup bonding during suckling. *Journal of Neuroscience* **25**, pp. 11637–11644.

384. **Fekete, E.,** and **Zorrilla, E.** (2007). Physiology, pharmacology, and therapeutic relevance of urocortins in mammals: ancient CRF paralogs. *Frontiers in Neuroendocrinology* **28**, pp. 1–27.

385. **Fellman, D., Bugnon, C., Bresson, J. L., Gouget, A., Cardot, J., Clavequin, M. C.,** and **Hadjiyiassemis, M.** (1984). The CRF neuron: immunocytochemical study. *Peptides* **5** Suppl 1, pp. 19–33.

386. **Fendt, M., Endres, T.,** and **Apfelbach, R.** (2003). Temporary inactivation of the bed nucleus of the stria terminalis but not of the amygdala blocks freezing induced by trimethylthiazoline, a component of fox feces. *Journal of Neuroscience* **23**, pp. 23–28.

387. **Fendt, M., Koch, M.,** and **Schnitzler, H. U.** (1997). Corticotropin-releasing factor in the caudal pontine reticular nucleus mediates the expression of fear-potentiated startle in the rat. *European Journal of Neuroscience* **9**, pp. 299–305.

388. **Fenoglio, K., Chen, Y.,** and **Baram, T.** (2006). Neuroplasticity of the hypothalamic-pituitary-adrenal axis early in life requires recurrent recruitment of stress-regulating brain regions. *Journal of Neuroscience* **26**, pp. 2434–2442.

389. **Fenoglio, K**, **Brunson, K. L.**,**Avishai-Eliner, S.**, **Chen, Y.**, and **Baram, T. Z.** (2004). Region-specific onset of handling-induced changes in corticotropin-releasing factor and glucocorticoid receptor expression. *Endocrinology* **145**, pp. 2702–2706.

390. **Ferguson, J. N.**, **Aldag, J. M.**, **Insel, T. R.**, and **Young, L. J.** (2001). Oxytocin in the medial amygdala is essential for social recognition in the mouse. *The Journal of Neuroscience* **21**, pp. 8278–8285.

391. **Ferin, M.** (1995). The antireproductive role of corticotropin releasing hormone and interleukin-1 in the female rhesus monkey. *Annales d'Endocrinologie* **56**, pp. 181–186.

392. **Fettweis, J. M.**, **Brooks, J. P.**, **Serrano, M. G.**, **Sheth, N. U.**, **Girerd, P. H.**, **Edwards, D. J.**, **Strauss, J. F.**, **the Vaginal Microbiome Consortium**, **Jefferson, K. K.**, **Buck, G. A.** (2014). Differences in vaginal microbiome in African American women versus women of European ancestry. *Microbiology* **160**, pp. 2272–2282.

393. **Figueiredo, H.**, **Bodie, B.**, **Tauchi, M.**, **Dolgas, C.**, and **Herman, J.** (2003). Stress integration aftera cute and chronic predator stress: differential activation of central stress circuitry and sensitization of the hypothalamo-pituitary-adrenocortical axis. *Endocrinology* **144**, pp. 5249–5258.

394. **Fink, G.** (1988). The G.W. Harris Lecture: Steroid Control of Brain and Pituitary Function. *Quarterly Journal of Experimental Physiology* **73**, pp. 257–293.

395. **Fink, G.** (1997). Mechanisms of negative and positive feedback of steroids in the hypothalamic–pituitary system. *Principals Med. Biol* **10A**, pp. 29–100.

396. **Fink, G.** (2015). Memoir: Harris' neuroendocrine revolution: of portal vessels and self-priming. *Journal of Endocrinology* **226**, pp. T13–T24

397. **Fitzsimons, J. T.** (1979). *The physiology of thirst and sodium appetite.* Cambridge: Cambridge University Press.

398. **Fitzsimons, J. T.** (1998). Angiotensin, thirst, and sodium appetite. *Physiological Reviews* **78**, pp. 583–686.

399. **Flandreau, E. I.**, **Ressler, K. J.**, **Owens, M. J.**, and **Nemeroff, C. B.** (2012). Chronic overexpression of corticotropin-releasing factor from the central amygdala produces HPA axis hyperactivity and behavioral anxiety associated with gene-expression changes in the hippocampus and paraventricular nucleus of the hypothalamus. *Psychoneuroendocrinology* **37**, pp. 27–38.

400. **Flandreau, E.**, **Risbrough, V.**, **Lu, A.**, **Ableitner, M.**, **Geyer, M. A.**, **Holsboer, F.**, and **Deussing, J. M.** (2015). Cell type-specific modifications of corticotropin-releasing factor (CRF) and its type 1 receptor (CRF 1) on startle behavior and sensorimotor gating. *Psychoneuroendocrinology* **53**, pp. 16–28.

401. **Florio, P.**, **Franchini, A.**, **Reis, F.**, **Pezzani, I.**, **Ottaviani, E.**, and **Petraglia, F.** (2000). Human placenta, chorion, amnion, and decidua express different variants of corticotropin-releasing factor receptor messenger RNA. *Placenta* **21**, pp. 32–37.

402. **Florio, P.**, **Imperatore, A.**, **Sanseverino, F. Torricelli, M.**, **Reis, F.M.**, **Lowry, P.J.**, and **Petraglia, F.** (2004). The measurement of maternal plasma corticotropin-releasing factor (CRF) and CRF-binding protein improves the early prediction of preeclampsia. *Journal of Clinical Endocrinology and Metabolism* **89**, pp. 4673–4677.

403. **Florio, P.**, **Severi, F.**, **Ciarmela, P.**, **Fiore, G.**, **Calonaci, G.**, **Merola, A.**, **De Felice, C.**, **Palumbo, M.**, and **Petraglia, F.** (2002). Placental stress factors and maternal-fetal adaptive response: the corticotropin releasing factor faily. *Endocrinology* **19**, pp. 91–102.

404. **Florio, P., Severi, F.M., Fiore, G., Micheli, L., Bocchi, C., Nencini, C., Pezzani, I., Giorgi, G.,** and **Petraglia, F.** (2003). Impaired uterine artery blood flow at mid gestation and low levels of maternal plasma corticotropin releasing factor. *Journal of the Society for Gynecologic Investigation* **10**, pp. 294–297.

405. **Fluharty, S. J.,** and **Epstein, A. N.** (1983). Sodium appetite elicited by intracerebroventricular infusion of angiotensin II in the rat: II. Synergistic interaction with systemic mineralocorticoids. *Behavioral Neuroscience* **97**, pp. 746–758.

406. **Forcelli, P. A., Orefice, L. L.,** and **Heinrichs, S. C.** (2007). Neural, endocrine and electroencephalographic hyperreactivity to human contact: a diathesis-stress model of seizure susceptibility in El mice. *Brain Research* **1144**, pp. 248–256.

407. **Fosque, B. F., Sun, Y., Dana, H., Yang, C. T., Ohyama, T., Tadross, M. R., Patel, R., Zlatic, M. Kim, D. S., Ahrens, M. B., Jajyaraman, V., Looger, L. L.,** and **Shreiter, E. R.** (2015). Labeling of active neural circuits in vivo with designed calcium integrators. *Science* **347**, pp. 755–760.

408. **Francis, D. D., Champagne, F. C.,** and **Meaney, M. J.** (2000). Variations in maternal behaviour are associated with differences in oxytocin receptor levels in the rat. *Journal of Neuroendocrinology* **12**, pp. 1145–1148.

409. **Francis, D. D., Diorio, J., Liu, D.,** and **Meaney, M. J.** (1999). Nongenomic transmission across generations of maternal behavior and stress responses in the rat. *Science* **286**, pp. 1155–1158.

410. **Francis, D. D., Young, L. J., Meaney, M. J.,** and **Insel, T. R.** (2002). Naturally occurring differences in maternal care are associated with the expression of oxytocin and vasopressin (V1a) receptors: gender differences. *Journal of Neuroendocrinology* **14**, pp. 349–353.

411. **Frankenne, F., Closset, J., Gomez, F., Scippo, M. L., Smal, J.,** and **Hennen, G.** (1988). The physiology of growth hormones in pregnant women and partial characterization of the placental GH variant. *Journal of Clinical Endocrinology and Metabolism* **66**, pp. 1171–1180.

412. **Frazier, C. R., Trainor, B. C., Cravens, C. J., Whitney, T. K.,** and **Marler, C. A.** (2006). Paternal behavior influences development of aggression and vasopressin expression in male California mouse offspring. *Hormones and Behavior* **50**, pp. 699–707.

413. **French, J. A., Fite, J. E., Jensen, H., Oparowski, K., Rukstalis, M. R., Fix, H. Jones, B., Maxwell, H., Pacer, M., Power, M. L.,** and **Schulkin, J.** (2007). Treatment with CRH-1 antagonist antalarmin reduced behavioral and endocrine responses to social stressors in marmosets. *American Journal of Primatology* **69**, pp. 877–889.

414. **Freud, S.** (1959). *Inhibitions, symptoms and anxiety*. Trans. Strachey, A. Revised and newly edited by James Strachey. New York: W.W. Norton.

415. **Frijda, N. H.** (1986). *The emotions*. Cambridge: Cambridge University Press.

416. **Frim, D. M., Emanuel, R. L., Robinson, B. G., Smas, C. M., Adler, G. K.,** and **Majzoub, J. A.** (1988). Characterization and gestational regulation of corticotropin-releasing hormone messenger RNA in human placenta. *Journal of Clinical Investigation* **82**, pp. 287–292.

417. **Frim, D. M., Robinson, B. G., Pasieka, K. B.,** and **Majzoub, J. A.** (1990). Differential regulation of corticotropin-releasing hormone mRNA in rat brain. *American Journal of Physiology* **258**, pp. E686–E692.

418. **Frye, C.**, and **Walf, A.** (2004). Estrogen and/or progesterone administered systematically or to the amygdala can have anxiety, fear, and pain reducing effects in ovariectomized rats. *Behavioral Neuroscience* **118**, pp. 306–313.

419. **Fuchs, T.**, **Iacobucci, P.**, **MacKinnon, K. M.**, and **Panksepp, J.** (2010). Infant-mother recognition in a social rodent (Octodon degus). *Journal of Comparative Psychology* **124**, pp. 166–175.

420. **Funabashi, T.**, **Kawaguchi, M.**, **Furuta, M.**, **Fukushima, A.**, and **Kimura, F.** (2004). Exposure to bisphenol A during gestation and lactation causes loss of sex difference in corticotropin-releasing hormone-immunoreactive neurons in the bed nucleus of the stria terminalis of rats. *Psychoneuroendocrinology* **29**, pp. 475–485.

421. **Funada, M.**, **Hara, C.**, and **Wada, K.** (2001). Involvement of corticotropin releasing factor receptor subtype 1 in morphine withdrawal regulation of the brain noradrenergic system. *European Journal of Pharmacology* **430**, pp. 277–281.

422. **Funder, J. W.**, **Pearce, P. T.**, **Smith, R.**, and **Smith, A. I.** (1988). Mineralocorticoid action: target tissue specificity is enzyme, not receptor, mediated. *Science* **242**, pp. 583–585.

423. **Funk, C.**, and **Koob, G.** (2007). A CRF(2) agonist administered into the central nucleus of the amygdala decreases ethanol self-administration in ethanol dependent rats. *Brain Research* **1155**, pp. 172–178.

424. **Funk, C.**, **O'Dell, L.**, **Crawford, E.**, and **Koob, G.** (2006). Corticotropin-releasing factor within the central nucleus of the amygdala mediates enhanced ethanol self-administration in withdrawn, ethanol-dependent rats. *Journal of Neuroscience* **26**, pp. 11324–11332.

425. **Funk, C. K.**, **Zorrilla, E. P.**, **Lee, M. J.**, **Rice, K. C.**, and **Koob, G. C.** (2007). Corticotropin- releasing factor 1 antagonists selectively reduce ethanol self-administration in ethanol-dependent rats. *Biological Psychiatry* **61**, pp. 78–86.

426. **Füzesi, T.**, **Daviu, N.**, **Cusulin, J.**, **Bonin, R.**, and **Bains, J.** (2016). Hypothalamic CRH neurons orchestrate complex behaviours after stress. *Nature Communications* **7**, pp. 1–13.

427. **Gafford, G. M.**, **Guo, J. D.**, **Flandreau, E. I.**, **Hazra, R.**, **Rainnie, D. G.**, and **Ressler, K. J.** (2012). Cell-type specific deletion of GABA(A)α1 in corticotropin-releasing factor-containing neurons enhances anxiety and disrupts fear extinction. *Proceedings of the National Academy of Sciences of the United States of America* **109**, pp. 16330–16335.

428. **Gafford, G. M.**, and **Ressler, K. J.** (2015). GABA and NMDA receptors in CRF neurons have opposing effects in fear acquisition and anxiety in central amygdala vs. bed nucleus of the stria terminalis. *Hormones and Behavior* **76**, pp. 136–142.

429. **Gaffrod, G.**, **Jasnow, A.**, and **Ressler, K.** (2014). Grin1 receptor deletion within CRF neurons enhances fear memory. *PLoS One* **9**, p. e111009.

430. **Gallagher, J. P.**, **Orozco-Cabal, L. F.**, **Liu, J.**, **Shinnick-Gallagher, P.** (2008). Synaptic physiology of central CRH system. *European Journal of Pharmacology* **583**, pp. 215–225.

431. **Gambarana, C.**, **Masi, F.**, **Tagliamonte, A.**, **Scheggi, S.**, **Ghiglieri, O.**, and **Graziella De Montis, M.** (1999). A chronic stress that impairs reactivity in rats also decreases dopaminergic transmission in the nucleus accumbens: a microdialysis study. *Journal of Neurochemistry* **72**, pp. 2039–2046.

432. **Gammie, S. C.**, and **Stevenson, S. A.** (2006). Intermale aggression in corticotropin-releasing factor receptor 1 deficient mice. *Behavioural Brain Research* **171**, pp. 63–69.

433. **Gammie, S., Hasen, N., Steveneson, S., Bale, T.,** and **Anna, K.** (2005). Elevated stress sensitivity in corticotropin releasing factor receptor 2 deficient mice decreases maternal, but not intermale aggression. *Behavioral Brain Research* **160**, pp. 169–177.

434. **Ganestad, S. W., Caldwell Hooper, A. E., Eaton, M. A.** (2012). On the function of placental corticotropin releasing hormone: a role in maternal-fetal conflicts over blood glucose concentrations. *Biological Reviews of the Cambridge Philosophical Society* **87**, pp. 856–873.

435. **Gangestad, S. W., Caldwell Hooper, A. E.,** and **Eaton, M. A.** (2012). On the function of placental corticotropin-releasing hormone: a role in maternal-fetal conflicts over blood glucose concentrations. *Biological Reviews* **87**, pp. 856–873.

436. **Gao, H. R., Zhuang, Q. X., Li, B., Li, H. Z., Chen, Z. P., Wang, J. J.,** and **Zhu, J. N.** (2016). Corticotropin releasing factor excites neurons of posterior hypothalamic nucleus to produce tachycardia in rats. *Scientific Reports* **6**, p. **20206**.

437. **Gao, L., Tao, Y., Hu, T., Liu, W., Xu, C., Liu, J., You, X., Gu, H.,** and **Ni, X.** (2012). Regulation of Estradiol and Progesterone Production by CRH-R1 and R2 is through divergent signaling pathways in cultured human placental tryphoblasts. *Endocrinology* **153**, pp. 4918–4928.

438. **Garcia, I., Quast, K. B., Huang, L., Herman, A. M., Selever, J., Deussing, J. M., Justice, N.J.** and **Arenkiel, B. R.** (2014). Local CRH signaling promotes synaptogenesis and circuit integration of adult-born neurons. *Developmental Cell* **30**, pp. 645–659.

439. **Garcia, J., Ervin, F. R.,** and **Koelling, R. A.** (1966). Learning with prolonged delay of reinforcement. *Psychonomic Science* **5**, pp. 121–122.

440. **Gavish, L., Carter, C. S.,** and **Getz, L. L.** (1983). Male-female interactions in prairie voles. *Animal Behaviour* **31**, pp. 511–517.

441. **Gay, J., Kokkotou, E., O'Brien, M., Pothoulakis, C.,** and **Karalis, K.** (2008). Corticotropin-releasing hormone deficiency is associated with reduced local inflammation in a mouse model of experimental colitis. *Endocrinology* **149**, pp. 3403–3409.

442. **George, O., Ghozland, S., Azar, M. R., Cottone, P., Zorrilla, E. P., Parsons, L. H., O'Dell, L.E., Richardson, H.N.,** and **Koob, G. F.** (2007). CRF–CRF1 system activation mediates withdrawal-induced increases in nicotine self-administration in nicotine-dependent rats. *Proceedings of the National Academy of Sciences* **104**, pp. 17198–17203.

443. **George, O., Sanders, C., Freilig, J., Grigoryan, E., Vu, S., Allen, C.D., Crawford, E., Mandyam, C.D.,** and **Koob, G.F.** (2012). Recruitment of medial prefrontal cortex neurons during alcohol withdrawal predicts cognitive impairment and excessive alcohol drinking. *Proceedings of the National Academy of Sciences of the United States of America* **109**, pp. 18156–18161.

444. **George, S., Khan, S., Briggs, H.,** and **Abelson, J.** (2010). CRH-stimulated cortisol release and food intake in healthy, non-obese adults. *Psychoneuroendocrinology* **35**, pp. 607–612.

445. **Georgiou, P., Zanos, P., Garcia-Carmona, J. A., Hourani, S., Kitchen, I., Laorden, M. L.,** and **Bailey, A.** (2016). Methamphetamine abstinence induces changes in μ-opioid receptor, oxytocin and CRF systems: association with an anxiogenic phenotype. *Neuropharmacology* **105**, pp. 520–532.

446. **Ghisleni, G., Capiotti, K. M., Da Silva, R. S., Oses, J. P., Piato, A. L., Soares, V., Bogo, M. R.,** and **Bonan, C. D.** (2012). The role of CRH in behavioral responses to

acute restrain stress in zebrafish. *Progress in Neuro-Psychopharmacology & Biological Psychiatry* **36**, pp. 176–182.

447. **Ghitza, U., Gray, S., Epstein, D., Rice, K.,** and **Shaham, Y.** (2006). The anxiogenic drug yohimbine reinstates palatable food seeking in a rat relapse model: a role of CRF1 receptors. *Neuropsychopahrmacology* **31**, pp. 2188–2196.
448. **Giardino, W. J., Pastor, R., Anacker, A. M., Spangler, E., Cote, D. M., Li, J., Stenzel-Poore, M. P., Phillips, T. J.,** and **Ryabinin, A. E.** (2011). Dissection of corticotropin-releasing factor system involvement in locomotor sensitivity to methamphetamine. *Genes Brain and Behavior* **10**, pp. 78–89.
449. **Giardino, W. J., Ryabinin, A. E.** (2013). CRF1 receptor signaling regulates food and fluid intake in the drinking-in-the-dark model of binge alcohol consumption. *Alcoholism Clinical and Experimental Research* **37**, pp. 1161–1170.
450. **Gibson, T. A., McKelvie, M., Schade, A., Schulkin, J., Pinnock, S. B.,** and **Herbert, J.** (2010). CRH regulates progenitor cell mitosis in the adult dentate gyrus, moderates the stimulating action of fluoxetine and represses Wnt3a expression. Society for Neuroscience Abstract.
451. **Gilhooley, M. J., Pinnock, S. B.,** and **Herbert, J.** (2011). Rhythmic expression of per1 in the dentate gyrus is suppressed by corticosterone: implications for neurogenesis. *Neuroscience Letters* **489**, pp. 177–181.
452. **Gillman, M., Rich-Edwards, J., Huh, S., Majzoub, J., Oken, E., Taveras, E.,** and **Rifas-Shiman, S.** (2006). Maternal corticotropin-releasing hormone levels during pregnancy and offspring adiposity. *Obesity* **14**, pp. 1647–1653.
453. **Gilman, T. L., DaMert, J. P., Meduri, J. D.,** and **Jasnow, A. M.** (2015). Grin1 deletion in CRF neurons sex-dependently enhances fear, sociability, and social stress responsivity. *Psychoneuroendocrinology* **58**, pp. 33–45.
454. **Gilpin, N. W., Richardson, H. N.,** and **Koob, G. F.** (2008). Effects of CRF1-receptor and opioid-receptor antagonists on dependence-induced increases in alcohol drinking by alcohol-preferring (P) rats. *Alcoholism: Clinical and Experimental Research* **32**, pp. 1535–1542.
455. **Ginsberg, A., Pecoraro, N., Warne, J., Horneman, H.,** and **Dallman, M.** (2010). Rapid alteration of stress-induced hypothalamic-pituitary-adrenal hormone secretion in the rat: a comparison of glucocorticoids and cannabinoids. *Stress* **13**, pp. 248–257.
456. **Glennemeier, K.,** and **Denver, R.** (2002). Small changes in whole-body corticosterone content affect larval Rana pipiens fitness components. *General and Comparative Endocrinology* **127**, pp. 16–25.
457. **Glowa, J.,** and **Gold, P.** (1991). Corticotropin releasing hormone produces profound anorexigenic effects in the rhesus monkey. *Neuropeptides* **18**, pp. 55–61.
458. **Gluckman, P. D., Hanson, M. A.,** and **Beedle, A. S.** (2007). Early life events and their consequences for later disease: a life history and evolutionary perspective. *American Journal of Human Biology* **19**, pp. 1–19.
459. **Glynn, L. M., Schetter, C. D., Chicz-DeMet, A., Hobel, C. J.,** and **Sandman, C. A.** (2007). Ethnic differences in adrenocorticotropic hormone, cortisol and corticotropin-releasing hormone during pregnancy. *Peptides* **28**, pp. 1155–1161.
460. **Glynn, L. M., Wadhwa, P. D., Dunkel-Schetter, C., Chicz-DeMet, A.,** and **Sandman, C. A.** (2001). When stress happens matters: effects of earthquake timing on stress responsivity in pregnancy. *American Journal of Obstetrics and Gynecology* **184**, pp. 637–642.

461. **Glynn, L.**, **Davis, E.**, and **Sandman, C.** (2013). New insights into the role of perinatal HPA-axis dysregulation in postpartum depression. *Neuropeptides* **47**, pp. 363–370.

462. **Glynn, L. M.**, **Wadhwa, P. D.**, **Dunkel-Schetter, S.**, **Chicz-Hammet, A.**, **Sandman, C. A.** (2001). When stress happens matters: the effects of earthquake timing on stress responsivity in pregnancy. *American Journal of Obstetrics and Gynecology* **184**, pp. 637–642.

463. **Goebel, M.**, **Fleming, S. M.**, **Million, M.**, **Stengel, A.**, **Taché, Y.**, **Wang, L.** (2010). Mice overexpressing corticotropin-releasing factor show brain atrophy and motor dysfunctions. *Neuroscience Letters* **473**, pp. 11–15.

464. **Goeders, N.** (2002). Stress and cocaine addiction. *Journal of Pharmacology and Experimental Therapeutics* **301**, pp. 785–789.

465. **Goeders, N.**, and **Guerin, G. F.** (1996). Role of corticosterone in intravenous cocaine self-administration in rats. *Neuroendocrinology* **64**, pp. 337–348.

466. **Goeders, N.**, and **Guerin, G. F.** (2000). Effects of the CRH receptor antagonist CP-154,526 on intravenous cocaine self-administration in rats. *Neuropsychopharmacology* **23**, pp. 577–586.

467. **Goeders, N. E.** (1998). Stress, the hypothalamic-pituitary-adrenal axis, and vulnerability to drug abuse. *NIDA Research Monograph* **169**, pp. 83–104.

468. **Goland, R.**, **Conwell, I. M.**, **Warren, W. B.**, and **Wardlaw, S. L.** (1992). Placental corticotropin-releasing hormone and pituitary-adrenal function during pregnancy. *Neuroendocrinology* **56**, pp. 742–749.

469. **Goland, R. S.**, **Wardlaw, S. L.**, **Blum, M.**, **Tropper, P. J.**, and **Stark, R. I.** (1988). Biologically active corticotropin-releasing hormone in maternal and fetal plasma during pregnancy. *American Journal of Obstetrics and Gynecology* **159**, pp. 884–890.

470. **Goland, R. S.**, **Wardlaw, S. L.**, **Stark, R. I.**, **Brown, L. S. Jr.**, and **Frantz, A. G.** (1986). High levels of corticotropin-releasing hormone immunactivity in maternal and fetal plasma during pregnancy. *Journal of Clinical Endocrinology and Metabolism* **63**, pp. 1199–1203.

471. **Goland, R. S.**, **Jozak, S.**, and **Conwell, I.** (1994). Placental corticotropin-releasing hormone and the hyptercortisolism of pregnancy. *American Journal of Obstetrics and Gynecology* **171**, pp. 1287–1291.

472. **Goland, R. S.**, **Jozak, S.**, **Warren, W. B.**, **Conwell, I. M.**, **Stark, R. I.**, and **Tropper, P. J.** (1993). Elevated levels of umbilical cord plasma corticotropin-releasing hormone in growth-retarded fetuses. *Journal of Clinical Endocrinology and Metabolism* **77**, pp. 1174–1179.

473. **Goland, R. S.**, **Tropper, P. J.**, **Warren, W. B.**, **Stark, R. I.**, **Jozak, S. M.**, and **Conwell, I. M.** (1995). Concentrations of corticotropin-releasing hormone in the umbilical-cord blood of pregnancies complicated by pre-eclampsia. *Reproduction Fertility and Development* 7, pp. 1227–1230.

474. **Gold, P. W.**, **Drevets, W. C.**, **Charney, D. S.**, and **Drevets, W. C.** (2002). New insights into the role of cortisol and the glucocorticoid receptor in severe depression. *Biological Psychiatry* **52**, pp. 381–385.

475. **Gold, P. W.**, **Goodwin, F. K.**, and **Chrousos, G. P.** (1988). Clinical and biochemical manifestations of depression. *New England Journal of Medicine* **319**, pp. 348–353.

476. **Goldenberg, R. L.** (2002). The management of preterm labor. *Obstetrics and Gynecology* **100**, pp. 1020–1037.

477. **Gonzales, X. F., Desmukh, A., Pulse, M., Johnson, K.**, and **Jones, H. P.** (2008). Stress induced differences in primary and secondary resistance against bacterial sepsis corresponds with diverse corticotropion releasing hormone receptor expression by pulmonary CD11c + MHCII + and CD11c- MHCII + APCs. *Brain Behavior and Immunity* **22**, pp. 552–564.

478. **Gonzalez, D. C.**, and **Valatx, J. L.** (1997). Effect of intracerebroventricular administration of α-helical CRH (9–41) on the sleep/waking cycle in rats under normal conditions or after subjection to an acute stressful stimulus. *Journal of Sleep Research* **6**, pp. 164–170.

479. **Goodson, J., Evans, A.**, and **Wang, Y.** (2006). Neuropeptide binding reflects convergent and divergent evolution in species typical group sizes. *Hormone and Behavior* **50**, pp. 223–236.

480. **Gos, T., Schulkin, J., Gos, A., Bock, J., Poeggel, G.**, and **Braun, K.** (2014). Paternal deprivation affects the functional maturation of corticotropin-releasing hormone (CRH)- and calbindin-D28k-expressing neurons in the bed nucleus of the stria terminalis (BNST) of the biparental Octodon degus. *Brain Structure & Function* **219**, pp. 1983–1990.

481. **Gould, E., Vall, N., Wagers, M.**, and **Gross, C. G.** (2001). Adult-generated hippocampal and neocortical neurons in macaques have a transient existence. *Proceedings of the National Academy of Sciences of the United States of America* **98**, pp. 10910–10916.

482. **Gourcerol, G., Wang, L., Wang, Y. H., Million, M.**, and **Taché, Y.** (2007). Urocortins and cholecystokinin-8 act synergistically to increase satiation in lean but not obese mice: involvement of corticotropin-releasing factor receptor-2 pathway. *Endocrinology* **148**, pp. 6115–6123.

483. **Gourley, S. L., Kedves, A. T., Olausson, P.**, and **Taylor, J. R.** (2008). A history of corticosterone exposure regulates fear extinction and cortical NR2B, GluR2/3 and BDNF. *Neuropsychopharmacology* **34**, pp. 707–716.

484. **Goy, R. W.**, and **McEwen, B. S.** (1980). *Sexual differentiation of the brain*. Cambridge, MA: MIT Press.

485. **Goymann, W.**, and **Wingfield, J. C.** (2004). Allostatic load, social status and stress hormones: the costs of social status matter. *Animal Behaviour* **67**, pp. 591–602.

486. **Grammatopoulos, D.** (2007). The role of CRH receptors and their agonists in myometrial contractility and quiescence during pregnancy and labor. Frontiers in Bioscience **12**, pp. 561–571.

487. **Grammatopoulos, D.** (2012). Insights into mechanisms of corticotropin-releasing hormone receptor signal transduction. *British Journal of Pharmacology* **166**, pp. 85–97.

488. **Grammatopoulos, D., Dai, Y., Chen, J., Karteris, E., Papadopoulos, N. Easton, A.J.**, and **Hillhouse, E.W.** (1998). Human corticotropin-releasing hormone receptor: differences in subtype expression between pregnant and nonpregnant myometrium. *Journal of Clinical Endocrinology and Metabolism* **83**, pp. 2539–2544.

489. **Gravett, M. G., Hummel, D., Eschenbach, D. A.**, and **Holmes, K. K.** (1986). Preterm labor associated with subclinical amniotic fluid infection and with bacterial vaginosis. *Obstetrics & Gynecology* **67**, pp. 229–237.

490. **Gray, J. D., Milner, T. A.,** and **McEwen, B. S.** (2013). Dynamic plasticity: the role of glucocorticoids, brain-derived neurotrophic factor and other trophic factors. *Neuroscience* **239**, pp. 214–227.

491. **Gray, J. M., Vecchiarelli, H. A., Morena, M., Lee, T. T., Hermanson, D. J., Kim, A. B., McLaughlin, R. J., Hassan, K. I., Kuhne, C., Wotjak, C.T., Deussing, J. M., Patel, S,** and **Hill, M. N.** (2015). Corticotropin-releasing hormone drives anandamide hydrolysis in the amygdala to promote anxiety. *The Journal of Neuroscience* **35**, pp. 3879–3892.

492. **Gray, J. M., Wilson, C. D., Lee, T. T., Pittman, Q. J., Deussing, J. M., Hillard, C. J., McEwen, B. S., Schulkin, J., Karatsoreos, I. N., Patel, S.,** and **Hill, M. N.** (2016). Sustained glucocorticoid exposure recruits cortico-limbic CRH signaling to modulate endocannabinoid function. *Psychoneuroendocrinology* **566**, pp. 151–158.

493. **Gray, T. S.** (1999). Functional and anatomical relationships among the amygdala, basal forebrain, central striatum and cortex. *Annals of the New York Academy of Sciences* **877**, pp. 439–444.

494. **Gray, T. S., Piechowski, R. A., Yracheta, J. M., Rittenhouse, P. A., Bethea, C. L.,** and **Van de Kar, L. D.** (1993). Ibotenic acid lesions in the bed nucleus of the stria terminalis attenuate conditioned stress induced increases in prolactin, ACTH and corticosterone. *Neuroendocrinology* **57**, pp. 517–524.

495. **Green, J. D.,** and **Harris, G. W.** (1946). The neurovascular link between the neurohypophysis and adenohypophysis. *Journal of Endocrinology* **5**, pp. 136–140.

496. **Greenwood-Van Meerveld, B., Gibson, M., Gunter, W., Shepard, J., Foreman, R.,** and **Myers, D.** (2001). Stereotaxic delivery of corticosterone to the amygdala modulates colonic sensitivity in rats. *Brain Research* **893**, pp. 135–142.

497. **Greenwood-Van Meerveld, B., Johnson, A. C., Cochrane, S., Schulkin, J.,** and **Myers, D. A.** (2005). Corticotropin-releasing factor 1 rceptor mediated mechanisms inhibit colonic hypersensitivity in rats. *Neurogastroenterol Motil* **17**(3), pp. 415–422.

498. **Greenwood-Van Meerveld, B., Johnson, A. C., Schulkin, J.,** and **Myers, D. A.** (2006). Long-term expression of corticotropin-releasing factor (CRF) in the paraventricular nucleus of the hypothalamus in response to an acute colonic inflammation. *Brain Research* **1071**, pp. 91–96.

499. **Grieder, T., Herman, M., Contet, C., Tan, L., Vargas-Perez, H., Cohen, A., Chwalek, M., Maal-Bared, G., Freiling, J., schlosburg, J., Clarke, L., Crawford, E., Koebel, P., Canonigo, V., Sanna, P., Tapper, A., Roberto, M., Kieffer, B., Sawchenko, P., Koob, G., Van der Kooy, D.,** and **George, O.** (2014). VTA CRF neurons mediate the aversive effects of nicotine withdrawal and promote intake escalation. *Nature Neuroscience* **17**, pp. 1751–1758.

500. **Griffin, D. R.** (1958). *Listening in the dark: the acoustic orientation of bats and men.* New Haven, CT: Yale University Press.

501. **Grigoriadis, D. E.** (2003). Corticotropin-releasing factor receptor antagonists: potential novel therapies for human diseases. *Sigma-Aldrich Corporation: Celltransmissions* **19**, pp. 3–10.

502. **Grigoriadis, D., Hoare, S., Lechner, S., Slee, D.,** and **Williams, J.** (2009). Drugability of extracellular targets: discovery of small molecule drugs targeting allosteric, functional, and subunit-selective sites on GPCRs and ion channels. *Neuropsychopharmacology* **34**, pp. 106–125.

503. **Grill, H. J.**, and **Kaplan, J. M.** (2002). The neuroanatomical axis for control of energy balance. *Frontiers in Neuroendocrinology* **23**, pp. 2–40.

504. **Grill, H. J., Markison, S., Ginsberg, A.**, and **Kaplan, J.** (2000). Long-term effects on feeding and body weight after stimulation of forebrain or hindbrain CRH receptors with urocortin. *Brain Research* **867**, pp. 19–28.

505. **Grillon, C., Hale, E., Lieberman, L., Davis, A., Pine, D.**, and **Ernst, M.** (2015). The CRH1 antagonist GSK561679 increases human fear but not anxiety as assessed by startle. *Neuropsychopharmacology* **40**, pp. 1064–1071.

506. **Grobman, W. A., Bailit, J. L., Rice, M. M., Wapner, R. J., Varner, M. W., Thorp, J. M. Jr., Leveno, K. J., Caritis, S. N., Iams, J. O., Tita, A. T., Saade, G., Sorokin, Y., Rouse, D., Tolosa, J., Dorsten, J.** (2014). Can differences in obstetric outcomes be explained by differences in the care provided? The MFMU Network APEX study. *AJOG* **211**, pp. 147.e1–147e16.

507. **Groeneweg, F. L., Karst, H., de Kloet, E. R.**, and **Joëls, M.** (2011). Rapid non-genomic effects of corticosteroids and their role in the central stress response. *Journal of Endocrinology* **209**, pp. 153–167.

508. **Grone, B. P.**, and **Maruska, K. P.** (2015). Divergent evolution of two corticotropin-releasing hormone (CRH) genes in teleost fishes. *Frontiers in Neuroscience* **9**, p. **365**.

509. **Grunwald, N.**, and **Brunton, P.** (2015). Prenatal stress programs neuroendocrine stress responses and affective behaviors in second generation rats in a sex-dependent manner. *Psychoneuronendocrinology* **62**, pp. 204–216.

510. **Guardino, C. M., Schetter, C. D., Saxbe, D. E., Adam, E. K., Ramey, S. L.**, and **Shalowitz, M. U.** (2016). Diurnal salivary cortisol patterns prior to pregnancy predict infant birth weight. *Health Psychology*, **35(6)**, 625–633.

511. **Guardiola-Diaz, H. M., Kolinske, J. S., Gates, L. H.**, and **Seasholtz, A. F.** (1996). Negative glucocorticoid regulation of cyclic adenosine 3', 5' monophosphate stimulated corticotropin-releasing hormone reporter expression in AtT-20 cells. *Molecular Endocrinlology* **10**, pp. 317–329.

512. **Guendelman, S., Kosa, J., Pearl, M., Graham, S.**, and **Kharrazi, M.** (2008). Exploring the relationship of second-trimester corticotropin releasing hormone, chronic stress and preterm delivery. *Journal of Maternal-Fetal and Neonatal Medicine* **21**, pp. 788–795.

513. **Guillemin, R.** (1977). Peptides in the brain. The new endocrinology of the neuron. Nobel Lecture. http://www.nobelprize.org/nobel_prizes/medicine/laureates/1977/guillemin-lecture.html (Accessed February 18, 2016).

514. **Guillemin, R.**, and **Schally, A. V.** (1963). Concentration of corticotropin releasing factor (CRF) in acid extracts of sheep hypothalamus by gel filtration and counter current distribution. *Texas Reports on Biology and Medicine* **21**, pp. 541–545.

515. **Guillemin, R.**, and **Rosenberg, B.** (1955). Humoral hypothalamic control of anterior pituitary: a study with combind tissuecultures. *Endocrinology* **57**, pp. 599–607.

516. **Gunduz-Cinar, O., Hill, M., McEwen, B.**, and **Holmes, A.** (2013). Amygdala FAAH and anandamide: mediating protection and recovery from stress? *Trends in Pharmacological Sciences* **34**, pp. 637–643.

517. **Gunduz-Cinar, O., Macpherson, K. P., Cinar, R., Gamble-George, J., Sugden, K., Williams, B., Godlewski, G., Ramikie, T. S., Gorka, A .X., Alapafuja, S. O., Nikas, S. P., Makriyannis, A., Poulton, R., Patel, S., Hariri, A. R., Caspi, A., Moffitt, T. E.,

**Kunos, G.**, and **Holmes, A.** (2013). Convergent translational evidence of a role for anandamide in amygdala-mediated fear extinction, threat processing and stress-reactivity. *Molecular Psychiatry* **18**, pp. 813–823.

518. **Gunnar, M. R.** (1998). Quality of early care and buffering of neuroendocrine stress reactions: potential effects on the developing human brain. *Preventive Medicine* **27**, pp. 208–211.
519. **Gunnar, M. R.**, and **Davis, E. P.** (2005). Stress and emotion in early childhood. In: Lerner, R. M., Easterbrooks, M. A., and Mistry, J. (eds.), *Comprehensive handbook of psychology, Vol. 6. Developmental psychology*. New York: Wiley. 113–134.
520. **Gunnar, M. R.**, **Mangelsdorf, S.**, **Larson, M.**, and **Hertsgaard, L.** (1989). Attachment, temperament, and adrenocortical activity in infancy: a study of psychoendocrine regulation. *Developmental Psychology* **25**, pp. 355–363.
521. **Gvilia, I.**, **Suntsova, N.**, **Kumar, S.**, **McGinty, D.**, and **Szymusiak, R.** (2015). Suppression of preoptic sleep-regulatory neuronal activity during corticotropin-releasing factor-induced sleep disturbance. *American Journal of Physiology-Regulatory, Integrative and Comparative Physiology* **309**, pp. R1092–R1100.
522. **Habib, K. E.**, **Weld, K. P.**, **Rice, K. C.**, **Pushkas, J.**, **Champoux, M.**, **Listwak, S.**, **Webster, E. L.**, **Atkinson, A. J.**, **Schulkin, J.**, **Contoreggic, C.**, **Chrousos, G. P.**, **McCann, S. M.**, **Suomi, S. J.**, **Higley, J. D.**, and **Gold, P. W.** (2000). Oral administration of a corticotropin-releasing hormone receptor antagonist significantly attenuates behavioral, neuroendocrine, and autonomic responses to stress in primates. *Proceedings of the National Academy of Sciences* **97**, pp. 6079–6084.
523. **Habib, K. E.**, **Weld, K. P.**, **Schulkin, J.**, **Listwak, S.**, **Champoux, M.**, **Shannon, C.**, **Chrousos, G. P.**, **Suomi, S.**, and **Gold, P. W.**, and **Higley, J. D.** (1999). Cerebrospinal fluid lvels of corticotropin-releasing hormone positively correlate with acute and chronice social stress in non-human primates. Paper presented at Society for Neuroscience Annual Meeting, Miami Beach, Florida.
524. **Haller, J.**, **Mikics, E.**, **Makara, G. B.** (2008). The effects of non-genomic glucocorticoid mechanisms on bodily functions and the central neural system. A critical evaluation of findings. *Frontiers in Neuroendocrinology* **29**, pp. 273–291.
525. **Hamilton, K. L.**, **Harris, A. C.**, and **Gewirtz, J. C.** (2013). Affective and neuroendocrine effects of withdrawal from chronic, long-acting opiate administration. *Brain Research* **1538**, pp. 73–82.
526. **Hamlin, A. S.**, **Buller, K. M.**, **Day, T. A.**, and **Osborne, P. B.** (2004). Effect of naloxone-precipitated morphine withdrawal on c-fos expression in rat corticotropin-releasing hormone neurons in the paraventricular hypothalamus and extended amygdala. *Neuroscience Letters* **362**, pp. 39–43.
527. **Hammock, E. A.**, and **Young, L. J.** (2004). Functional microsatellite polymorphism associated with divergent social structure in vole species. *Molecular Biology and Evolution* **21**, pp. 1057–1063.
528. **Hannerfors, A. K.**, **Hellgren, C.**, **Schijven, D.**, **Iliadis, S. I.**, **Comasco, E.**, **Skalkidou, A.**, **Olivier, J. D. A.**, and **Sundström-Poromaa, I.** (2015). Treatment with serotonin reuptake inhibitors during pregnancy is associated with elevated corticotropin-releasing hormone levels. *Psychoneuroendocrinology* **58**, pp. 104–113.
529. **Hansson, A. C.**, **Cipptielli, A.**, **Sommer, W. H.**, **Fedeli, A.**, **Bjőrk, K.**, **Soverchia, L.**, **Terasmaa, A.**, **Massi, M.**, **Heilig, M.**, and **Ciccocioppo, R.** (2006). Variation at the rat Crhr1 locus and sensitivity to relapse into alcohol seeking induced by environmental

stress. *Proceedings of the National Academy of Sciences of the United States of America* **103**, pp. 15236–15241.

530. **Haramati, S., Navon, I., Issler, O., Ezra-Nevo, G., Gil, S., Zwang, R., Hornstein, E.,** and **Chen, A.** (2011). MicroRNA as repressors of stress-induced anxiety: the case of amygdalar miR-34. *The Journal of Neuroscience* **31**, pp. 14191–14203.
531. **Harlow, H. F.,** and **Soumi, S. J.** (1971). Social recovery by isolation-reared monkeys. *Proceedings of the National Academy of Sciences of the United States of America* **68**, pp. 1534–1538.
532. **Harris, A. P., Holmes, M. C., De Kloet, E. R., Chapman, K. E.,** and **Seckl, J. R.** (2013). Mineralocorticoid and glucocorticoid receptor balance in control of HPA axis and behaviour. *Psychoneuroendocrinology* **38**, pp. 648–658.
533. **Harris, G. W.** (1937). The induction of ovulation in the rabbit, by electrical stimulation of the hypothalamo-hypophysial mechanism. Proceedings of the Royal Society of London. *Series B, Biological Sciences* **122**, pp. 374–394.
534. **Hauger, R. L., Grigoriadis, D. E., Dallman, M. F., Plotsky, P. M., Vale, W. W.,** and **Dautzenberg, F. M.** (2003). International Union of Pharmacology. XXXVI. Current status of the nomenclature for receptors for corticotropin-releasing factor and their ligands. *Pharmacological Reviews*, pp. **55**: 21–26.
535. **Hebb, D. O.** (1946). Emotion in man and animal: an analysis of the intuitive processes of recognition. *Psychological Review* **53**, p. **88**.
536. **Heilig, M.** (2015). *The thirteenth step: addiction in the age of brain science.* New York: Columbia University Press.
537. **Heim, C., Bradley, B., Mletzko, T. C., Deveau, T. C., Musselman, D. L., Nemeroff, C. B., Ressler, K. J.,** and **Binder, E. B.** (2009). Effect of childhood trauma on adult depression and neuroendocrine function: sex-specific moderation by CRH receptor 1 gene. *Frontiers in Behavioral Neuroscience* **3**, p. **41**.
538. **Heinrichs, M., Baumgartner, T., Kirschbaum, C.,** and **Ehlert, U.** (2003). Social support and oxytocin interact to suppress cortisol and subjective responses to psychosocial stress. *Biological Psychiatry* **54**, pp. 1389–1398.
539. **Heinrichs, M., Domes, G.** (2008). Neuropeptides and social behaviour: effects of oxytocin and vasopressin in humans. *Progress in Brain Research* **170**, pp. 337–350.
540. **Heinrichs, S.** (2003). Modulation of social learning in rats by brain corticotropin-releasing factor. Brain Research **994**, pp. 107–114.
541. **Heinrichs, S. C., Menzaghi, F., Pich, E. M., Britton, K. T.,** and **Koob, G. F.** (1995). The role of CRF in behavioral aspects of stress. *Annals of the New York Academy of Sciences* **771**, pp. 92–104.
542. **Held, K., Künzel, H., Ising, M., Schmid, D. A., Zobel, A., Murck, H., Holsboer, F.,** and **Steiger, A.** (2004). Treatment with the CRH 1-receptor-antagonist R121919 improves sleep-EEG in patients with depression. *Journal of Psychiatric Research* **38**, pp. 129–136.
543. **Helmeke, C., Seidel, K., Poeggel, G., Bredy, T. W., Abraham, A.,** and **Braun, K.** (2009). Paternal deprivation during infancy results in dendrite-and time-specific changes of dendritic development and spine formation in the orbitofrontal cortex of the biparental rodent Octodon degus. *Neuroscience* **163**, pp. 790–798.

544. **Henckens, M.J., Deussing, J.M., Chen, A.** (2016). Region-specific roles of the corticotropin-releasing factor-urocortin system in stress. Nature Reviews Neuroscience 17(10), 636–651.

545. **Hennessy, M. B., Jacobs, S., Schiml, P. A., Hawk, K., Stafford, N.,** and **Deak, T.** (2013). Maternal inhibition of infant behavioral response following isolation in novel surroundings and inflammatory challenge. *Devlopmental Psychobiology* **55**, pp. 395–403.

546. **Henry, B., Vale, W.,** and **Markou, A.** (2006). The effect of lateral septum corticotropin-releasing factor receptor 2 activation on anxiety is modulated by stress. *Journal of Neuroscience* **26**, pp. 9142–9152.

547. **Herbert, J.** (1993). Peptides in the limbic system: neurochemical codes for co-ordinated adaptive responses to behavioural and physiological demand. Progress in Neurobiology **41**, pp. 723–791.

548. **Herbert, J.,** and **Schulkin, J.** (2002). Neurochemical coding of adaptive responses in the limbic system. In: Pfaff, D. et al. (eds.), *Hormones, brain and behavior*. San Diego: Academic Press; New York: Elsevier Press. pp. 659–689.

549. **Herget, U.,** and **Ryu, S.** (2015). Coexpression analysis of nine neuropeptides in the neurosecretory preoptic area of larval zebrafish. Frontiers in Neuroanatomy **9**: 2.

550. **Herkenham, M.** (1987). Mismatches between neurotransmitter and receptor localizations in brain: observations and implications. *Neuroscience* **23**, pp. 1–38.

551. **Herman, J. P.** (2013). Neural control of chronic stress adaptation. *Frontiers in Behavioral Neuroscience* 7, pp. 1–12.

552. **Herman, J. P., Figueiredo, H., Mueller, N. K., Ulrich-Lai, Y., Ostrander, M. M., Choi, D. C.,** and **Cullinan, W. E.** (2003). Central mechanisms of stress integration: hierarchical circuitry controlling hypothalamo–pituitary–adrenocortical responsiveness. *Frontiers in Neuroendocrinology* **24**, pp. 151–180.

553. **Herman, J. P., Dolgas, C. M.,** and **Carlson, S. L.** (1998). Ventral subiculum regulates hypothalamo–pituitary–adrenocortical and behavioural responses to cognitive stressors. *Neuroscience* **86**, pp. 449–459.

554. **Herman, J.P., Ostrander, M. M., Mueller, N. K.,** and **Figueiredo, H.** (2005). Limbic system mechanisms of stress regulation: hypothalamo-pituitary-adrenocortical axis. *Progress in Neuro-Psychopharmacology and Biological Psychiatry* **29**, pp. 1201–1213.

555. **Herman, M. A., Contet, C., Justice, N. J., Vale, W.,** and **Roberto, M.** (2013). Novel subunit-specific tonic GABA currents and differential effects of ethanol in the central amygdala of CRF receptor-1 reporter mice. *Journal of Neuroscience* **33**, pp. 3284–3298.

556. **Herrick, C. J.** (1905). The central gustatory pathway in the brain of body fishes. Journal of *Comparative Neurology and Psychology* **15**, pp. 375–456.

557. **Herrick, C J.** (1926/1963). *Brain in rats and men*. New York, NY: Hafner Publishers.

558. **Higley, J. D.,** and **Bennett, A. J.** (1999). Central nervous system serotonin and personality as variables contributing to excessive alcohol consumption in non-human primates. *Alcohol and Alcoholism* **34**: 402–418.

559. **Higley, J. D., Hasert, M. F., Suomi, S. J.,** and **Linnoila, M.** (1991). Nonhuman primate model of alcohol abuse: effects of early experience, personality, and stress on alcohol consumption. *Proceedings of the National Academy of Sciences* **88**, pp. 7261–7265.

560. **Higley, J. D., Suomi, S. J.**, and **Linnoila, M.** (1992). A longitudinal assessment of CSF monoamine metabolite and plasma cortisol concentrations in young rhesus monkeys. *Biological Psychiatry* **32**, pp. 127–145.

561. **Higley, J. D., Suomi, S. J.**, and **Linnoila, M.** (1996). A nonhuman primate model of type ii excessive alcohol consumption? part 1. low cerebrospinal fluid 5-hydroxyindoleacetic acid concentrations and diminished social competence correlate with excessive alcohol consumption. *Alcoholism: Clinical and Experimental Research* **20**, pp. 629–642.

562. **Higley, J. D., Thompson, W. W., Champoux, M., Goldman, D., Hasert, M. F., Kraemer, G. W., Scanlan, G. W., Suomi, S. J.** and **Linnoila, M.** (1993). Paternal and maternal genetic and environmental contributions to cerebrospinal fluid monoamine metabolites in rhesus monkeys (Macaca mulatta). *Archives of General Psychiatry* **50**, pp. 615–623.

563. **Hill, M. N.**, and **McEwen, B. S.** (2009). Endocannabinoids: The silent partner of glucocorticoids in the synapse. *Proceedings of the National Academy of Sciences* **106**, pp. 4579–4580.

564. **Hill, M., Bierer, L., Makotkine, I., Golier, J., Galea, S., McEwen, B., Hillard, C.**, and **Yehuda, R.** (2013). Reductions in circulating endocannabinoid levels in individuals with post-traumatic stress disorder following exposure to the world trade center attacks. Psychoneuroendocrinolgy **38**, pp. 2952–2961.

565. **Hill, M. N., Ho, W. S., Meier, S. E., Gorzalka, B. B.**, and **Hillard, C. J.** (2005). Chronic corticosterone treatment increases the endocannabinoid 2-arachidonylglycerol in the rate amygdala. *European Journal of Pharmacology* **528**, pp. 99–102.

566. **Hill, M. N., Hunter, R. G.**, and **McEwen, B. S.**, (2009). Chronic stress differentially regulates cannabinoid CB1 receptor binding in distinct hippocampal subfields. European *Journal of Pharmacology* **614**, pp. 66–69.

567. **Hillhouse, E. W.**, and **Grammatopoulos, D. K.** (2002). Role of stress peptides during human pregnancy and labour. *Reproduction* **124**: , pp. 323–329.

568. **Hillhouse, E. W.**, and **Grammatopoulos, D. K.** (2006). The molecular mechanisms underlying the regulation of the biological activity of corticotropin-releasing hormone receptors: implications for physiology and pathophysiology. *Endocrine Review* **27**, pp. 260–286.

569. **Hobel, C. J., Goldstein, A.**, and **Barrett, E. S.** (2008). Psychosocial stress and pregnancy outcome. *Clinical Obstetrics and Gynecology* **51**, pp. 333–348.

570. **Hobel, C. J., Duinkel-Schetter, C., Roesch, S. C., Castro, L. C.**, and **Arora, C. P.** (1999). Maternal plasma corticotropin-releasing hormone associated with stress at 20 weeks' gestation in pregnancies ending in preterm delivery. *American Journal of Obstetrics and Gynecology* **180**, pp. S257–S263.

571. **Hodos, W.**, and **Butler, A. B.** (1997). Evolution of sensory pathways in vertebrates. *Brain, Behavior and Evolution* **50**, pp. 189–197.

572. **Hofer, M. A.** (1973). The role of nutrition in the physiological and behavioral effects of early maternal separation on infant rats. *Psychosomatic Medicine* **35**, pp. 350–359.

573. **Hofer, M. A.**, and **Sullivan, R. M.** (2001). Toward a neurobiology of attachment. In: *Handbook of developmental cognitive neuroscience*. Cambridge, MA: MIT Press. 599–616.

574. **Hökfelt, T.**, **Fahrenkrug, J.**, **Tatemoto, K.**, **Mutt, V.**, **Werner, S.**, **Hulting, A.**, **Terenius, L.**, and **Chang, K. J.** (1983). The PHI (PHI-27)/corticotropin-releasing factor/enkephalin immunoreactive hypothalamic neuron: Possible morphological basis for integrated control of prolactin, corticotropin, and growth hormone secretion. *Proceedings of the national academy of sciences of the United States of America* **80**, pp. 895–898.

575. **Hökfelt, T.**, **Johansson, O.**, **Ljungdahl, Å.**, **Lundberg, J. M.**, and **Schultzberg, M.** (1980). Peptidergic neurones. *Nature* **284**, pp. 515–521.

576. **Holland, L. Z.**, and **Short, S.** (2008). Gene duplication, co-option and recruitment during the origin of the vertebrate brain from the invertebrate chordate brain. *Brain, Behavior and Evolution* **72**, pp. 91–105.

577. **Holliday, R.** (2002). Epigenetics comes of age in the twentyfirst century. *Journal of Genetics* **81**, pp. 1–4.

578. **Holliday, R.** and **Ho, T.** (1998). Evidence for gene silencing by endogenous DNA methylation. *Proceedings of the National Academy of Sciences* **95**, pp. 8727–8732.

579. **Hollon, N.**, **Burgeno, L.**, and **Phillips, P.** (2015). Stress effects on the neural substrates of motivated behavior. *Nature Neuroscience* **18**, pp. 1405–1412.

580. **Holly, E. N.**, **Boyson, C. O.**, **Montagud-Romero, S.**, **Stein, D. J.**, **Gobrogge, K. L.**, **DeBold, J. F.**, and **Miczek, K. A.** (2016). Episodic social stress-escalated cocaine self-administration: role of phasic and tonic corticotropin releasing factor in the anterior and posterior ventral tegmental area. *Journal of Neuroscience* **36**, pp. 4093–4105.

581. **Holly, E. N.**, **DeBold, J. F.**, and **Miczek, K. A.** (2015). Increased mesocorticolimbic dopamine during acute and repeated social defeat stress: modulation by corticotropin releasing factor receptors in the ventral tegmental area. *Psychopharmacology* **232**, pp. 4469–4479.

582. **Holly, E. N.**, **DeBold, J. F.**, and **Miczek, K. A.** (2015). Increased mesocorticolimbic dopamine during acute and repeated social defeat stress: modulation by corticotropin releasing factor receptors in the ventral tegmental area. *Psychopharmacology* **232**, pp. 4469–4479.

583. **Holsboer, F.** (1999). The rationale for corticotropin-releasing hormone receptor (CRH-R) antagonists to treat depression and anxiety. *Journal of Psychiatric Research* **33**, pp. 181–214.

584. **Holsboer, F.**, and **Ising, M.** (2008). Central CRH systemin depression and anxiety—evidence from clinical studies with CRH1 receptor antagonists. *European Journal of Pharmacology* **583**, pp. 350–357.

585. **Hope, P. J.**, **Turnbull, H.**, **Farr, S.**, **Morley, J. E.**, **Rice, K. C.**, **Chrousos, G. P.**, **Torpy, D. J.**, and **Wittert, G. A.** (2000). Peripheral administration of CRF and urocortin: effects on food intake and the HPA axis in the marsupial Sminthopsis crassicaudata. *Peptides* **21**, pp. 669–677.

586. **Horst, J.**, **Mark, M.**, **Kentrop, J.**, **Arp, M.**, **Veen, R.**, **Kloet, E.**, and **Oitzl, M.** (2014). Deletion of the forebrain mineralcorticoid receptor impairs social discrimination and decision-making in male, but not in female mice. *Frontiers in Behavioral Neuroscience* **8**:, pp. 1–10.

587. **Hosono, K.**, **Kikuchi, Y.**, **Miyanishi, H.**, **Hiraki-Kajiyama, T.**, **Takeuchi, A.**, **Nakasone, K.**, **Maehiro, S.**, and **Okubo, K.** (2015). Teleocortin: a novel member of the CRH family in teleost fish. *Endocrinology* **156**, pp. 2949–2957.

588. **Hostetler, C. M.**, and **Ryabinin, A. E.** (2013). The CRF system and social behavior: a review. *Frontiers in Neuroscience* 7, p. **92**.

589. **Howerton, A. R., Roland, A. V., Fluharty, J. M., Marshall, A., Chen, A., Daniels, D., Beck, S. G.**, and **Bale, T.** (2014). Sex differences in corticotropin-releasing factor receptor-1 action within the dorsal raphe nucleus in stress responsivity. *Biological Psychiatry* 75, pp. 873–883.

590. **Howerton, C. L., Morgan, C. P., Fischer, D. B.**, and **Bale, T. L.** (2013). O-GlcNAc transferase (OGT) as a placental biomarker of maternal stress and reprogramming of CNS gene transcription in development. *Proceedings of the National Academy of Sciences* **110**, pp. 5169–5174.

591. **Howland, M. A., Sandman, C. A., Glynn, L. M., Crippen, C.**, and **Davis, E. P.** (2016). Fetal exposure to placental corticotropin-releasing hormone is associated with child self-reported internalizing symptoms. *Psychoneuroendocrinology* **67**, pp. 10–17.

592. **Hoyle, C. H.** (1998). Neuropeptide families: evolutionary perspectives. *Regulatory Peptides* 73, pp. 1–33.

593. **Hsu, D. T., Mickey, B. J., Langenecker, S. A., Heitzeg, M. M., Love, T. M., Wang, H., Kennedy, S.E., Pecina, M., Shafir, T., Hodgkinson, C.A., Enoch, M. A., Goldman, D.**, and **Zubiera, J-K.** (2012). Variation in the corticotropin-releasing hormone receptor 1 (CRHR1) gene influences fMRI signal responses during emotional stimulus processing. *Journal of Neuroscience* **32**, pp. 3253–3260.

594. **Hu, X.**, and **Funder, J.** (2006). The evolution of mineralocorticoid receptors. *Molecular Endocrinology* **20**:, pp. 1471–1478.

595. **Huang, H., Li, L., Wang, Y., Tang, L., Wang, C.**, and **Leung, L.** (2011). Geinstein upregulates placental corticotropin-releasing hormone expression in lipopolysaccharide-sensitized mice. *Placenta* **32**, pp. 757–762.

596. **Hubbard, C. S., Labus, J. S., Bueller, J., Stains, J., Suyenobu, B., Dukes, G.E., Kelleher, D. L., Tillisch, K., Naliboff, B. D.**, and **Mayer, E. A.** (2011). Corticotropin-releasing factor receptor 1 antagonist alters regional activation and effective connectivity in an emotional-arousal circuit during expectation of abdmoninal pain. *Journal of Neuroscience* **31**, pp. 12491–12500.

597. **Hubbard, D. T., Nakashima, B. R., Lee, I.**, and **Takahashi, L. K.** (2007). Activation of basolateral amygdala corticotropin-releasing factor 1 receptors modulates the consolidation of contextual fear. *Neuroscience* **150**, pp. 818–828.

598. **Hughes, A.**, and **Friedman, R.** (2013). Parallel evolution by gene duplication in the genomes of two unicellular fungi. *Genome Research* **13**, pp. 794–799.

599. **Hui, D., Liu, G., Kavuma, E,. Hewson, S. A., McKay, D.**, and **Hannah, M. E.** (2007). Preterm labour and birth: a survey of clinical practice regarding use of tocolytics, antenatal corticosteroids, and progesterone. *Journal of Obstetrics and Gynaecology Canada* **29**:, pp. 117–130.

600. **Hui, G. K., Figueroa, I. R., Poytress, B. S., Roozendaal, B., McGaugh, J. L.**, and **Weinberger, N. M.** (2004). Memory enhancement of classical fear conditioning by post-training injections of corticosterone in rats. *Neurobiology of Learning and Memory* **81**, pp. 67–74.

601. **Huising, M. O., Van Schooten, C., Taverne-Thiele, A. J., Hermsen, T., Verburg-van Kemenade, B. M.**, and **Flik, G.** (2004). Structural characterisation of a cyprinid

(Cyprinus carpio L.) CRH, CRH-BP and CRH-R1, and the role of these proteins in the acute stress response. *Journal of Molecular Endocrinology* **32**, pp. 627–648.

602. **Huising, M.**, and **Flik, G.** (2005). The Remarkable conservation of corticotropin releasing hormone (CRH)-binding protein in the honeybee (Apis mellifera) dates the CRH system to a common ancestor of insects and vertebrates. *Endocrinology* **146**, pp. 2165–2710.
603. **Huising, M. O.**, **van der Aa, L. M.**, **Metz, J. R.**, **de Fátima Mazon, A.**, **Kemenade, B. M.**, and **Flik, G.** (2007). Corticotropin-releasing factor (CRF) and CRF-binding protein expression in and release from the head kidney of common carp: evolutionary conservation of the adrenal CRF system. *Journal of Endocrinology* **193**, pp. 349–357.
604. **Huitinga, I.**, **van der Cammen, M.**, **Salm, L.**, **Erkut, Z.**, **van Dam, A. M.**, **Tilders, F.**, and **Swaab, D.** (2000). IL-1β immunoreactive neurons in the human hypothalamus: reduced numbers in multiple sclerosis. *Journal of Neuroimmunology* **107**, pp. 8–20.
605. **Hwa, L. S.**, **Holly, E. N.**, **DeBold, J. F.**, and **Miczek, K. A.** (2015). Social stress-escalated intermittent alcohol drinking: modulation by CRF-R1 in the ventral tegmental area and accumbal dopamine in mice. *Psychopharmacology* **14**, pp. 1–10.
606. **Hwa, L.**, **DeBold, J.**, and **Miczek, K.** (2014). Alcohol in excess: CRF1 receptors in the rat and mouse VTA and DRN. *Psychopharmacology* **225**, pp. 313–327.
607. **Hyman, R. W.**, **Fukushima, M.**, **Diamond, L.**, **Kumm, J.**, **Giudice, L. C.**, and **Davis, R. W.** (2005). Microbes on the human vaginal epithelium. *Proceedings of the National Academy of Sciences* **102**, pp. 7952–7957.
608. **Iams, J. D.** (2014). Clinical practice: prevention of preterm parturition. *New England Journal of Medicine* **370**, pp. 254–261.
609. **Iemolo, A.**, **Blasio, A.**, **St Cyr, S. A.**, **Jiang, F.**, **Rice, K. C.**, **Sabino, V.**, and **Cottone, P.** (2013). CRF–CRF1 receptor system in the central and basolateral nuclei of the amygdala differentially mediates excessive eating of palatable food. *Neuropsychopharmacology* **38**, pp. 2456–2466.
610. **Ikemoto, S.**, and **Panksepp, J.** (1999). The role of nucleus accumbens dopamine in motivated behavior: a unifying interpretation with special reference to reward-seeking. *Brain Research Reviews* **31**, pp. 6–41.
611. **Ilias, I.**, **Mastorakis, G.** (2003). The merging role of peripheral corticotropin-releasing hormone (CRH). *Journal of Endocrinological Investigation* **26**, pp. 364–371.
612. **Im, E.**, **Rhee, S. H.**, **Park, Y. S.**, **Fiocchi, C.**, **Taché, Y.**, and **Pothoulakis, C.** (2010). Corticotropin-releasing hormone family of peptides regulates intestinal angiogenesis. *Gastroenterology* **138**, pp. 2457–2467.
613. **Imaki, T.**, **Nahan, J. L.**, **Rivier, C.**, **Sawchenko, P. E.**, and **Vale, W.** (1991). Differential regulation of corticotropin-releasing factor mRNA in rat brain regions by glucocorticoids and stress. *Journal of Neuroscience* **11**, pp. 585–598.
614. **Insel, T. R.** (1992). Oxytocin—a neuropeptide for affiliation: evidence from behavioral, receptor autoradiographic, and comparative studies. *Psychoneuroendocrinology* **17**, pp. 3–35.
615. **Insel, T. R.**, **O'Brien, D. J.**, and **Leckman, J. F.** (1999). Oxytocin, vasopressin, and autism: is there a connection? *Biological Psychiatry* **45**, pp. 145–157.
616. **Ising, M.**, and **Holsboer, F.** (2007). CRH1 receptor antagonists for the treatment of depression and anxiety. *Experimental and Clinical Psychopharmacology* **15**: , pp. 519–528.

617. **Ising, M., Zimmerman, U., Kunzel, H., Uhr, M., Foster, A., Learned-Coughlin, S., Holsboer, F.,** and **Grigoriadis, D.** (2007). High-affinity CRF1 receptor antagonist NBI-34041: preclinical and clinical data suggest safety and efficacy in attenuating elevated stress response. *Neuropsychopharmacology* **32**, pp. 1941–1949.

618. **Isogawa, K., Bush, D. E.,** and **LeDoux, J. E.** (2013). Contrasting effects of pretraining, posttraining, and pretesting infusions of corticotropin-releasing factor into the lateral amygdala: attenuation of fear memory formation but facilitation of its expression. *Biological Psychiatry* **73**, pp. 353–359.

619. **Issler, O., Carter, R. N., Paul, E. D., Kelly P. A., Olverman, H. J., Neufeld-Cohen, A., Kuperman, Y., Lowry, C. A., Seckl, J. R., Chen, A.,** and **Jamieson, P. M.** (2014). Increased anxiety in corticotropin releasing factor type 2 receptor-null mice requires recent acute stress exposure and is associated with dysregulated serotonergic activity in limbic brain areas. *Biology of Mood and Anxiety Disorders* **4**, pp. 1–14.

620. **Ito, R., Dalley, J. W., Howes, S. R., Robbins, T. W.,** and **Everitt, B. J.** (2000). Dissociation in conditioned dopamine release in the nucleus accumbens core and shell in response to cocaine cues and during cocaine-seeking behavior in rats. *Journal of Neuroscience* **20**, pp. 7489–7495.

621. **Ivy, A., Rex, C., Chen, Y., Dubé, C., Maras, P., Grigoriadis, D., Gall, C., Lynch, G.,** and **Baram, T.** (2010). Hippocampal dysfunction and cognitive impairments provoked by chronic early-life stress involve excessive activation of CRH receptors. *Journal of Neuroscience* **30**, pp. 13005–13015.

622. **Iwasaki-Sekno, A., Mano-Otagiri, A., Ohata, H., Ymauchi, N.,** and **Shibasaki, T.** (2009). Gender differences in corticotropin and corticosterone secretion and corticotropin releasing factor mrna expression in the paraventricular nucleus of the hypothalamus and the central nucleus of the amygdala in response to footshock stress or psychological stress in rats. *Psychoneuorendocrinology* **34**, pp. 226–237.

623. **Jacobson-Pick, S.,** and **Richter-Levin, G.** (2010). Differential impact of juvenile stress and corticosterone in juvenility and in adulthood, in male and female rats. *Behavioral Brain Research* **214**, pp. 268–276.

624. **Jaferi, A. M.,** and **Bhatnagar, S.** (2007). Corticotropin-releasing hormone receptors in the medial prefrontal cortex regulate hypothalamic–pituitary–adrenal activity and anxiety-related behavior regardless of prior stress experience. *Brain Research* **1186**, pp. 212–223.

625. **Jasnow, A. M., Banks, M., Owens, E.,** and **Huhman, K.** (1999). Differential effects of two corticotropin-releasing factor antagonists on conditioned defeat in male Syrian hamsters (mesocricetus auratus). *Brain Research* **846**, pp. 122–128.

626. **Jasnow, A. M., Schulkin, J.,** and **Pfaff, D.** (2006). Estrogen facilitates fear conditioning and increases corticotropin-releasing hormone mRNA expression in the central amygdala in female mice. *Hormones and Behavior* **49**, pp. 197–205.

627. **Jasnow, A M., Davis, M.,** and **Huhman, K. L.** (2004). Involvement of central amygdala and bed nucleus of the stria terminalis corticotropin-releasing factor in behavioral responses to social defeat. *Behavioral Neuroscience* **118**, pp. 1052–1061.

628. **Jaspers, K.** (1913, 1997). *General psychopathology, vol. I and II*. Trans. Hoenig, J. and M. Hamilton, M. W. Baltimore: Johns Hopkins University Press.

629. **Jaspers, K.** (1951, 1954). *Way to wisdom*. New Haven: Yale University Press.

630. **Jee, C., Lee, J., Lim, J. P., Parry, D., Messing, R. O.,** and **McIntire, S. L.** (2013). SEB-3, a CRF receptor-like GPCR, regulates locomotor activity states, stress responses and ethanol tolerance in Caenorhabditis elegans. *Genes, Brain and Behavior* **12**, pp. 250–262.

631. **Jeanneteau, F. D., Lambert, W. M., Ismaili, N., Bath, K. G., Lee, F. S., Garabedian, M. J.,** and **Chao, M.** (2012). BDNF and glucocorticoids regulate corticotrophin-releasing hormone (CRH) homeostasis in the hypothalamus. *Proceedings of the National Academy of Sciences of the United States of America* **109**, pp. 1305–1310.

632. **Jeong, K., Jacobson, L., Widmaier, E.,** and **Majzoub, J.** (1999). Normal suppression of the reproductive axis following stress in corticotropin-releasing-hormone-deficient mice. *Endocrinology* **140**, pp. 1702–1708.

633. **Jeschke, U., Mylonas, I., Richter, D. U., Höcker, I., Briese, V., Makrigiannakis, A.,** and **Friese, K.** 2005 Regulation of progesterone production in human term trophoblasts in vitro by CRH, ACTH, and cortisol (prednisolone). *Arch Gynecol Obstet* **272**, pp. 7–12.

634. **Ji, G.,** and **Neugebauer, V.** (2007). Differential effects of CRF1 and CRF2 receptor antagonists on pain-related sensitization of neurons in the central nucleus of the amygdala. *Journal of Neurophysiology* **97**: , pp. 3893–3904.

635. **Ji, G., Fu, Y., Ruppert, K. A.,** and **Neugebauer, V.** (2007). Pain-related anxiety-like behavior requires CRF1 receptors in the amygdala. *Molecular Pain* **3**, p. **13**.

636. **Jin, D., He, P., You, X., Zhu, X., Dai, L., He, Q., Liu, C., Hui, N., Sha, J., Ni, X.** (2007). Expression of corticotrophin-releasing hormone receptor type 1 and 2 in human pregnant myometrium. *Reproductive Sciences* **14**: , pp. 568–577.

637. **Joëls, M.** (1997). Steroid hormones and excitability in the mammalian brain. Frontiers in *Neuroendocrinology* **18**, pp. 2–48.

638. **Joëls, M.,** and **Baram, T. Z.** (2009). The neuro-symphony of stress. *Nature Review* **10**, pp. 459–466.

639. **Johnson, A. C., Tran, L., Schulkin, J.,** and **Greenwood-Van Meerveld, B.** (2012). Importance of stress receptor-mediated mechanisms in the amygdala on visceral pain perception in an intrinsically anxious rat. *Neurogastroenterology and Motility* **24**, pp. 479–486.

640. **Johnston, J. B.** (1923). Further contributions to the study of the evolution of the forebrain. *Journal of Comparative Neurology* **35**, pp. 337–481.

641. **Jones, S. A., Brooks, A. N.,** and **Challis, J. R.** (1989). Steroids modulate corticotropin-releasing hormone production in human fetal membranes and placenta. *Journal of Clinical Endocrinology and Metabolism* **68**, pp. 825–830.

642. **Jorgensen, H., Knigge, U., Kjaer, A., Moller, M.,** and **Warberg, J.** (2002). Serotonergic stimulation of corticotropin-releasing hormone and pro-opiomelanocortin gene expression. *Journal of Neuroendocrinology* **14**: , pp. 788–795.

643. **Jurek, B., Slattery, D. A., Hiraoka, Y., Liu, Y., Nishimori, K., Aguilera, G., Neumann, I. D.,** and **van den Burg, E. H.** (2015). Oxytocin regulates stress-induced Crf gene transcription through CREB-regulated transcription coactivator 3. *Journal of Neuroscience* **35**, pp. 12248–12260.

644. **Kafka, F.** (1915). *The metamorphosis*. New York: Bantam.

645. **Kagan, J.** (2002). *Surprise, uncertainty and mental structures*. Cambridge, MA: Harvard University Press.

646. **Kageyama, K., Gaudriault, G., Bradbury, M.**, and **Vale, W.** (2000). Regulation of corticotropin-releasing factor receptor type 2B messenger ribonucleuic acid in the rat cardiovascular system by urocortin, glucocorticoids, and cytokines. *Endocrinology* **141**, pp. 2285–2293.

647. **Kalantaridou, S., Makrigiannakis, A., Zoumakis, E.**, and **Chrousos, G.** (2007). Peripheral corticotropin releasing hormone is produced in the immune and reproductive systems: actions, potential roles and clinical implications. *Frontiers in Bioscience* **12**, pp. 572–580.

648. **Kalantaridou, S., Makrigiannakis, A., Zoumakis, E.**, and **Chrousos, G.** (2004). Reproductive functions of corticotropoin-releasing hormone. Research and potential clinical utility of antalarmins (CRH receptor type 1 antagonists). *American Journal of Reproductive Immunology* **51**, pp. 269–274.

649. **Kalin, N. H., Shelton, S. E.**, and **Barksdale, C. M.** (1989). Behavioral and physiologic effects of CRH administered to infant primates undergoing maternal separation. *Neuropsychopharmacology* **2**, pp. 97–104.

650. **Kalin, N. H., Shelton, S. E.**, and **Davidson, R. J.** (2000). Cerebrospinal fluid corticotropin-releasing hormone levels are elevated in monkeys with patterns of brain activity associated with fearful temperament. *Biological Psychiatry* **47**, pp. 579–585.

651. **Kalin, N. H., Takahashi, L. K.**, and **Chen, F. L.** (1994). Restraint stress increases corticotropin-releasing hormone mRNA content in the amygdala and paraventricular nucleus. *Brain Research* **656**, pp. 182–186.

652. **Kalin, N. H.**, (1985). Behavioral effects of ovine corticotropin-releasing factor administered to rhesus monkeys. *Federation Proceedings* **44**, pp. 249–253.

653. **Kalin, N. H., Larson, C., Shelton, S. E.**, and **Davidson, R. J.** (1998). Asymmetric frontal brain activity, cortisol, and behavior associated with fearful temperament in rhesus monkeys. *Behavioral Neuroscience* **112**, pp. 286–292.

654. **Kalin, N. H.**, and **Takahashi, L. K.** (1990). Fear-motivated behavior by prior shock experience is mediated by corticotropin-releasing hormone. *Brain Research* **509**, pp. 80–84.

655. **Kalin, N. H., Fox, A. S., Kovner, R., Riedel, M. K., Fekete, E. M., Roseboom, P. H., Tromp do, P. M., Grabow, B. P., Olsen, M. E., Brodsky, E. K., McFarlin, D. R., Alexander, A. L., Emborg, M.E., Block, W. F., Fudge, J. L.**, and **Oler, J. A.** 2016 Overexpressing corticotropin-releasing factor in the primate amygdala increase anxious temperament and alters its neural circuit. *Biol Psychiatry* **80**(5), pp. 345–355.

656. **Kalra, S. P.** (1993). Mandatory neuropeptide-steroid signaling for the preovulatory luteinizing hormone-releasing hormone discharge. *Endocrine Reviews* **14**, pp. 507–538.

657. **Kalynchuk, L., Gregus, A., Bourdreau, D.**, and **Perrot-Sinal, T.** (2004). Corticosterone incrases depression-like behavior, with some effects on predator odor-indcued defensive behavior, in male and female rats. *Behavioral Neuroscience* **118**, pp. 1365–1377.

658. **Kaminski, K. L.**, and **Watts, A. G.** (2012). Intact catecholamine inputs to the forebrain are required for appropriate regulation of corticotrophin-releasing hormone and vasopressin gene expression by corticosterone in the rat paraventricular nucleus. *Journal of Neuroendocrinology* **24**, pp. 1517–1526.

659. **Kamprath, K., Plendl, W., Marscicano, G., Deussing, J., Wurst, W., Lutz, B.,** and **Wotjak, C.** (2009). Endocannabinoids mediate acute fear adaptation via glutamatergic neurons independently of corticotropin releasing hormone signaling. *Genes, Brain, Behavior* **8**, pp. 203–211.

660. **Kang-Park, M., Kieffer, B. L., Roberts, A. J., Siggins, G. R.,** and **Moore, S. D.** (2015). Interaction of CRF and kappa opioid systems on GABAergic neurotransmission in the mouse central amygdala. *Journal of Pharmacology and Experimental Therapeutics* **355**, pp. 206–211.

661. **Kanwal, J. S.,** and **Rao, P. P.** (2002). Oxytocin within auditory nuclei: a neuromodulatory function in sensory processing? *Neuroreport* **13**, pp. 2193–2197.

662. **Kanwal, J. S., and Ehret, G., eds. (2006).** *Behavior and neurodynamics for auditory communication*. Cambridge: Cambridge University Press.

663. **Karalis, K., Goodwin, G.,** and **Majzoub, J. A.** (1996). Cortisol blockade of progesterone: a possible molecular mechanism involve in the initation of human labor. *Nature Medicine* **2**, pp. 556–560.

664. **Karalis, K., Muglia, L., Bae, D., Hilderband, H.,** and **Majzoub, J.** (1997). CRH and the immune system. *Journal of Neuroimmunology* **72**: , pp. 131–136.

665. **Karanikas, C., Lu, Y.,** and **Richardson, H.** (2013). Adolescent drinking targets corticotropin-releasing factor (CRF) peptide labeled cells in the central amygdala of male and female rats. *Neuroscience* **249**, pp. 98–105.

666. **Karteris, E., Markovic, D., Chen, J., Hillhouse, E.,** and **Grammatopoulos, D.** (2010). Identification of a novel corticotropin-releasing hormone type 1B-Like receptor variant lacking exon 13 in human pregnant myometrium regulated by estradiol- 17B and progesterone. *Endocrinology* **151**: , pp. 4959–4968.

667. **Karteris, E., Papadopoulou, N., Grammatopoulos, D. K.,** and **Hillhouse, E. W.** (2004). Expression and signaling characteristics of the corticotrophin-releasing hormone receptors during the implantation phase in the human endometrium. *Journal of Molecular Endocrinology* **32**, pp. 21–32.

668. **Kasahara, M., Groenink, L., Kas, M. J., Bijlsma, E. Y., Olivier, B.,** and **Sarnyai, Z.** (2011). Influence of transgenic corticotropin-releasing factor (CRF) over-expression on social recognition memory in mice. *Behavioural Brain Research* **218**, pp. 357–362.

669. **Kasckow, J. W., Aguilera, G., Mulchahey, J. J., Sheriff, S.,** and **Herman, J. P.** (2003). In vitro regulation of corticotropin-releasing hormone. *Life Sciences* **73**, pp. 769–781.

670. **Kassahn, K., Ragan, M.,** and **Funder, J.** (2011). Mineralocorticoid receptors: evolutionary and pathophysiological considerations. *Endocrinology* **152**, pp. 1883–1890.

671. **Kaur, S., Li, J., Stenzel-Poore, M.,** and **Ryabinin, A.** (2012). Corticotropin releasing factor acting on corticotropin releasing factor receptor type 1 is critical for binge alcohol drinking in mice. *Alcoholism Clinical and Experimental Research* **36**, pp. 369–376.

672. **Kavushansky, A.,** and **Richter-Levin, G.** (2006). Effects of stress and corticosterone on activity and plasticity in the amygdala. *Journal of Neuroscience Research* **84**, pp. 1580–1587.

673. **Kawahito, Y., Sano, H., Mukai, S., Asai, K., Kimura, S., Yamamura, Y., Kato, H., Chrousos, G. P., Wilder, R. L.,** and **Kondo, M.** (1995). Corticotropin releasing hormone in colonic mucosa in patients with ulcerative colitis. *Gut* **37**, pp. 544–551.

674. **Keller-Wood, M. E.**, and **Dallman, M. F.** (1984). Corticosteroid inhibition of ACTH secretion. *Endocrine Reviews* **5**, pp. 1–24.

675. **Keller-Wood, M. E.**, **Shinsako, J.**, and **Dallman, M. F.**(1983). Feedback inhibition of adrenocorticotropin hormone by physioloigical increases in plasma coritcosteroids in conscious dogs. *Journal of Clinical Investigation* **71**, pp. 859–866.

676. **Kendrick, K. M.** (2004). The neurobiology of social bonds. *Journal of Neuroendocrinology* **16**, pp. 1007–1008.

677. **Keverne, E. B.** (1997). Genomic imprinting in the brain. *Current Opinion in Neurobiology* **7**, pp. 463–468.

678. **Keverne, E. B.**, and **Curley, J. P.** (2004). Vasopressin, oxytocin and social behaviour. *Current Opinion in Neurobiology* **14**, pp. 777–783.

679. **Keverne, E. B.**, and **Curley, J. P.** (2008). Epigenetics, brain evolution and behavior. *Frontiers in Neuroendocrinology* **29**, pp. 398–412.

680. **Khan, B. Z.** (2009). *Democratizaion of invention: patents and copyrights in american economic development, 1790-1920.* Cambridge: Cambridge University Press.

681. **Khor, B. S.**, **Jamil, M. F.**, **Adenan, M. I.**, and **Shu-Chien, A. C.** (2011). Mitragynine attenuates withdrawal syndrome in morphine-withdrawn zebrafish. *PloS One* **6**, p. e28340.

682. **Kikusui, T.**, **Takeuchi, Y.**, and **Mori, Y.** (2000). Involvement of corticotropin-releasing factor in the retrieval process of fear-conditioned ultrasonic vocalization in rats. *Physiology and Behavior* **71**, pp. 323–328.

683. **Kim, A.**, **Gray, J.**, **Hermanson, D.**, **McLaughlin, R.**, **Wilson, C.**, **Vecchiarelli, H.**, **McEwen, B.**, **Schulkin, J.**, **Karatsoreos, I.**, **Patel, S.**, and **Hill, M.** (2015). Corticotropin releasing hormone enhances 2-arachidonoylglycerol in the medial prefrontal cortex. *Society for Neuroscience* 2014.

684. **Kim, J. J.**, **Lee, H. J.**, **Han, J. S.**, and **Packard, M. G.** (2001) Amygdala is critical for stress-induced modulation of hippocampal long-term potentiation and learning. *Journal of Neuroscience* **21**, pp. 5222–5228.

685. **Kim, S. H.**, **Bennett, P. R.**, and **Terzidou, V.** (2015). Diverse roles of oxytocin. *Inflammation and Cell Signaling* **2**, pp. 1–9.

686. **Kimura, M.**, **Müller-Preuss, P.**, **Lu, A.**, **Wiesner, E.**, **Flachskamm, C.**, **Wurst, W.**, **Holsboer, F.** and **Deussing, J. M.** (2010). Conditional corticotropin-releasing hormone overexpression in the mouse forebrain enhances rapid eye movement sleep. *Molecular Psychiatry* **15**, pp. 154–165.

687. **King, B. R.**, and **Nicholson, R. C.** (2007). Advances in understanding corticotrophin-releasing hormone gene expression. *Frontiers in Bioscience* **12**, pp. 581–590.

688. **King, B. R.**, **Smith, R.**, and **Nicholson, R. C.** (2001). The regulation of human corticotropin releasing hormone gene expression in the placenta. *Peptides* **22**, pp. 795–801.

689. **King, B. R.**, **Smith, R.**, and **Nicholson, R. C.** (2002). Novel glucocorticoid and cAMP interactionso n the CRH gene promoter. *Molecular and Cellular Endocrinology* **194**, pp. 19–28.

690. **Kiss, J. Z. Mezey, E.**, and **Skirboll, L.** (1984). Corticotropin-releasing factor-immunoreactive neurons of the paraventricular nucleus become vasopression positive after adrenalectomy. *Proceedings of the National Academy of Sciences of the United States of America* **81**, pp. 1854–1858.

691. **Kitamura, O., Wee, S., Specio, S. E., Koob, G.F.,** and **Pulvirenti, L.** Escalation of methamphetamine self-administration in rats: a dose-effect function. *Psychopharmacology (Berl)* (2006) **186**, pp. 48–53. doi:10.1007/s00213-006-0353-z

692. **Klampfl, S. M., Brunton, P. J., Bayerl, D. S.,** and **Bosch, O. J.** (2014). Hypoactivation of CRF receptors, predominantly type 2, in the medial-posterior BNST is vital for adequate maternal behavior in lactating rats. *Journal of Neuroscience* **34**, pp. 9665–9676.

693. **Klampfl, S. M., Brunton, P. J., Bayerl, D. S.,** and **Bosch, O. J.** (2016). CRF-R1 activation in the anterior-dorsal BNST induces maternal neglect in lactating rats via an HPA axis-independent central mechanism. *Psychoneuroendocrinology* **64**, pp. 89–98.

694. **Klampfl, S. M., Neumann, I. D.,** and **Bosch, O. J.** (2013). Reduced brain corticotropin-releasing factor receptor activation is required for adequate maternal care and maternal aggression in lactating rats. *European Journal of Neuroscience* **38**, pp. 2742–2750.

695. **Kling, M. A., Rubinow, D. R., Doran, A. R., Roy, A., Davis, C. L., Calabrese, J. R., Nieman, L. K., Post, R.M., Chrousos, G. P.,** and **Gold, P. W.** (1993). Cerebrospinal fluid immunoreactive somatostatin concentrations in patients with Cushing's disease and major depression: relationship to indices of corticotropin-releasing hormone and cortisol secretion. *Neuroendocrinology* **57**, pp. 79–88.

696. **Kling, M. A., Smith, M. A., Glowa, J. R., Pluznik, D., Demas, J., DeBellis, M. D., Gold, P. W.,** and **Schulkin, J.** (1993). Facilitation of cocaine kindling by glucocorticoids in rats. *Brain Research* **629**, pp. 163–166.

697. **Knapp, D. J., Whitman, B. A., Wills, T. A., Angel, R. A., Overstreet, D. H., Criswell, H. E., Ming, Z.,** and **Breese, G. R.** (2011a). Cytokine involvement in stress may depend on corticotropin releasing factor to sensitize ethanol withdrawal anxiety. *Brain Behav Immun Suppl* **1**, pp. S146–154.

698. **Knapp, D., Overstreet, D., Huang, M., Wills, T., Whitman, B., Angel, R., Sinnett, S.,** and **Breese, G.** (2011a). Effects of a stressor and corticotrophin releasing factor on ethanol deprivation-induced ethanol intake and anxiety-like behavior in alcohol-preferring P rats. *Psychopharmacology* **218**:, pp. 179–189.

699. **Knapp, D. J., Whitman, B. A., Wills, T. A., Angel, R. A., Overstreet, D. H., Criswell, H. E., Ming, Z.,** and **Breese, G. R.** (2011b). Cytokine involvement in stress may depend on corticotrophin releasing factor to sensitize ethanol withdrawal anxiety. *Brain Behavior Immunity* **25**, pp. S146–S154.

700. **Koehl, M., Meerlo, P., Gonzales, D., Rontal, A., Turek, F.,** and **Abrous, D.** (2008). Exercise-induced promotion of hippocampal cell proliferation requires beta-endorphin. *FASEB Journal* **22**:, pp. 2253–2262.

701. **Kokkotou, E., Torres, D., Moss, A. C., O'Brien, M., Grigoriadis, D. E., Karalis, K.,** and **Pothoulakis, C.** (2006). Corticotropin-releasing hormone receptor 2-deficient mice have reduced intestinal inflammatory responses. *Journal of Immunology* **177**, pp. 3355–3361.

702. **Kolber, B. J., Boyle, M. P., Wieczorek, L., Kelley, C. L., Onwuzurikie, C. C., Nettles, S. A., Vogt, S. K.,** and **Muglia, L. J.** (2010). Transient early-life forebrain corticotropin-releasing hormone elevation causes long lasting anxiogenic and despair-like changes mice. *Journal of Neuroscience* **30**, pp. 2571–2581.

703. **Kolber, B. J., Roberts, M. S., Howell, M. P., Wozniak, D. F., Sands, M. S.,** and **Muglia, L. J.** (2008). Central amygdala glucocorticoid receptor action promotes fear-associated CRH activation and conditioning. *Proceedings of the national academy of sciences of the United States of America* **105**: 12004–12009.

704. **Komuro, H., Sato, N., Sasaki, A., Suzuki, N., Kano, M., Tanaka, Y., Yamaguchi-Kabata, Y., Kanazawa, M., Warita, H., Aoki, M.,** and **Fukudo, S.** (2016). Corticotropin-releasing hormone receptor 2 gene variants in irritable bowel syndrome. *PloS One* **11**, p. e0147817.

705. **Kondoh, K., Lu, Z., Ye, X., Olson, D. P., Lowell, B. B.,** and **Buck, L. B.** (2016). A specific area of olfactory cortex involved in stress hormone responses to predator odours. *Nature* **532**, pp. 103–106.

706. **Koob, G. F.** (1998). Neuroscience of addiction. *Neuron* **21**, pp. 467–476.

707. **Koob, G. F.** (1999). Corticotropin-releasing factor, norepinephrine, and stress. *Biological Psychiatry* **46**, pp. 1167–1180.

708. **Koob, G. F.** (2009). Neurobiological substrates for the dark side of compulsivity in addiction. *Neuropharmacology* **56**, pp. 18–31.

709. **Koob, G. F.,** and **Volkow, N.** (2010). Neurocircuitry of addiction. *Neuropsychopharmacology* **35**, pp. 217–238.

710. **Koob, G. F.** (2013). Negative reinforcement in drug addiction: the darkness within. *Current Opinion in Neurobiology* **23**, pp. 559–563.

711. **Koob, G. F., Heinrichs, S. C., Pich, E. M., Menzaghi, F., Baldwin, H., Miczek, K.,** and **Britton, K. T.** (1993). The role of corticotropin-releasing factor in behavioural responses to stress. *Corticotropin-Releasing Factor* **172**, pp. 277–295.

712. **Koob, G. F.** and **Le Moal, M.** (2002). *Neurobiology of drug addiction. Stages and pathways of drug involvement: examining the gateway hypothesis*. New York: Cambridge University Press.

713. **Koob, G. F,** and **Le Moal, M.** (2004). Drug addiction and allostasis. In: Schulkin J. (ed.), *Allostasis, homeostasis, and the costs of physiological adaptation*, pp. 150–163.

714. **Koob, G. F.,** and **Le Moal, M.** (2005). *Neurobiology of addiction*. New York: Elsevier

715. **Koob, G. F.,** and **Zorrilla, E. P.** (2010). Neurobiological mechanisms of addiction: focus on corticotropin-releasing factor. *Current Opinion in Investigational Drugs* **11**, p. **63**.

716. **Koob, G. F.,** and **Bloom, F. E.** (1985). Corticotropin-releasing hormone and behavior. *Federation Proceedings* **44**, pp. 259–263.

717. **Koob, G. F.,** and **Kreek, M. J.** (2007). Stress, dysregulation of drug reward pathways, and the transition to drug dependence. *American Journal of Psychiatry* **164**: , pp. 1149–1159.

718. **Koob, G. F.,** and **Le Moal, M.** (2001). Drug addiction, dysregulation of reward, and allostasis. *Neuropsychopharmacology* **24**, pp. 97–129.

719. **Koob, G. F.** (2009). Brain stress system in the amygdala and addiction. *Brain Research* **1293**, pp. 61–75.

720. **Koob, G. F.** (2010). The role of CRF and CRF-related peptides in the dark side of addiction. *Brain Research* **1314**, pp. 3–14.

721. **Koob, G. F.** (2015). The dark side of emotion. *European Journal of Pharmacology* **753**, pp. 73–87.

722. **Koob, G. F.**, and **Le Moal, M.** (2008). Addiction and the brain antireward system. *Annual Review of Psychology* **59**, pp. 29–53.

723. **Koob, G. F.**, **Markou, A.**, **Weiss, F.**, and **Schulteis, G.** (1993). Opponent process and drug dependence: neurobiological mechanisms. *Seminars in Neuroscience* **5**, pp. 351–358.

724. **Korebrits, C.**, **Ramirez, M. M.**, **Watson, L.**, **Brinkman, E.**, **Bocking, A. D.**, and **Challis, J. R. G.** (1998). Maternal corticotropin-releasing hormone is increased with impending preterm birth. *Journal of Clinical Endocrinology and Metabolism* **83**, pp. 1585–1591.

725. **Korosi, A.**, **Shanabrough, M.**, **McClelland, S.**, **Liu, Z. W.**, **Borok, E.**, **Gao, X. B.**, **Horvath, T. L.**, and **Baram, T. Z.** (2010). Early-life experience reduces excitation to stress-responsive hypothalamic neurons and reprograms the expression of corticotropin-releasing hormone. *Journal of Neuroscience* **30**, pp. 703–713.

726. **Korte, S.**, and **Boer, S.** (2003). A robust animal model of state anxiety: fear potentiated behavior in the elevated plus-maze. *European Journal of Pharmacology* **463**, pp. 163–175.

727. **Korte, S. M** (2001). Corticosteroids in relation to fear, anxiety and psychopathology. *Neuroscience and Biobehavioral Reviews* **25**, pp. 117–142.

728. **Koutmani, Y.**, **Politis, P.**, **Elkouris, M.**, **Agrogiannis, G.**, **Kemerli, M.**, **Patsouris, E.**, **Remboutsika, E.**, and **Karalis, K.** (2013). Corticotropin releasing hormone exerts direct effects on neuronal progenitor cells: implications for neuroprotection. *Molecular Psychiatry* **18**, pp. 300–307.

729. **Kovo, M.**, **Vitner, D.**, **Holtmzan, D.**, **Krophach, N.**, **Golan, A.**, and **Chen, A.** (2013). Expression of the placental corticotropin releasing factor (CRF) and CRF binding protein in perfused cotyledon with lipopolysaccharide and magnesium sulphate. *AJOG* **208**, pp. S284–S285.

730. **Kozicz, T.** (2007). On the role of urocortin 1 in the non-preganglionic Edinger-Westphal nucleus in stress adaptation. *General and Comparative Endocrinology* **153**:, pp. 235–240.

731. **Kraemer, G. W.**, and **McKinney, W. T.** (1985). Social separation increases alcohol consumption in rhesus monkeys. *Psychopharmacology* **86**, pp. 182–189.

732. **Kramer, K. M.**, **Choe, C.**, **Carter, C. S.**, and **Cushing, B. S.** (2006). Developmental effects of oxytocin on neural activation and neuropeptide release in response to social stimuli. *Hormones and Behavior* **49**, pp. 206–214.

733. **Kramer, M.**, **Lydon, J.**, **Goulet, L.**, **Kahn, S.**, **Dahhou, M.**, **Platt, R.**, **Sharma, S.**, **Meaney, M.**, and **Seguin, L.** (2013). Maternal stress/distress, hormonal pathways and spontaneous preterm birth. *Pediatric and Perinatal Epidemiology* **27**, pp. 237–246.

734. **Kramer, M.**, **Lydon, J.**, **Seguin, L.**, **Goulet, L.**, **Kahn, S.**, **McNamara, H.**, **Genest, J.**, **Dassa, C.**, **Chen, M.**, **Sharma, S.**, **Meaney, M.**, **Thomson, S.**, **Van Uum, S.**, and **Koren, G.** (2009). Stress pathways to spontaneous preterm birth: the role of stressors, psychological distress, and stress hormones. *American Journal of Epidemiology* **169**, pp. 1319–1326.

735. **Kranzler, H. R.**, **Feinn, R.**, **Nelson, E. C.**, **Covault, J.**, **Anton, R. F.**, **Farrer, L.**, and **Gelernter, J.** (2011). A CRHR1 haplotype moderates the effect of adverse childhood experiences on lifetime risk of major depressive episode in African-American women. *American Journal of Medical Genetics Part B: Neuropsychiatric Genetics* **156**, pp. 960–968.

736. **Kreek, M. J.** (1997). Opiate and cocaine addictions: challenge for pharmacotherapies. *Pharmacology Biochemistry and Behavior* **57**, pp. 551–569.

737. **Kreek, M.J., Zhou, Y., Butelman, E.,** and **Levran, O.** (2009). Opiate and cocaine addiction: from bench to clinic and back to the bench. *Current oOpinion in Ppharmagology* **9**: , pp. 74–80.

738. **Krettek, J. E.,** and **Price, J. L.** (1977a). Projections from the amygdaloid complex and adjacent olfactory structures to the entorhinal cortex and to the subiculum in the rat and cat. *Journal of Comparative Neurology* **172**, pp. 723–752.

739. **Krettek, J. E.,** and **Price, J. L.** (1977b). Projections from the amygdaloid complex to the cerebral cortex and thalamus in the rat and cat. *Journal of Comparative Neurology* **172**, pp. 687–722.

740. **Kropach, N., Holtzman, D., Vitner, D., Bar, J., Chen, A.,** and **Kovo, M.** (2016). The effect of magnesium sulfate on the placental corticotropin-releasing factor (CRF) and CRF binding protein mRNA expression in perfused human placental cotyledon. *Journal of Maternal-Fetal and Neonatal Medicine* **29**, pp. 376–379.

741. **Kuhar, M.** (1984). Corticotropin-releasing factor receptors in rate forebrain: autoradiographic identification. *Science* **224**, pp. 14491–1451.

742. **Kuhn, C. M., Pauk, J.,** and **Schanberg, S. M.** (1990). Endocrine responses to mother-infant separation in developing rats. *Developmental Psychobiology* **23**, pp. 395–410.

743. **Kühn, E. R., De Groef, B., Grommen, S. V., Van der Geyten, S.,** and **Darras, V. M.** (2004). Low submetamorphic doses of dexamethasone and thyroxine induce complete metamorphosis in the axolotl (Ambystoma mexicanum) when injected together. *General and Comparative Endocrinology* **137**, pp. 141–147.

744. **Kühn, E., De Groef, B., Van der Geyten, S.,** and **Darras, V.** (2005). Corticotropin-releasing hormone-mediated metamorphosis in the neotenic axolotl Ambystoma mexicanum: synergistic involvement of thyroxine and corticoids on brain type II deiodinase. *General and Comparative Endocrinology* **143**, pp. 75–81.

745. **Kulkarni, S. S., Singamsetty, S.,** and **Buchholz, D. R.** (2010). Corticotropin-releasing factor regulates the development in the direct developing frog, Eleutherodactylus coqui. *General and Comparative Endocrinology* **169**, pp. 225–230.

746. **Kumar, D., Springel, E., Moore, R.M., Mercer, B.M., Philipson, E., Mansour, J.M., Mesiano, S., Schatz, F., Lockwood, C.J.,** and **Moore, J.J.** (2015). Progesterone inhibits in vitro fetal membrane weakening. *American Journal of Obstetrics and Gynecology* **213**: , pp. 520, e1–e9.

747. **Kunzel, H., Ising, M., Zobel, A., Nickel, T., Ackl, N., Sonntag, A., Holsboer, F.,** and **Uhr, M.** (2005). Treatment with a CRH-1-receptor antagonist (R121919) does not affect weight or plasma leptin concentration in patients with major depression. *Journal of Psychiatric Research* **39**: , pp. 173–177.

748. **Kunzel, H.E., Zobel, A.W., Nickel, T., Ackl, N., Uhr, M., Sonntag, A., Ising, M.,** and **Holsboer, F.** (2003). Treatment of depression with the CRH-1-receptor antagonist R121919: endocrine changes and side effects. *Journal of Psychiatric Research* **37**, pp. 525–533.

749. **Kupferschmidt, D., Klas, P.,** and **Erb, S.** (2012). Cannabinoid CB1 receptors mediate the effects of corticotropin-releasing factoron the reinstatement of cocaine seeking

and expression of cocaine-induced behavioural sensitization. *British Journal of Pharmacology* **167**, pp. 196–206.

750. **Kupferschmidt, D., Newman, A., Boonstra, R.,** and **Erb, S.** (2012). Antagonism of cannabinoid 1 receptors reverses the anxiety-like behavior induced by central injections of corticotropin releasing factor and cocaine withdrawal. *Neuroscience* **204**, pp. 125–133.
751. **Kusnecov, A., Liang, R.,** and **Shurin, G.** (1999). T-Lymphocyte activation increases hypothalamic and amygdaloid expression of CRH mRNA and emotional reactivity to novelty. *Journal of Neuroscience* **19**, pp. 4533–4543.
752. **Kwako, L., Spagnolo, P., Schwandt, M., Thorsell, A., George, D., Momenan, R., Rio, D., Huestis, M., Anizan, S., Concheiro, M., Sinha, R.,** and **Heilig, M.** (2015). The corticotropin releasing hormone-1 (CRH1) receptor antagonist pexacerfont in alcohol dependence: a randomized controlled experimental medicine study. *Neuropsychopharmacology* **40**, pp. 1053–1063.
753. **La, J. H., Sung, T. S., Kim, H. J., Kim, T. W., Kang, T. M.,** and **Yang, I. S.** (2008). Peripheral corticotropin releasing hormone mediates post-inflammatory visceral hypersensitivity in rats. *World Journal of Gastroenterology* **14**: , pp. 731–736.
754. **Laatikainen, T., Virtanen, T., Kaaja, R.,** and **Salminen-Lappalainen, K.** (1991). Corticotropin-releasing hormone in maternal and cord plasma in pre-eclampsia. *European Journal of Obstetrics and Gynecology and Reproductive Biology* **39**(1), pp. 19–24.
755. **Laatikainen, T., Virtanen, T., Raisanen, I.,** and **Salminen, K.** (1987). Immunoreactive corticotropin releasing factor and corticotropin during pregnancy and puerperium. *Neuropeptides* **10**, pp. 343–353.
756. **Laberge, F., Mühlenbrock-Lenter, S., Grunwald, W.,** and **Roth, G.** (2006). Evolution of the amygdala: new insights from studies in amphibians. *Brain Behavior Evolution* **67**, pp. 177–187.
757. **Labermaier, C., Kohl, C., Hartmann, J., Devigny, C., Altmann, A., Weber, P., Arloth, J., Quast, C., Wagner, K.V., Scharf, S.H., Czibere, L., Widner-Andra, R., Brenndorfer, J., Landraf, R., Hausch, F., Jones, K.A., Muller, M.B., Uhr, M., Holsboer, F., Binder, E.B.** and **Schmidt, M.V.** (2014). A Polymorphism in the crhr1 gene determines stress vulnerability in male mice. *Endocrinology* **155**, pp. 2500–2510.
758. **Lachize, S., Apostolakis, E. M., van der Laan, S., Tijssen, A. M. I., Xu, J., de Kloet, E. R.,** and **Meijer, O. C.** (2009). Steroid receptor coactivation-1 is necessary for regulation of corticotropin-releasing hormone by chronic stress and glucocorticoids. *Proceedings of the National Academy of Sciences of the United States of America* **106**, pp. 8038–8042.
759. **Laflamme, N., Nappi, R. E., Drolet, G., Labrie, C.,** and **Rivest, S.** (1998). Expression and neuropeptidergic characterization of estrogen receptors (ERa and ERb) throughout the rat brain: anatomical evidence of distinct roles of each subtype. *Journal of Neurobiology* **36**, pp. 357–378.
760. **Landgraf, R.,** and **Wigger, A.** (2003). Born to be anxious: neuroendocrine and genetic correlates of trait anxiety in HAB rats. Stress: The International Journal on the Biology of Stress **6**, pp. 111–119.
761. **Landys, M. M., Ramenofsky, M.,** and **Wingfield, J. C.** (2006). Actions of glucocorticoids at a seasonal baseline as compared to stress-related levels in the regulation of periodic life processes. *General and Comparative Endocrinology* **148**, pp. 132–149.

762. **Lang, P. J.**, **Bradley, M. M.**, and **Cuthbert, B. N.** (1998). Emotion, motivation, and anxiety: brain mechanisms and psychophysiology. *Biological Psychiatry* **44**, pp. 1248–1263.

763. **Lannon, S. M.**, **Vanderhoeven, J. P.**, **Eschenbach, D. A.**, **Gravett, M. G.**, and **Waldorf, K. M. A.** (2014). Synergy and interactions among biological pathways leading to preterm premature rupture of membranes. *Reproductive Sciences* **21**, pp. 1215–1227.

764. **Larhammar, D.**, **Dreborg, S.**, **Larssson, T.A.**, and **Sundstrom, G.** (2009). Early duplications of opioid receptor and peptide genes in vertebrate evolution. *Annals of the New York Academy of Sciences* **1163:**, pp. 451–453.

765. **Laryea, G.**, **Arnett, M. G.**, and **Muglia, L. J.** (2012). Behavioral studies and genetic alterations in corticotropin-releasing hormone (CRH) neurocircuitry: insights into human psychiatric disorders. *Behavioral Sciences* **2**, pp. 135–171.

766. **Lashley, K. S.** (1938). An experimental analysis of instinctive behavior. *Psychological Review* **45**, pp. 445–471.

767. **Laskey, W.** and **Polosa, C.** (1988) Characteristices of the sympathetic preganglionic neuron and its synaptic input. *Progress in Neurobiology* **31**, pp. 47–84.

768. **Lastein, S.**, **Höglund, E.**, **Overli, O.**, and **Døving, K.B.** (2008). Effects of antalarmin, a CRF receptor 1 antagonist, on fright reaction and endocrine stress response in crucian carp. *Journal of Comparative Physiology* **194**, pp. 1007–1012.

769. **Laugero, K. D.**, **Gomez, F.**, **Manalo, S.**, and **Dallman, M. F.** (2002). Corticosterone infused intracerebroventricularly inhibits energy storage and stimulates the hypothalamo-pituitary axis in adrenalectomized rats drinking sucrose. *Endocrinology* **143**, pp. 4552–4562.

770. **Le, M. H.**, **Weissmiller, A. M.**, **Monte, L.**, **Lin, P. H.**, **Hexom, T. C.**, **Natera, O.**, **Wu, C.** and **Rissman, R. A.** (2016). Functional impact of corticotropin-releasing factor exposure on tau phosphorylation and axon transport. *PloS One* **11**, pp. 1–15.

771. **Lebow, M.A.**, **Chen, A.** (2016). Overshadowed by the amygdala: the bed nucleus of the stria terminalis emerges as key to psychiatric disorders. Molecular Psychiatry 21(4), pp. 450–463.

772. **LeDoux, J.** (1996). *The emotional brain*. New York: Simon and Shuster.

773. **LeDoux, J.** (2012). Rethinking the emotional brain. *Neuron* **73**, pp. 653–676.

774. **LeDoux, J.** (2015). Feelings: What are they and how does the brain make them? *Daedalus* **144**, pp. 96–111.

775. **LeDoux, J.E.** (2000). Emotion circuits in the brain. *Annual Review of Neuroscience* **23**, pp. 155–184.

776. **LeDoux, J.E.** (2014). Coming to terms with fear. *Proceedings of the national academy of sciences of the United States of America* **111**, pp. 2871–2878.

777. **LeDoux, J.**, **Iwata, J.**, **Cicchetti, P.**, and **Reis, D.** (1988). Different projections of the central amygdaloid nucleus mediate autonomic and behavioral correlates of conditioned fear. *Journal of Neuroscience* **8**, pp. 2517–2529. **Lee, E. H. Y.**, **Hung, H. C.**, **Lu, K. T.**, **Chen, W. H.**, and **Chen, H. Y.** (1992). Protein synthesis in the hippocampus associated with memory facilitation by corticotropin-releasing factor in rats. *Peptides* **13**, pp. 927–937.

778. **Lee, P. H. K.**, **Grimes, L.**, and **Hong, J. S.** (1989). Glucocorticoids potentiate kainic acid-induced seizures and wet dog shakes. *Brain Research* **480**, pp. 322–325.

779. **Lee, S.**, and **Rivier, C.** (1996). Gender differences in the effect of prenatal alcohol exposure on the hypothalamic pituitary adrenal axis response to immune signals. *Psychoneuroendocrinology* **21**, pp. 145–155.

780. **Lee, Y.**, and **Davis, M.** (1997). Role of the hippocampus, the bed nucleus of the stria terminalis, and the amygdala in the excitatory effect of corticotropin-releasing hormone on the acoustic startle reflex. *Journal of Neuroscience* **17**, pp. 6434–6446.

781. **Lee, Y.**, **Fitz, S.**, **Johnson, P.**, and **Shekhar, A.** (2008). Repeated stimulation of CRH receptors in the BNST of rats selectively induces social but not panic-like anxiety. *Neuropsychopharmacology* **33**, pp. 2586–2594.

782. **Lee, Y.**, **Schulkin, J.**, and **Davis, M.** (1994). Effect of corticosterone on the enhancement of the acoustic startle reflex by corticotropin releasing factor (CRF). *Brain Research* **666**, pp. 93–98.

783. **Lehner, M.**, **Taracha, E.**, **Skorzewska, A.**, **Turzynska, D.**, **Sobolewska, A.**, **Maciejak, P.**, **Szyndler, J.**, **Hamed, A.**, **Bidzinski, A.**, **Wislowska-Stanek, A.**, and **Plaznik, A.** (2008). Expression of c-Fos and CRF in the brains of rats differing in the strength of a fear response. *Behavioral Brain Research* **188**, pp. 154–167.

784. **Lehrman, D.** (1958). Induction of broodiness by participation in courtship and nest-building in the ring dove (Streptopella risoria). *Journal of Comparative and Physiological Psychology* **51**, pp. 32–36.

785. **Lehrman, D. S.** (1965). Interaction between internal and external environments in the regulation of the reproductive cycle of the ring dove. In: F. A. Beach (ed.), *Sex and behavior*. New York: Wiley, pp. 43–58.

786. **Lei, K.**, **Chen, L.**, **Georgiou, E.X.**, **Sooranna, S.R.**, **Khanjani, S.**, **Brosens, J.J.**, **Bennett, P.R.**, and **Johnson, M.R.** (2012). Progesterone acts via the nuclear glucocorticoid receptor to suppress IL-1β-Induced COX-2 expression in human term myometrial cells. *PLoS One* 7, p. e50167.

787. **Lemos, J. C.**, **Wanat, M. J.**, **Smith, J. S.**, **Reyes, B. A.**, **Hollon, N. G.**, **Van Bockstaele, E. J.**, **Chavkin, C.**, and **Phillips, P. E.** (2012). Severe stress switches CRF action in the nucleus accumbens from appetitive to aversive. *Nature* **490**, pp. 402–406.

788. **Lenglos, C.**, **Mitra, A.**, **Guèvremont, G.**, and **Timofeeva, E.** (2013). Sex differences in the effects of chronic stress and food restriction on body weight gain and brain expression of CRF and relaxin-3 in rats. *Genes, Brain and Behavior* **12**, pp. 370–387.

789. **Leonhardt, S.**, **Boonyaratnakornkit, V.**, and **Edwards, D.** (2003). Progesterone receptor transcription and non-transcription signaling mechanisms. *Steroids* **68**, pp. 761–770.

790. **Leshem, M.**, and **Schulkin, J.** (2012). Transgenerational effects of infantile adversity and enrichment in male and female rats. *Developmental Psychobiology* **54**, pp. 169–186.

791. **Leuner, B.**, and **Shors, T. J.** (2013). Stress, anxiety, and dendritic spines: what are the connections? *Neuroscience* **251**, pp. 108–119.

792. **Leung, E.**, **Tasker, S.**, **Atkinson, L.**, **Vaillancourt, T.**, **Schulkin, J.**, and **Schmidt, L.** (2010). Perceived maternal stress during pregnancy and its relation to infant stress reactivity at 2 days and 10 months of postnatal life. *Clinical Pediatrics* **49**, pp. 158–165.

793. **Leung, T.N.**, **Chung, T.K.**, **Madsen, G.**, **Lam, P.L.**, **Sahota, D.**, and **Smith, R.** (2001). Rate of rise in maternal plasma corticotropin-releasing hormone and its relation to gestational length. *BJOG* **108**, pp. 527–532.

794. **Levine, S.** (1967). Maternal and environmental influences on the adrenocortical response to stress in weanling rats. *Science* **256:** , pp. 258–260.

795. **Levine, S.** (1975). Psychosocial factors in growth and development. In: Levi, L. (ed.), *Society, stress, and disease*. London: Oxford University Press. 43–50.

796. **Levine, S.** (2000). Modulation of CRF gene expression by early experience. *Neuropsychopharmacology* **23**, pp. S72–S74.

797. **Levine, S., Haltmeyer, G. C., Karas, G. G.,** and **Denenberg, V. H.** (1967). Physiological and behavioral effects of infantile stimulation. *Physiology and Behavior* **2**, pp. 55–59.

798. **Leviton, A., Allred, E. N., Kuban, K. C., O'Shea, T. M., Paneth, N.,** and **Majzoub, J.** (2016). ELGAN Study Investigators. Brain disorders associated with corticotropin-releasing hormone expression in the placenta among children born before the 28th week of gestation. *Acta Paediatrica* **105**, pp. e7–e11.

799. **Leviton, A., Allred, E. N., Kuban, K. C., O'Shea, T. M., Paneth, N.,** and **Majzoub, J.** (2016). Brain disorders associated with corticotropin-releasing hormone expression in the placenta among children born before the 28th week of gestation. *Acta Paediatrica* **105**, pp. e7–e11.

800. **Lewis, K., Li, C., Perrin, M. H., Blount, A., Kunitake, K., Donaldson, C., Vaughan, J., Reyes, T. M., Gulyas, J., Fischer, W., Bilezikjian, L., Rivier, J., Sawchenko, P. E.,** and **Vale, W. W.** (2001). Identification of urocortin III, an additional member of the corticotropin releasing factor (CRF) family with high affinity for the CRF2 receptor. *Proceedings of the National Academy of Sciences of the United States of America* **98**, pp. 7570–7575.

801. **Lezak, K. R., Roman, C. W., Braas, K. M., Schutz, K. C., Falls, W. A., Schulkin, J., May, V.,** and **Hammack, S.** (2014). Regulation of bed nucleus of the stria terminalis PACAP expression by stress and corticosterone. *Journal of Molecular Neuroscience* **54:** , pp. 477–484.

802. **Li, H. C., Penzo, M., Taniguchi, H., Kopec, C., Huang, Z.,** and **Li, B.** (2013). Experience dependent modification of a central amygdala fear circuit. *Nature Neuroscience* **16**, pp. 332–339.

803. **Liang, K. C., Melia, K. R., Miserendino, M. J., Falls, W. A., Campeau, S.,** and **Davis, M.** (1992b). Corticotropin-releasing factor: long-lasting facilitation of the acoustic startle reflex. *Journal of Neuroscience* **12**, pp. 2303–2312.

804. **Liang, K. C., Melia, K. R., Campeau, S., Falls, W. A., Miserendino, M. J.,** and **Davis, M.** (1992a). Lesions of the central nucleus of the amygdala but not the paraventricular nucleus of the hypothalamus, block the excitatory effects of corticotropin-releasing factor on the acoustic startle reflex. *Journal of Neuroscience* **12**, pp. 2313–2320.

805. **Lieberman, M. D.** (2000) Intuition: a social cognitive neuroscience approach. *Psychol Bull* **126**(1), pp. 109–37.

806. **Liebsch, G., Landgraf, R., Gerstberger, R., Probst, C., Wotjak, C., Engelmann, M., Holsboer, F.,** and **Motkowski, A.** (1995). Chronic infusion of a CRH1 receptor antisense oligodeoxynucleotide into the central nucleus of the amygdala reduced anxiety-related behavior in socially defeated rats. *Regulatory Peptides* **59**, pp. 229–239.

807. **Lightman, S.** (2008). The neuroendocrinology of stress: a never ending story. *Journal of Neuroendocrinology* **20**, pp. 880–884.

808. **Lim, A. C., Goossens, A., Ravelli, A. C., Boer, K., Bruinise, H. W.,** and **Mol, B. W.** (2010). Use of progesterone treatment for the prevention of recurrent preterm

birth: identification of obstacles to change. *American Journal of Perinatology* **27**: , pp. 241–250.

809. **Lim, M. M., Liu, Y., Ryabinin, A. E., Bai, Y., Wang, Z.,** and **Young, L. J.** (2007). CRF receptors in the nucleus accumbens modulate partner preference in prairie voles. *Hormones and Behavior* **51**, pp. 508–515.
810. **Lim, M. M., Nair, H. P.,** and **Young, L. J.** (2005). Species and sex differences in brain distribution of corticotropin-releasing factor receptor subtypes 1 and 2 in monogamous and promiscuous vole species. *Journal of Comparative Neurology* **487**, pp. 75–92.
811. **Lim, M. M., Liu, Y., Ryabinin, A. E., Bai, Y., Wang, Z.,** and **Young, L.** (2007). CRF receptors in the nucleus accumbens modulate partner preference in prairie voles. *Hormones and Behavior* **51**, pp. 508–515.
812. **Lim, M. M., Tsivkovskaia, N. O., Bai, Y., Young, L. J.,** and **Ryabinin, A. E.** (2006). Distribution of corticotropin-releasing factor and urocortin 1 in the vole brain. *Brain, Behavior, and Evolution* **68**, pp. 229–240.
813. **Lind, R. W., Swanson, L. W.,** and **Ganten, D.** (1984). Angiotensin II immunoreactivity in the neural afferents and efferents of the subfornical organ of the rat. *Brain Research* **321**, pp. 209–215.
814. **Ling, L., Faraco, J., Li, R., Kadotani, H., Rogers, W., Lin, X., Qiu, X., de Johng, P., Nishino, S.,** and **Mignot, E.** (1999). The sleep disorder canine narcolepsy is caused by a mutation in the Hypocretin (Orexin) Receptor 2 gene. *Cell* **98**: , pp. 365–376.
815. **Linthorst, A. C., Flachskamm, C., Hopkins, S. J., Hoadley, M. E., Labeur, M. S., Holsboer, F.,** and **Paul, J. M.** (1997). Long-term intracerebroventricular infusion of corticotropin-releasing hormone alters neuroendocrine, neurochemical, autonomic, behavioral and cytokine responses to a systemic inflammatory challenge. *Journal of Neuroscience* **17**, pp. 4448–4460.
816. **Linton, E. A., Perkins, A. V., Woods, R. J., Eben, F., Wolfe, C. D., Behan, D.P., Potter, E., Vale, W. W.,** and **Lowry, P. J.** (1993). Corticotropin releasing hormone binding protein (CRH-BP): plasma levels decrease during the third trimester of normal human pregnancy. *Journal of Clinical Endocrinology and Metabolism* **76**, pp. 260–262.
817. **Liu, C. X., Takahashi, S., Murata, T., Hashimoto, K., Agatsuma, T., Matsukawa, S.,** and **Higuchi, T.** (1996). Changes in oxytocin receptor mRNA in the rat uterus measured by competitive reverse transcription-polymerase chain reaction. *Journal of Endocrinology* **150**, pp. 479–486.
818. **Liu, D., Diorio, J., Tannenbaum, B., Caldji, C., Francis, D., Freedman, A., Sharma, S., Pearson, D., Plotsky, P. M.,** and **Meaney, M. J.** (1997). Maternal care, hippocampal glucocorticoid receptors, and hypothalamic–pituitary–adrenal responses to stress. *Science* **277**, pp. 1659–1662.
819. .**Liu, S., Chang, J., Long, N., Beckwith, K., Talhouarne, G., Brooks, J. J., Qu, M. H., Ren, W., Wood, J.D., Cooper, S.** & and **Bhargava, A.** (2015). Endogenous CRF in rat large intestine mediates motor and secretory responses to stress. *Neurogastroenterology and Motility* **28**, pp. 281–290.
820. **Liu, X., Wellman, L., Yang, L., Ambrozewicz, M., Tang, X.,** and **Sanford, L.** (2011). Antagonizing corticotropin releasing factor in the central nucleus of the amygdala attenuates fear-induced reductions in sleep but not freezing. *Sleep* **34**: , pp. 1539–1549.
821. **Liu, Y., Fang, X., Yuan, J., Sun, Z., Li, C., Li, R., Li, L., Zhu, C., Wan, R., Guo, R., Jin, L.,** and **Li, S.** (2014). The role of corticotropin-releasing hormone receptor 1 in the

development of colitis-associated cancer in mouse model. *Endocrine-Related Cancer* **21**, pp. 639–651.

822. **Liu, Y.**, and **Wang, Z. X.** (2003). Nucleus accumbens oxytocin and dopamine interact to regulate pair bond formation in female prairie voles. *Neuroscience* **121**, pp. 537–544.
823. **Lockwood, C.** (2004). The initation of parturition at term. *Obstetrics and Gynecology Clinics of North America* **31**, pp. 935–937.
824. **Lodge, D.**, and **Lawrence, A.** (2003). The CRF1 receptor antagonist antalarmin reduces volitional ethanol consumption in isolation-reared fawn-hooded rats. *Neuroscience* **117**, pp. 243–247.
825. **Loite, U.**, **Kingo, K.**,**Rreimann, E.**, **Reemann, P.**, **Vasar, E.**, **Silm, H.**, and **Köks, S.** (2013). Gene expression analysis of the corticotropin-releasing hormone-proopiomelanocortin system in psoriasis skin biopsies. *Acta Dermato-Venereologica* **93**, pp. 400–405.
826. **Lovejoy, D. A.** (2009). Structural evolution of urotensin-I: Reflections of life before corticotropin releasing factor. *General and Comparative Endocrinology* **164**, pp. 15–19.
827. **Lovejoy, D. A.**, and **de Lannoy, L.** (2013). Evolution and phylogeny of the corticotropin-releasing factor (CRF) family of peptides: Expansion and specialization in the vertebrates. *Journal of Chemical Neuroanatomy* **54**, pp. 50–56.
828. **Lovejoy, D. A.**, and **Balment, R. J.** (1999). Evolution and physiology of the corticotropin-releasing factor (CRF) family and neuropeptides in vertebrates. *General and Comparative Endocrinology* **115**, pp. 1–22.
829. **Lovejoy, D. A.**, **Chang, B. S.**, **Lovejoy, N. R.**, and **del Castillo, J.** (2014). Molecular evolution of GPCRS: CRH/CRH receptors. *Journal of Molecular Endocrinology* **52**, pp. T43–T60.
830. **Lovejoy, D. A.**, and **Jahan, S.** (2006). Phylogeny of corticotropin-releasing factor family of peptides in the metazoan. *General and Comparative Endocrinology* **146**, pp. 1–8.
831. **Lovejoy, D. A.**, **Rotzinger, S.**, and **Barsyte-Lovejoy, D.** (2009). Evolution of complementary peptide systems: teneurin C-terminal-associated peptides and corticotropin-releasing factor superfamilies. *Annals of the New York Academy of Sciences* **1163**, pp. 215–220.
832. **Lovenberg, T. W.**, **Chalmers, D. T.**, **Liu, C.**, and **De Souza, E. B.** (1995a). CRF2 alpha and CRF2 beta receptor mRNAs are differentially distributed between the rat central nervous system and peripheral tissues. Endocrinology **136**, pp. 4139–4142.
833. **Lovenberg, T. W.**, **Liaw, C. W.**, **Grigoriadis, D. E.**, **Clevenger, W.**, **Chalmers, D. T.**, **De Souza, E. B.**, and **Oltersdorf, T.** (1995b). Cloning and characterization of a functionally distinct corticotropin-releasing factor receptor subtype from rat brain. *Proceedings of the National Academy of Sciences* **92**, pp. 836–840
834. **Lowry, C. A.**, and **Moore, F. L.** (1991). Corticotropin-releasing factor (CRF) antagonist suppresses stress-induced locomotor activity in an amphibian. *Hormones and Behavior* **25**, pp. 84–96.
835. **Lucas, L. R.**, **Angulo, J. A.**, **Moal, M. L.**, **McEwen, B. S.**, and **Piazza, P. V.** (1998). Neurochemical characterization of individual vulnerability to addictive drugs in rats. *European Journal of Neuroscience* **10**, pp. 3153–3163.
836. **Lucas, M.**, **Chen, A.**, and **Richter-Levin, G.** (2013). Hypothalamic corticotropin-releasing factor is centrally involved in learning under moderate stress. *Neuropsychopharmacology* **38**, pp. 1825–1832.

837. **Lukkes, J. L., Mokin, M. V., Scholl, J. L.,** and **Forster, G. L.** (2009). Adult rats exposed to early-life social isolation exhibit increased anxiety and conditioned fear behavior, and altered hormonal stress responses. *Hormones and Behavior* **55**, pp. 248–256.

838. **Lukkes, J. L., Norman, K. J., Meda, S.,** and **Andersen, S. L.** (2016). Sex differences in the ontogeny of CRF receptors during adolescent development in the dorsal raphe nucleus and ventral tegmental area. *Synapse* **70**, pp. 125–132.

839. **Lupien, S.J., Fioco, A., Wan, N., Maheu, F., Lord, C., Schramek, T.,** and **Tu, T.** (2005). Stress hormones and human memory function across the lifespan. *Psychoneuroendocrinology* **30**, pp. 225–242.

840. **Lutfy, K., Aimiuwu, O., Mangubat, M., Shin, C. S., Nerio, N., Gomez, R., Liu, Y.** and **Friedman, T. C.** (2012). Nicotine stimulates secretion of corticosterone via both CRH and AVP receptors. *Journal of Neurochemistry* **120**, pp. 1108–1116.

841. **Ma, S., Blasiak, A., Olucha-Bordonau, F.E., Verberne, A.J., Gundlach, A.L.** (2013). Heterogeneous responses of nucleus incertus neurons to corticotrophin-releasing factor and coherent activity with hippocampal theta rhythm in the rat. *Journal of Physiology* 591(16), pp. 3981–4001.

842. **Ma, X. M., Camacho, C.,** and **Aguilera, G.** (2001). Regulation of corticotropin-releasing hormone (CRH) transcription and CRH mRNA stability by glucocorticoids. *Cellular and Molecular Neurology* **21**, pp. 465–475.

843. **Maccari, S., Piazza, P. V., Deminiere, J. M., Lemaire, V., Mormede, P., Angelucci, L.,** and **Le Moal, M.** (1991). Life events-induced decrease of corticosteroid type I receptors is associated with reduced corticosterone feedback and enhanced vulnerability to amphetamine self-administration. *Brain Rresearch* **547**, pp. 7–20.

844. **Machado, R. B., Tufik, S.,** and **Suchecki, D.** (2010). Modulation of sleep homeostasis by corticotropin releasing hormone in REM sleep-deprived rats. *International Journal of Endocrinology* **147**, pp. 1–12.

845. **Madhappan, B., Kempuraj, D., Christodoulou, S., Tsapikidis, S., Boucher, W., Karagiannis, V., Athanassiou, A.,** and **Theoharides, T.** (2003). High levels of intrauterine corticotropin releasing hormone, urocortin, tryptase, and interleukin-8 in spontaneous abortions. *Endocrinology* **144**, pp. 2285–2290.

846. **Magalhaes, A. C., Homes, K. D., Dale, L. B., Comps-Agrar, L., Lee, D., Yadav, P. N., Drysdale, L., Poulter, M. O., Roth, B. L., Pin, J. P., Anisman, H.,** and **Ferguson, S. S.** (2010). CRF receptor 1 regulates anxiety behavior via sensitization of 5-HT2 receptor signaling. *Nature Neuroscience* **13**, pp. 622–630.

847. **Magdeldin, S., Blaser, R., Yamamoto, T.,** and **Yates, J.** (2015). Behavioral and proteomic analysis of stress response in zebrafish (Danio rerio). *Journal of Proteome* **14**, pp. 943–952.

848. **Magiakou, M. A., Mastorakos, G., Rabin, D., Dubbert, B., Gold, P. W.,** and **Chrousos, G. P.** (1996). Hypothalamic corticotropin-releasing hormone suppression during the postpartum period: implications for the increase in psychiatric manifestations at this time. *Journal of Clinical Endocrinology and Metabolism* **81**, pp. 1912–1917.

849. **Majzoub, J. A.** (2006). Corticotropin-releasing hormone physiology. European Journal of *Endocrinology* **155**, pp. S71–S76.

850. **Majzoub, J. A.,** and **Karlis, K. P.** (1999). Placental corticotropin-releasing hormone: function and regulation. *American Journal of Obstetrics and Gynecology* **180**, pp. S242–246.

851. **Majzoub, J.A., McGregor, J.A., Lockwood, C.J., Smith, R., Taggart, M.S.,** and **Schulkin, J.** (1999). A central theory of preterm and term labor: putative role for corticotropin-releasing hormone. *American Journal of Obstetrics and Gynecology* **180**, pp. S232–S241.

852. **Makino, S., Gold, P. W.,** and **Schulkin, J.** (1994a). Corticosterone effects on corticotropin-releasing hormone mRNA in the central nucleus of the amygdala and the parvocellular region of the paraventricular nucleus of the hypothalamus. *Brain Research* **640**, pp. 105–112.

853. **Makino, S., Gold, P. W.,** and **Schulkin, J.** (1994b). Effects of corticosterone on CRH mRNA and content in the bed nucleus of the amygdala and the paraventricular nucleus of the hypothalamus. *Brain Research* **657**, pp. 141–149.

854. **Makino, S., Hashimoto, K.,** and **Gold, P. W.** (2002). Multiple feedback mechanisms activating corticotropin-releasing hormone system in the brain during stress. *Pharmacology, Biochemistry, and Behavior* **73**, pp. 147–158.

855. **Makino, S., Schulkin, J., Smith, M. A., Pacak, K., Palkorits, M.,** and **Gold, P. W.** (1995). Regulation of corticotropin-releasing hormone receptor messenger ribonucleic acid in the rat brain and pituitary by glucocorticoids and stress. *Endocrinology* **136**, pp. 4517–4525.

856. **Makino, S., Shibasaki, T., Yamauchi, N., Nishioka, T., Mimoto, T., Wakabayashi, I., Gold, P. W.,** and **Hashimoto, K.** (1999). Psychological stress increased corticotropin-releasing hormone mRNA and content in the central nucleus of the amygdala but not in the hypothalamic paraventricular nucleus in the rat. *Brain Research* **850**, pp. 136–143.

857. **Makrigiannakis, A., Zoumakis, E., Kalantaridou, S.,** and **Chrousos, G.** (2004). Endometrial and placental CRH as regulators of human embryo implantation. *Journal of Reproductive Immunology* **62**, pp. 53–59.

858. **Malagoli, D., Di Paolo, I.,** and **Ottaviani, E.** (2007). Presence of and stress-related changes in urocortin-like molecules in neurons and immune cells from the mussel Mytilus galloprovincialis. *Peptides* **28**, pp. 1545–1552.

859. **Malisch, J. L., Breuner, C. W., Kolb, E. M., Wada, H., Hannon, R. M., Chappell, M.A., Middleton, K. M.,** and **Garland T. Jr.** (2009). Behavioral despair and home-cage activity in mice with chronically elevated baseline corticosterone concentrations. *Behavior Genetics* **39**, pp. 192–201.

860. **Malkoski, S. P., Handanos, C. M.,** & and **Dorin, R. I.** (1997). Localization of a negative glucocorticoid response element of the human corticotropin releasing hormone gene. *Molecular and Cellular Endocrinology* **127**, pp. 189–199.

861. **Mansbach, R., Brooks, E. N.,** and **Chen, Y. L.** (1997). Antidepressant like effects of CP-154,526, a selective CRF1 receptor antagonist. *European Journal of Pharmacology* **323**, pp. 21–26.

862. **Mantsch, J. R., Taves, S., Khan, T., Katz, E. S., Sajan, T., Tang, L. C., Cullinan, W. E.,** and **Ziegler, D. R.** (2007). Restraint-induced corticosterone secretion and hypothalamic CRH mRNA expression are augmented during acute withdrawal from chronic cocaine administration. *Neurosciencel Letters* **415**, pp. 268–273.

863. **Manzon, L. A.** (2002). The role of prolactin in fish osmoregulation: a review. *General and Comparative Endocrinology* **125**, pp. 291–310.

864. **Marchal, C., Schramm, F., Kern, A., Luft, B., Yang, X., Schujit, T., Hovius, J., Jaulhac, B.,** and **Boulanger, N.** (2011). Antilarmin effect of tick saliva during the transmission of lyme disease. *Infection and Immunity* **79**, pp. 774–4785.

865. **Marchalant, Y., Rosi, S.,** and **Wenk, G.** (2007). Anti-inflammatory property of the cannabinoid agonist WIN-55212-2 in a rodent model of chronic brain inflammation. *Neuroscience* **144**, pp. 1516–1522.

866. **Marcinkiewcz, C., Prado, M., Isaac, S., Marshall, A., Rylkova, D.,** and **Bruijnzeel, A.** (2009). Corticotropin-releasing factor within the central nucleus of the amygdala and the nucleus accumbens shell mediates the negative affective state of nicotine withdrawal in rats. *Neuropsychopharmacology* **34**, pp. 1743–1752.

867. **Marcinkiewcz, C. A., Mazzone, C. M., D'Agostino, G., Halladay, L. R., Hardaway, J. A., DiBerto, J. F., Navarro, M., Burnham, N., Cristiano, C., Dorrier, C. E., Tipton, G. J., Ramakrishnan, C., Kozicz, T., Deisseroth, K., Thiele, T. E., McElligott, Z. A., Holmes, A., Heisler, L. K.,** and **Kash TL.** 2016 Serotonin engages an anxiety and fear-promoting circuit in the extended amygdala. *Nature* **537**(7618), pp. 97–101.

868. **Marinelli, M., Le Moal, M.,** and **Piazza, P. V.** (1996). Acute pharmacological blockade of corticosterone secretion reverses food restriction-induced sensitization of the locomotor response to cocaine. *Brain Research* **724**, pp. 251–255.

869. **Marinoni, E., Korebritz, C., Di Iorio, R., Cosmi, E. V.,** & and **Challis, J. R. G.** (1998). Influence of glucocorticoids on placental CRH. In: *Labor and Deliverydelivery: The the Proceedings proceedings of the 2nd World Congress on Labor and Delivery*. Rome, Italy: Taylor & and Francis, pp. 48–62.

870. **Marks, I.,** and **Tobena, A.** (1990) Learning and unlearning fear: A clinical and evolutionary perspective. *Neuroscience and Biobehavioral Reviews* **14**, pp. 365–384.

871. **Marsiglio, W., Amato, P., Day, R. D.,** and **Lamb, M. E.** (2000). Scholarship on fatherhood in the 1990s and beyond. *Journal of Marriage and Family* **62**, pp. 1173–1191.

872. **Marsiglio, W.,** and **Cohan, M.** (2000). Contextualizing father involvement and paternal influence: Sociological and qualitative themes. *Marriage & and Family Review* **29**, pp. 75–95.

873. **Marsiglio, W., Day, R. D.,** and **Lamb, M. E.** (2000). Exploring fatherhood diversity: Implications for conceptualizing father involvement. *Marriage and Family Review* **29**, pp. 269–293.

874. **Martínez-García, F., Martínez-Marcos, A.,** and **Lanuza, E.** (2002). The pallial amygdala of amniote vertebrates: evolution of the concept, evolution of the structure. *Brain Research Bulletin* **1**, pp. 463–469.

875. **Martinez, V., Wang, L., Rivier, J., Vale, E. W.,** and **Taché, Y.** (2002). Differential actions of peripheral corticotropin-releasing factor (CRF), urocortin II, and urocortin III on gastric emptying and colonic transit in mice: role of CRF receptor suptypes 1 and 2. *Journal of Pharmacology and Experimental Therapeutics* **301**, pp. 611–617.

876. **Martin-Fardon, R., Zorrilla, E., Ciccocioppo, R.,** and **Weiss, F.** (2010). Role of innate and drug-induced dysregulation of brain stress and arousal systems in addiction: focus on corticotropin releasing factor, nociception/orphanin FQ, and orexin/hypocretin. *Brain Research* **1314**, pp. 145–161.

877. **Martins, A. P., Marras, R. A.,** and **Guirnarães, F. S.** (1997a). Anxiogenic effect of corticotropin-releasing hormone in the dorsal periaqueductal grey. *Neuroreport* **8**, pp. 3601–3604.

878. **Martins, J. M., Banks, W. A.,** and **Kastin, A. J.** (1997b). Transport of CRH from mouse brain directly affects peripheral production of β-endorphin by the spleen. *American Journal of Physiology* **273**, pp. E1083–E1089.

879. **Masini, C. V., Sauer, S.,** and **Campeau, S.** (2005). Ferret odor as a processive stress model in rats: neurochemical, behavioral, and endocrine evidence. *Behavioral Neuroscience* **119**, pp. 280.

880. **Mastorakos, G., Scopa, C., Vryondiou, A., Friedman, T., Kattis, D., Phenekos, C., Merino, M.,** and **Chrousos, G.** (1994). Presence of ummunoreactiv corticotropin-releasing hormone in normal and polycystic ovaries. *Journal of Clinical Endocrinolology & Metabolism* **79**, pp. 934–939.

881. **Mastorakos, G., Webster, E. L., Friedman, T. C.,** and **Chrousos, G. P.** (1993). Immunoreactive corticotropin-releasing hormone (CRH) and its binding sites in the rat ovary. *Journal of Clinical Investigation* **92**, pp. 961–968.

882. **Matsuda, K., Kojima, K., Wada, K., Maruyama, K., Leprince, J., Tonon, M., Uchiyama, M.,** and **Vaudry, H.** (2010). The anorexigenic action of the octadecaneuropeptide (ODN) in goldfish is mediated through the MC4R- and subsequently the CRH receptor-signaling pathways. *Journal of Molecular Neuroscience* **42**, pp. 74–79.

883. **Matsuda, R.** (2013). *Morphology and evolution of the insect abdomen: with special reference to developmental patterns and their bearings upon systematics.* New York: Elsevier.

884. **Matthews, S.,** and **Owen, D.** (2006). Prenatal glucocorticoid exposure alters hypothalamic-pituitary-adrenal function in juvenile guinea pigs. *Journal of Neuroendocrinology* **19**, pp. 172–180.

885. **Matys, T., Pawlak, R., Matys, E., Pavlides, C., McEwen, B. S.,** and **Strickland, S.** (2004). Tissue plasminogen activator promotes the effects of corticotropin-releasing factor on the amygdala and anxiety-like behavior. *Proceedings of the National Academy of Sciences of the United States of America* **101**, pp. 16345–16350.

886. **Matz, S. P.,** and **Hofeldt, G. T.** (1999). Immunohistochemical localization of corticotropin-releasing factor in the brain and corticotropin-releasing factor and thyrotropin-stimulating hormone in the pituitary of chinook salmon (Oncorhynchus tshawytscha). *General and Comparative Endocrinology* **114**, pp. 151–160.

887. **McCall, J. G., Al-Hasani, R., Siuda, E. R., Hong, D. Y., Norris, A. J., Ford, C. P.,** and **Bruchas, M. R.** (2015). CRH engagement of the locus coeruleus noradrenergic system mediates stress-induced anxiety. *Neuron* **87**, pp. 605–620.

888. **McCarthy, M. M.** (2008). Estradiol and the developing brain. *Physiology Reviews* **88**, pp. 91–134.

889. **McCarthy, M. M., Curran, G. H.,** and **Siegel, H. I.** (1994). Evidence for the involvement of prolactin in the maternal behavior of the hamster. *Physiology and Behavior* **55**, pp. 181–184.

890. **McCarthy, M. M., Auger, A. P., Bale, T. L., De Vries, G. J., Dunn, G. A., Forger, N. G., Murray, E. K., Nugen, B. M., Schwarz, J. M.,** and **Wilson, M. C.** (2009). The

epigenetics of sex differences in the brain. *Journal of Neuroscience* **29**, pp. 12815–12823.

891. **McDonald, T., Le, W.,** and **Hoffman, G.** (2000). Brainstem catecholaminergic neurons activated by hypoxemia express GR and are coordinately activated with fetal sheep hypothalamic paraventricular CRH neurons. *Brain Research* **885**, pp. 70–78.
892. **McEuen, J. G., Semsar, K. A., Lim, M. A.,** & and **Bale, T. L.** (2009). Influence of sex and corticotropin-releasing factor pathways as determinants in serotonin sensitivity. *Endocrinology* **150**, pp. **3709–3716**.
893. **McEwen, B. S.** (1995). Steroid actions on neuronal signalling. *Ernst Schering Research Foundation Lecture Series* **27**, pp. 1–45.
894. **McEwen, B. S.** (1998). Protective and damaging effects of stress mediators. *New England Journal of Medicine* **338**, pp. 171–179.
895. **McEwen, B. S.** (2005). Glucocorticoids, depression, and mood disorders: structural remodeling in the brain. *Metabolism* **54**, pp. 20–23.
896. **McEwen, B. S.** (2009). Introduction: the end of sex as we once knew it. *Physiol Behav* **97**(2), pp. 143–145.
897. **McEwen, B. S.** (2007). Physiology and neurobiology of stress and adaptation: central role of the brain. *Physiology review Review* **87**, pp. 873–904.
898. **McEwen, B. S.,** and **Stellar, E.** (1993). Stress and the individual: mechanisms leading to disease. *Archives of Internal Medicine* **153**, pp. 2093–3101.
899. **McEwen, B. S.,** and **Wingfield, J. C.** (2003). The concept of allostasis in biology and medicine. *Hormones and Behavior* **43**, pp. 2–15.
900. **McEwen, B., Bowles, N., Gray, J., Hill, M., Hunter, R., Karatsoeros, I.,** and **Nasca, C.** (2015). Mechanisms of stress in the brain. *Nature Neuroscience* **18**, pp. 1353–1362.
901. **McEwen, B., Gray, J.,** and **Nasca, J.** (2014). Recognizing resilience: learning from the effects of stress on the brain. *Neurobiology of Stress* **1**, pp. 1–11.
902. **McFadden, K., Griffen, T. A., Levy, V., Wolfe, J. H.,** and **Valentino, R. J.** (2012). Overexpression of corticotropin-releasing factor in Barrington's nucleus neurons by adeno-associated viral transduction: effects on bladder function and behavior. *European Journal of Neuroscience* **36**, pp. 3356–3364.
903. **McGaugh, J. L.** (2000). Memory—a century of consolidation. *Science* **287**, pp. 248–251.
904. **McGaugh, J. L.** (2003). *Memory and emotion*. New York: Columbia University Pres.
905. **McGill, B. E., Bundle, S. F., Yaylaoglu, M. B., Carson, J. P., Thaller, C.,** and **Zoghbi, H. Y.** (2006). Enhanced anxiety and stress-induced corticosterone release are associated with increased Crh expression in a mouse model of Rett syndrome. *Proceedings of the National Academy of Sciences of the United States of America* **103**, pp. 18267–18272.
906. **McGinnis, J. M.,** and **Foege, W. H.** (1999). Mortality and morbidity attributable to use of addictive substances in the United States. *Proceedings of the Association of American Physicians* **111**, pp. 109–118.
907. **McInturf, S.,** and **Hennessy, M.** (1995). Peripheral administration of a corticotropin-releasing factor antagonist increases the vocalizing and locomotor activity of isolated guinea pig pups. *Physiology and Behavior* **60**, pp. 707–710.
908. **McKeown, K. J.,** and **Challis, J. R. G.** (2003). Regulation of 15-hydroxy prostaglandin dehydrogenase by corticotrophin-releasing hormone through a calcium-dependent

pathway in human chorion trophoblast cells. *Journal of Clinical Endocrinology & Metabolism* **88**, pp. 1737–1741.

909. **McLaughlin, R.**, **Hill, M.**, and **Gorzalka, B.** (2014). A critical role for prefrontocortical endocannabinoid signaling in the regulation of stress and emotional behavior. *Neuroscience and Biobehavioral Reviews* **42**, pp. 116–131.
910. **McLean, M.**, **Bists, A.**, **Davies, J.**, **Woods, R.**, **Lowry, P.**, and **Smith, R.** (1995). A placental clock controlling the length of human pregnancy. *Nature Medicine* **1**, pp. 460–463.
911. **McLean, S. A.**, **Williams, D. A.**, **Stein, P. K.**, **Harris, R. E.**, **Lyden, A. K.**, **Whalen, G.**, **Park, K. M.**, **Liberzon, I.**, **Sen, A.**, **Gracely, R. H.**, **Baraniuk, J. N.**, and **Clauw, D. J.** (2006). Cerebrospinal fluid corticotropin-releasing factor concentration is associated with pain but not fatigue symptoms in patients with fibromyalgia. *Neuropsychopharmacology* **31**, pp. 2776–2782.
912. **McNally, G. P.**, and **Akil, H.** (2002). Role of corticotropin-releasing hormone in the amygdala and bed nucleus of the stria terminalis in the behavioral, pain modulatory, and endocrine consequences of opiate withdrawal. *Neuroscience* **112**, pp. 605–617.
913. **McReynolds, J.**, **Pena, D.**, **Blacktop, J.**, and **Mantsch, J.** (2014). Neurobiological mechanisms underlying relapse to cocaine use: contributions of CRF and noradrenergic systems and regulation by glucocorticoids. *Stress* **17**, pp. 22–38.
914. **Meaney, M. J.**, **Aitken, D. H.**, **Van Berkel, C.**, **Bhatnagar, S.**, and **Sapolsky, R. M.** (1988). Effect of neonatal handling on age-related impairments associated with the hippocampus. *Science* **239**, pp. 766–768.
915. **Meaney, M. J.**, **Aitken, D. H.**, **Viau, V.**, **Sharma, S. H. A. K. T. I.**, and **Sarrieau, A.** (1989). Neonatal handling alters adrenocortical negative feedback sensitivity and hippocampal type II glucocorticoid receptor binding in the rat. *Neuroendocrinology* **50**, pp. 597–604.
916. **Meaney, M. J.**, **Diorio, J.**, **Francis, D.**, **Widdowson, J.**, **LaPlante, P.**, **Caldji, C.**, **Sharma, S.**, **Seckl, J. R.** and **Plotsky, P. M.** (1996). Early environmental regulation of forebrain glucocorticoid receptor gene expression: implications for adrenocortical responses to stress. *Developmental Neuroscience* **18**, pp. 61–72.
917. **Meaney, M. J.** (2001). Maternal care, gene expression, and the transmission of individual differences in stress reactivity across generations. *Annual Review of Neuroscience* **24**, pp. 1161–1192.
918. **Mecenas, C. A.**, **Giussani, D. A.**, **Owiny, J. R.**, **Jenkins, S. L.**, **Wu, W. X.**, **Honnebier, B. O. M.**, **Lockwood, C. J.**, **Kong, L.**, **Guller, S.**, and **Nathanielsz, P. W.** (1996). Production of premature delivery in pregnant rhesus monkeys by androstenedione infusion. *Nature Medicine* **2**, pp. 443–448.
919. **Meis, P.**, **Klebanoff, M.**, **Thom, E.**, **Dombroski, M.**, **Sibai, B.**, **Moawad, A.**, **Spong, C. Y.**, **Hauth, J. C.**, **Miodovnik, M.**, **Varner, M. W.**, **Leveno, K. J.**, **Caritis, S.N.**, **Iams, J. D.**, **Wapner, R. J.**, **Conway, D.**, **O'Sullivan, M. J.**, **Carpenter, M.**, **Mercer, B.**, **Ramin, S. M.**, **Thorp, J. M.**, **Peaceman, A. M.**, **Gabbe, S.N**, and **the National Institute of Child Health and Human Development Maternal-Fetal Medicine Units Network.** (2003). Prevention of recurrent preterm delivery by 17 alpha-hydroxyprogesterone caproate. *New England Journal of Medicine* **348**, pp. 2379–2385.
920. **Meltzer-Brody, S.** (2011). New insights into perinatal depression: pathogenesis and treatment during pregnancy and postpartum. *Dialogues in Clinical Neuroscience* **13**, pp. 89–100.

921. **Meltzer-Brody, S., Stuebe, A., Dole, N., Savitz, D., Rubinow, D.,** and **Thorp, J.** (2011). Elevated corticotropin releasing hormone (CRH) during pregnancy and risk of postpartum depression (PPD). *Journal of Clinical Endocrinology & Metabolism* **96**, pp. E40–E47.

922. **Meng, L., Lu, Z., Xiaoteng, W., Yue, H., Bin, L., Lina, M.,** and **Zhe, C.** (2015). Corticotropin-releasing factor changes the phenotype and function of dendritic cells in mouse mesenteric lymph nodes. *Journal of Neurogastroenterology and Motility* **21**, pp. 571–580.

923. **Menon, R., Arora, C., Hobel, C.,** and **Fortunato, S.** (2008). Corticotropin-releasing hormone in lipopolysaccharide-stimulated term fetal membranes and amniotic fluid from term and preterm birth in African Americans and Caucasians. *Reproductive Sciences* **15**, pp. 477–483.

924. **Menzaghi, F., Howard, R. L., Heinrichs, S. C., Vale, W., Rivier, J.,** and **Koob, G. F.** (1994). Characterization of a novel and potent corticotropin releasing factor antagonist in rats. *Journal of Pharmacology and Experimental Therapeutics* **269**, pp. 564–572.

925. **Merali, Z., Anisman, H., James, J. S., Kent, P.,** and **Schulkin, J.** (2008). Effects of corticosterone on corticotropin-releasing hormone and gastrin-releasing peptide release in response to an aversive stimulus in two regions of the forebrain (central nucleus of the amygdala and prefrontal cortex). *European Journal of Neuroscience* **28**, pp. 165–172.

926. **Merali, Z., Khan, S., Michaud, D. S., Shippy, D. A.,** and **Anisman, H.** (2004). Does amygdaloid corticotropin-releasing hormone (CRH) mediate anxiety-like behaviors? Dissociation of anxiogenic effects and CRH release. *European Journal of Neuroscience* **20**, pp. 229–239.

927. **Merali, Z., McIntosh, J., Kent, P., Michaud, D.,** and **Anisman, H.** (1998). Aversive and appetitive events evoke the release of corticotropin-releasing hormone and bombesin-like peptides at the central nucleus of the amygdala. *Journal of Neuroscience* **18**, pp. 4758–4766.

928. **Merali, Z., Michaud, D., McIntosh, J., Kent, P.,** and **Anisman, H.** (2003). Differential involvement of amygdaloid CRH system(s) in the salience and valence of stimuli. *Progress in Neuro-Psychopharmacology and Biological Psychiatry* **27**, pp. 1201–1212.

929. **Merali, Z., McIntosh, J.,** and **Anisman, H.** (2004b). Anticipatory cues differentially provoke in vivo peptidergic and monoaminergic release at the medial prefrontal cortex. *Neuropsychopharmacology* **29**, pp. 1409–1418.

930. **Merchenthaler, I., Lane, M. V., Numan, S.,** and **Dellovade, T. L.** (2004). Distribution of estrogen receptor α and β in the mouse central nervous system: in vivo autoradiographic and immunocytochemical analyses. *Journal of Comparative Neurology* **473**, pp. 270–291.

931. **Merlino, A., Welsh, T., Erdonmez, T., Madsen, G., Zakar, T., Smith, R., Mercer, B.,** and **Mesiano, S.** (2009). Nuclear progesterone receptor expression in the human fetal membranes and decidua at term before and after labor. *Reproductive Sciences* **16**, pp. 357–363.

932. **Merlo-Pich, E. M., Lorang, M., Yeganeh, M., Rodriguez de Fonseca, F., Raber, J., Koob, G. F.,** and **Weiss, F.** (1995). Increase of extracellular corticotropin-releasing factor-like immunoreactivity levels in the amygdala of awake rats during restraint stress and ethanol withdrawal as measured by microdialysis. *Journal of Neuroscience* **15**, pp. 5439–5447.

933. **Mesiano, S., Chan, E., Fitter, J., Kwek, K., Yeo, G.,** and **Smith, R.** (2002). Progesterone withdrawal and estrogen activation in human parturition are coordinated by progesterone receptor a expression in the myometrium. *Journal of Clinical Endocrinology & Metabolism* **87**, pp. 2924–2930.

934. **Mesiano, S., Wang, Y.,** and **Norwitz, E.** (2011). Progesterone receptors in the human pregnancy: do they hold the key to birth timing? *Reproductive Sciences* **18**, pp. 6–19.

935. **Michel, C., Dunn-Meynell, A.,** and **Levin, B.** (2004). Reduced brain CRH and GR mRNA expression preceds obesity in juvenile rats bred for diet-induced obesity. *Behavioural Brain Research* **154**, pp. 511–517.

936. **Michelson, D., Stratakis, C., Hill, L., Reynolds, J., Galliven, E., Chrousos, G.,** and **Gold, P.** (1996). Bone mineral density in women with depression. *New England Journal of Medicine* **335**, pp. 1176–1181.

937. **Micioni Di Bonaventura, M.V., Ciccocioppo, R., Romano, A., Bossert, J.M., Rice, K.C, Ubaldi, M., St Laurent, R., Gaetani, S., Massi, M., Shaham, Y., Cifani, C.** (2014). Role of bed nucleus of the stria terminalis corticotrophin-releasing factor receptors in frustration stress-induced binge-like palatable food consumption in female rats with a history of food restriction. *Journal of Neuroscience* 34(34), pp. 11316–11324.

938. **Miguel, T.,** and **Nunes-de-Souza, R.** (2011). Anxiogenic and intinociceptive effects induced by corticotropin-releasing factor (CRF) injections into the periaqueductal gray are modulated by CRF1 receptor in mice. *Hormones and Behavior* **60**, pp. 292–300.

939. **Millan, M. A., Jacobowitz, D. M., Hauger, R. L., Catt, K. J.,** and **Aguilera, G.** (1986). Distribution of corticotropin-releasing factor receptors in primate brain. *Proceedings of the National Academy of Sciences* **83**, pp. 1921–1925.

940. **Millan, M.A, Samra, A.B.A., Wynn, P.C., Catt, K.J.,** and **Aguilera, G.** (1987). Receptors and actions of corticotropin-releasing hormone in the primate pituitary gland. *Journal of Clinical Endocrinology & Metabolism* **64**, pp. 1036–1041.

941. **Miller, W.J., Suzuki, S., Miller, L.K., Handa, R.,** and **Uht, R. M.** (2004). Estrogen receptor (ER) beta isoforms rather than ER alpha regulate corticotropin-releasing hormone promoter activity through an alternate pathway. *Journal of Neuroscience* **24**, pp. 10628–10635.

942. **Missig, G., Ayers, L., Schulkin, J.,** and **Rosen, J.** (2010). Oxytocin reduces background anxiety in a fear-potentiated startle paradigm. *Neuropsychopharmacology* **35**, pp. 2607–2616.

943. **Mitra, R., Jadhav, S., McEwen, B. S., Vyas, A.,** and **Chattarji, S.** (2005). Stress duration modulates the spatiotemporal patterns of spine formation in the basolateral amygdala. *Proceedings of the National Academy of Sciences of the United States of America* **102**, pp. 9371–9376.

944. **Mitra, R.,** and **Sapolsky, R. M.** (2008). Acute corticosterone treatment is sufficient to induce anxiety and amygdaloid dendritic hypertrophy. *Proceedings of the National Academy of Sciences of the United States of America* **105**, pp. 5573–5578.

945. **Moffett, M.,** and **Goeders, N.** (2007). CP-154, 526, a CRF type-1 receptor antagonist, attenuates the cue- and methamphetamine-induced reinstatement of extinguished methamphetamine-seeking behavior in rats. *Psychopharmacology* **190**, pp. 171–180.

946. **Mogenson, G. J.**, and **Yim, C. Y.** (1981). Electrophysiological and neuropharmacological-behavioral studies of the nucleus accumbens: implications for its role as a limbic-motor interface. *Neurobiology of the Nucleus Accumbens* pp. 210–229.

947. **Mogenson, G. J., Jones, D. L.**, and **Yim, C. Y.** (1980). From motivation to action: functional interface between the limbic system and the motor system. *Progress in Neurobiology* **14**, pp. 69–97.

948. **Molander, A., Vengeliene, V., Heilig, M., Wurst, W., Deussing, J.**, and **Spanagel, R.** (2012). Brain-specific inactivation of the Crhr1 gene inhibits post-dependent and stress-induced alcohol intake, but does not affect relapse-like drinking. *Neuropsychopharmacology* **37**, pp. 1047–1056.

949. **Montecucchi, P. C.**, and **Henschen, A.** (1981). Amino acid composition and sequence analysis of sauvagine, a new active peptide from the skin of Phyllomedusa sawagei. *International Journal of Peptide and Protein Research* **18**, pp. 113–120.

950. **Moog, N. K., Buss, C., Entringer, S., Shahbaba, B., Gillen, D. L., Hobel, C. J.**, and **Wadhwa, P. D.** (2016). Maternal exposure to childhood trauma is associated during pregnancy with placental-fetal stress physiology. *Biological Psychiatry* **79**, pp. 831–839.

951. **Moore, C., Johnson, Z., Higgins, M., Toufexis, D.**, and **Wilson, M.** (2015). Antagonism of CRF Type 1 receptors attenuates caloric intake of free feeding subordinate female rhesus moneksy in a rich dietary environment. *Journal of Neuroendocrinology* **27**, pp. 33–43.

952. **Moore, F. L.**, and **Rose, J. D.** (2002). Sensorimotor processing model. In: Pfaff, D.W. et al. (eds.), *Hormones, brain and behavior*. New York: Academic Press.

953. **Moore-Ede, M. C., Sulzman, F. M.**, and **Fuller, C. A.** (1992). *The clocks that time us*. Cambridge, MA: Harvard University Press.

954. **Morales-Prieto, D., Ospina-Prieto, S., Chaiwangyen, W., Schoenleben, M.**, and **Markert, U.** (2013). Pregnancy-associated miRNA-clusters. *Journal of Reproductive Immunology* **97**, pp. 51–61.

955. **Morgan, M., Schulkin, J.**, and **Pfaff, D.** (2004). Estrogens and non-preductive behaviors related to activity and fear. *Neuroscience and Biobehavioral Reviews* pp. 55–63.

956. **Morgan, M. A., Schulkin, J.**, and **LeDoux, J. E.** (2003). Ventral medial prefrontal cortex and emotional perseveration: the memory for prior extinction training. *Behavioural Brain Research* **146**, pp. 121–130.

957. **Morgan, M. A.**, and **LeDoux, J. E.** (1995). Differential contribution of dorsal and ventral medial prefrontal cortex to the acquisition and extinction of conditioned fear. *Behavioral Neurology* **109**, pp. 681–688.

958. **Morgan, M. A.**, and **Pfaff, D. W.** (2001). Effects of estrogen on activity and fear-related behaviors in mice. *Hormones and Behavior* **40**, pp. 472–482.

959. **Morgan, M. A., Romanski, L. M.**, and **LeDoux, J. E.** (1993). Extinction of emotional learning: contribution of medial prefrontal cortex. *Neuroscience Letters* **163**, pp. 109–113.

960. **Moriceau, S., Roth, T. L., Okotoghaide, T.**, and **Sullivan, R. M.** (2004). Corticosterone controls and the developmental emergence of fear and amygdala function to predator odors in infant rat pups. *International Journal of Developmental Neuroscience* **22**, pp. 415–422.

961. **Moriceau, S., Shionoya, K., Jakubs, K.,** and **Sullivan, R. M.** (2009). Early-life stress disrupts attachment learning: the role of amygdala corticosterone, locus ceruleus corticotropin releasing hormone, and olfactory bulb norepinephrine. *Journal of Neuroscience* **29**, pp. 15745–15755.

962. **Morimoto, M., Morita, N., Ozawa, H., Yokoyama, K.,** and **Kawata, M.** (1996). Distribution of glucocorticoid receptor immunoreactivity and mRNA in the rat brain: an immunohistochemical and in situ hybridization study. *Neuroscience Research* **26**, pp. 235–269.

963. **Morisot, N.,** and **Contarino, A.** (2016). The CRF 1 and the CRF 2 receptor mediate recognition memory deficits and vulnerability induced by opiate withdrawal. *Neuropharmacology* **105**, pp. 500–507.

964. **Morrow, B. A., Elsworth, J. D., Rasmusson, A. M.,** and **Roth, R. H.** (1999). The role of mesoprefrontal dopamine neurons in the acquisition and expression of conditioned fear in the rat. *Neuroscience* **92**, pp. 553–564.

965. **Moussatche, P.,** and **Lyons, T.** (2012). Nongenomic progesterone signaling and its non canonical receptor. *Biochemical Society Transactions* **40**, pp. 200–204.

966. **Mueller, B.R.,** and **Bale, T. L.**(2008). Sex-specific programming of offspring emotionality after stress early in pregnancy. *Journal of Neuroscience* **28**, pp. 9055–9065.

967. **Muglia, L. J.** (2000). Genetic analysis of fetal development and parturition control in the mouse. *Pediatric Research* **47**, pp. 437–443.

968. **Muglia, L. J., Jacobson, L., Weninger, S. C., Karalis, K. P., Jeong, K. H.,** and **Majzoub, J. A.** (2001). The physiology of corticotropin-releasing hormone deficiency in mice. *Peptides* **22**, pp. 725–731.

969. **Muglia, L. J., Jacobson, L., Dikkes, P.,** and **Majzoub, J. A.** (1995). Corticotropin-releasing hormone deficiency reveals major fetal but not adult glucocorticoid need. *Nature* **373**, pp. 427–432.

970. **Muller, M., Keck, M., Zimmerman, S., Holsboer, F., Worst, W.** (2000). Disruption of feeding behavior in CRH receptor 1-deficient mice is dependent on glucocorticoids. *Neuroreport* **11**, pp. 1963–1966.

971. **Müller, M. B., Zimmermann, S., Sillaber, I., Hagemeyer, T. P., Deussing, J. M., Timpl, P., Kormann, M. S., Droste, S. K., Kuhn, R., Reul, J. M., Holsboer, F.,** and **Wurst, W.** (2003). Limbic corticotropin-releasing hormone receptor 1 medicates anxiety-related behavior and hormonal adaptation to stress. *Nature Neuroscience* **6**, pp. 1100–1107.

972. **Munck, A., Guyre, P. M.,** and **Holbrook, N. J.** (1984) Physiological regulation of glucocorticoids in stress and their regulation to pharmacological actions. *Endocrine Reviews* **5**, pp. 25–44.

973. **Muramatsu, Y., Sugino, N., Suzuki, T., Totsune, K., Takahashi, K., Tashiro, A., Hongo, M., Oki, Y.,** and **Sasano, H.** (2001). Urocortin and corticotropin-releasing factor receptor expression in normal cycling human ovaries. *Journal of Clinical Endocrinology & Metabolism* **86**, pp. 1362–1369.

974. **Muráni, E., Murániová, M., Ponsuksili, S., Schellander, K.,** and **Wimmers, K.** (2006). Molecular characterization and evidencing of the porcine CRH gene as a functional-positional candidate for growth and body composition. *Biochemical and Biophysical Research Communications* **342**, pp. 394–405.

975. **Murphy, W. J., Pringle, T. H., Crider, T. A., Springer, M. S.,** and **Miller, W.** (2007). Using genomic data to unravel the root of the placental mammal phylogeny. *Genome Research* **17**, pp. 413–421.

976. **Murray, L., Halligan, S., Goodyer, I.,** and **Herbert, J.** (2010). Disturbances in early parenting of depressed mothers and cortisol secretion in offspring: a preliminary study. *Journal of Affective Disorders* **1122;**, pp. 218–223.

977. **Myers, D., Gibson, M., Schulkin, J.,** and **Van Meerveld, B.** (2005). Corticosterone implants to the amygdala and type 1 CRH receptor regulation: effects on behavior and colonic sensitivity. *Behavioral Brain Research* **161**, pp. 39–44.

978. **Myers, B.,** and **Greenwood-Van Meerveld, B.** (2007). Corticosteroid receptor-mediated mechanisms in the amygdala regulate anxiety and colonic sensitivity. *AM J Physiol Gastrointest Liver Physiol* **292**(6), pp. G1622–1629.

979. **Myers, B.,** and **Greenwood-Van Meerveld, B.** (2010). Elevated corticosterone in the amygdala leads to persistent increases in anxiety-like behavior and pain sensitivity. *Behav Brain Res* **214**(2), pp. 465–469.

980. **Myers, B., Schulkin, J.,** and **Greenwood-Van Meerveld, B.** (2011). Sex steroid localized to the amygdala increase pain responses to visceral stimulation in rats. *Journal of Pain* **12**(4), pp. 486–494.

981. **Nagy, F. Z.,** and **Paré, D.** (2008). Timing of impulses from the central amygdala and bed nucleus of the stria terminalis to the brain stem. *Journal of Neurophysiology* **100**, pp. 3429–3436.

982. **Namburi, P., Beyeler, A., Yorozu, S., Calhoon, G., Halbert, S., Wichmann, R., Holden, S., Mertens, K., Anahator, M., Ortiz, A., Wickersham, I., Gray, J.,** and **Tye, K.** (2015). A circuit mechanism for differentiating positive and negative associations. *Nature Letters* **520**, pp. 675.

983. **Narita, K., Takei, Y., Suda, M., Aoyama, Y., Uehara, T., Kosaka, H., Amanuma, M., Fukuda, M.,** and **Mikuni, M.** (2010). Relationship of parental bonding styles with gray matter volume of dorsolateral prefrontal cortex in young adults. *Progress in Neuro-Psychopharmacology and Biological Psychiatry* **34**, pp. 624–631.

984. **Narla, C., Scidmore, T., Jeong, J., Everest, M., Chidiac, P.,** and **Poulter, M.** (2016). A switch in G protein coupling for type 1 corticotropin-releasing factor receptors promotes excitability in epileptic brains. *Neuroscience* **9**, pp. 1–9.

985. **Narla, C., Dunn, S., Ferguson, M.,** and **Poulter, M.** Suppression of piriform cortex activity in rat by corticotropin-releasing factor 1 and serotonin 2A/C receptors. *Front Cell Neuroscience* **9**, pp. 200.

986. **Nauta, W. J.** (1972). Neural associations of the frontal cortex. *Acta Neurobiologiae Experimentalis* **32**, pp. 125–140.

987. **Navarro-Zaragoza, J., Martinez-Laorden, E. Mora, L., Hidalgo, J. Milanės, M., Laorden, M. L.** (2014). Cardiac adverse effects of naloxone-precipitated morphine withdrawal on right ventricle: role of corticotropin-releasing factor (CRF) 1 receptor. *Toxicology and Applied Pharmacology* **275**, pp. 28–35.

988. **Nawata, Y., Kitaichi, K.,** and **Yamamoto, T.** (2012). Increases of CRF in the amygdala are responsible for reinstatement of methamphetamine-seeking behavior induced by footshock. *Pharmacology, Biochemistry, and Behvaior* **101**, pp. 297–302.

989. **Neckameyer, W. S.,** and **Leal, S. M.** (2002). Biogenic amines as circulating hormones in insects. In D.W. Pfaff, et al. (eds.), *Hormones, brain and behavior*. New York: Academic Press. 17–32.

990. **Negus, S.,** and **Rice, K.** (2009). Mechanisms of withdrawal associated increases in heroin self-administration: pharmacologic modulation of heroin vs. food choice in heroin dependent rhesus monkeys. *Neuropsychopharmacology* **34**, pp. 899–911.

991. **Nejati, R.**, **Kovacic, D.**, and **Slominski, A.** (2014). Neuro-immune-endocrine functions of the skin: an overview. *Expert review Review of dermatology Dermatology* **8**, pp. 581–583.

992. **Nelson, E.**, and **Panksepp, J.** (1996). Oxytocin mediates acquisition of maternally associated odor preferences in preweanling rat pups. *Behavioral Neuroscience* **110**, pp. 583.

993. **Nelson, E.**, and **Panksepp, J.** (1998). Brain substrates of infant–mother attachment: contributions of opioids, oxytocin, and norepinephrine. *Neuroscience and Biobehavioral Reviews* **22**, pp. 437–452.

994. **Nelson, R. J.**, **Demas, G. E.**, **Klein, S. L.**, and **Kriegsfeld, L. J.** (2002). *Seasonal patterns of stress, immune function, and disease*. Cambridge: Cambridge University Press.

995. **Nemeroff, C. B.**, **Krishnan, K. R. R.**, **Reed, D.**, **Leder, R.**, **Beam, C.**, and **Dunnick, N. R.** (1992). Adrenal gland enlargement in major depression: a computed tomographic study. *Archives of General Psychiatry* **49**, pp. 384–387.

996. **Nemeroff, C.**, **Widerlov, E.**, **Bissette, G.**, **Walleus, H.**, **Karlsson, I.**, **Eklund, K.**, **Kilts, C.**, **Loosen, P.**, and **Vale, W.** (1984). Elevated concentrations of CSF corticotropin-releasing-factor-like immunoreactivity in depressed patients. *Science* **22**, pp. 1342–1344.

997. **Nemeroff, C. B.**, **Owens, M. J.** (2004). Pharmacologic differences among the SSRIs: focus on monamine transporters and the HPA axis. *CNS Spectrums* **9**, pp. 23–31.

998. **Neumann, I. D.**, and **Landgraf, R.** (2012). Balance of brain oxytocin and vasopressin: implications for anxiety, depression, and social behaviors. *Trends in Neurosciences* **35**, pp. 649–659.

999. **Neumann, I.**, **Wigger, A.**, **Liebsch, G.**, **Holsboer, F.**, **Landgraf, R.** (1998). Increased basal activity of the hypothalamo-pituitary-adrenal axis during pregnancy in rats bred for high anxiety-related behavior. *Psychoneuroendocrinology* **23**, pp. 449–463.

1000. **Newman, S. W.** (2002) Phermonal signals access the medial extended amygdala: one node in a proposed social behavior network. In D.W. Pfaff, et al. (eds.), *Hormones, brain and behavior*. New York: Academic Press. 17–32.

1001. **Ni, X.**, **Hou, Y.**, **King, B.**, **Tang, X.**, **Read, M.**, **Smith, R.**, and **Nicholson, R.** (2004). Estrogen receptor-mediated down-regulatio of corticotropin-releasing hormone gene expression is dependent on a cyclic adenosine 3',5'-monophosphate regulatory element in human placental syncytiotrophoblast cells. *Journal of Clinical Endocrinology & and Metabolism* **89**, pp. 2312–2318.

1002. **Nicholson, R. C.**, **King, B. R.**, and **Smith, R.** (2004). Complex regulatory interactions control CRH gene expression. *Frontiers in Bioscience* **9**, pp. 32–39.

1003. **Nielsen, D.** (2006). Corticotropin releasing factor type 1 receptor antagonists: the next class of antidepressants? *Life Sciences* **78**, pp. 909–919.

1004. **Nijsen, M. J.**, **Croiset, G.**, **Diamant, M.**, **De Wied, D.**, and **Wiegant, V. M.** (2001). CRH signalling in the bed nucleus of the stria terminalis is involved in stress-induced cardiac vagal activation in conscious rats. *Neuropsychopharmacology* **24**, pp. 1–10.

1005. **Nijsen, M. J.**, **Croiset, G.**, **Diamant, M.**, **Stam, R.**, **Kamphuis, P. J.**, **Bruijnzeel, A.**, **de Wied, D.** and **Wiegant, V. M.** (2000). Endogenous corticotropin-releasing hormone inhibits conditioned-fear-induced vagal activation in the rat. *European Journal of Pharmacology* **389**, pp. 89–98.

1006. **Nitabach, M. N.**, **Schulkin, J.**, and **Epstein, A. N.** (1989). The medial amygdala is part of a mineralocorticoid-sensitive circuit controlling NaCl intake in the rat. *Behavioural Brain Research* **35**, pp. 127–134.

1007. **Nold, C.**, **Maubert, M.**, **Anton, L.**, **Yellon, S.**, and **Elovitz, M.** (2013). Prevention of preterm birth by progestational agents: what are the molecular mechanisms. *AJOG* **208**, pp. 223.e1–e7.

1008. **Noll, J. G.**, **Schulkin, J.**, **Trickett, P. K.**, **Susman, E. J.**, **Breech, L.**, and **Putnam, F. W.** (2007). Differential pathways to preterm delivery for sexually abused and comparison women. *Journal of Pediatric Psychology* **32**, pp. 1238–1248.

1009. **Norwitz, E. R.**, and **Caughey, A. B.** (2011). Progesterone supplementation and the prevention of preterm birth. *Reviews in Obstetrics and Gynecology* **4**, pp. 60–72.

1010. **Novembri, R.**, **Carrarelli, P.**, **Toti, P.**, **Rocha, A.**, **Borges, L.**, **Reis, Piomboni, P.**, **Florio, P.**, and **Petaglia, F.** (2011). Urocortin 2 and urocortin 3 in endometriosis: evidence for a possible role in inflammatory response. *Molecular Human Reproduction* **17**, pp. 587–593.

1011. **Novembri, R.**, **De Clemente, C.**, **Funghi, L.**, **Torricelli, M.**, **Voltolini, C.**, **Challis, J. R.**, and **Petraglia, F.** (2015). Corticotropin releasing hormone and Urocortin 2 activate inflammatory pathways in cultured trophoblast cell lines. *European Journal of Obstetrics and Gynecology and Reproductive Biology* **195**, pp. 200–205.

1012. **O'Donnell, K.**, **Jensen, A.**, **Freeman, L.**, **Khalife, N.**, **O'Connor, T.**, and **Glover, V.** (2011). Maternal prenatal anxiety and downregulation of placental 11B-HSD2. *Psychoneuroendocrinology* **37**, pp. 818–826.

1013. **O'Donovan, A.**, **Chao, L.**, **Paulson, J.**, **Samuelson, K.**, **Shigenaag, J.**, **Grunfeld, C.**, **Weiner, M.**, and **Neylan, T.** (2015). Altered inflammatory activity associated with reduced hippocampal volume and more severe posttraumatic stress symptoms in Gulf War Veterans. *Psychoneuroendocrinology* **41**, pp. 557–566.

1014. **O'Kane, M.**, **Murphy, E.P.**, and **Kirby, B.** (2006). The role of corticotropin-releasing hormone in immunie-mediated cutaneous inflammatory disease. *Experimental Dermatology* **15**, pp. 143–153.

1015. **Oberlander, J.**, and **Henderson, L.** (2012a). Corticotropin-releasing factor modulation of forebrain GABAergic transmission has a pivotal role in the expression of anabolic steroid-induced anxiety in the female mouse. *Neuropsychopharmacology* **37**, pp. 1483–1499.

1016. **Oberlander, J. G.**, and **Henderson, L. P.** (2012b). The Sturm und Drang of anabolic steroid use: angst, anxiety, and aggression. *Trends in Neuroscience* **35**, pp. 382–392.

1017. **Ock, J.**, **Lee, H.**, **Kim, S.**, **Lee, W.**, **Choi, D. Park, E. Kim, S.**, **Kim, I.**, and **Suk, K.** (2006). Induction of microglial apoptosis by corticotropin-releasing hormone. *Journal of Neurochemistry* **98**, pp. 962–972.

1018. **Oftedal, O. T.** (2002). The mammary gland and its origin during synapsid evolution. Journal of Mammary *Gland Biology and Neoplasia* 7, pp. 225–252.

1019. **Ohmura, Y.**, **Izumi, T.**, **Yamaguchi, T.**, **Tsutsui-Kimura, I.**, **Yoshida, T.**, and **Yoshioka, M.** (2010). The serotonergic projection from the median raphe nucleus to the ventral hippocampus is involved in the retrieval of fear memory through the corticotropin-releasing factor type 2 receptor. *Neuropsychopharmacology* **35**, pp. 1271–1278.

1020. **Ohmura, Y.**, **Yamaguchi, T.**, **Futami, Y.**, **Togashi, H.**, **Izumi, T.**, **Matsumoto, M.**, **Yoshida, T.**, and **Yoshioka, M.** (2009). Corticotropin releasing factor enhances attentional function as assessed by the five-choice serial reaction time task in rats. *Behavioural Brain Research* **198**, pp. 429–433.

1021. **Ojasoo, T.**, **Dore, J. C.**, **Gilbert, J.**, and **Raynaud, J. P.** (1988). Binding of steroids to the progestin and glucocorticoid receptors analyzed by correspondence analysis. *Journal of Medicinal Chemistry* **31**, pp. 1160–1169.

1022. **Okada, R.**, **Miller, M. F.**, **Yamamoto, K.**, **De Groef, B.**, **Denver, R. J.**, and **Kikuyama, S.** (2007). Involvement of the corticotropin-releasing factor (CRF) type 2 receptor in CRF-induced thyrotropin release by the amphibian pituitary gland. *General and Comparative Endocrinology* **150**, pp. 437–444.

1023. **Okawara, Y.**, **Morley, S. D.**, **Burzio, L. O.**, **Zwiers, H.**, **Lederis, K.**, and **Richter, D.** (1988). Cloning and sequence analysis of cDNA for corticotropin-releasing factor precursor from the teleost fish Catostomus commersoni. *Proceedings of the National Academy of Sciences* **85**, pp. 8439–8443.

1024. **Okuda, S.**, **Roozendaal, B.**, and **McGaugh, J. L.** (2004). Glucocorticoid effects on object recognition memory require training-associated emotional arousal. *Proceedings of the National Academy of Sciences of the United States of America* **101**, pp. 853–858.

1025. **Olazabal, D. E.**, and **Young, L. J.** (2006). Oxytocin receptors in the nucleus accumbens facilitate "spontaneous" maternal behavior in adult female prairie voles. *Neuroscience* **141**, pp. 559–568.

1026. **Olschowka, J. A.**, **O'donohue, T. L.**, **Mueller, G. P.**, and **Jacobowitz, D. M.** (1982a). Hypothalamic and extrahypothalamic distribution of CRF-like immunoreactive neurons in the rat brain. *Neuroendocrinology* **35**, pp. 305–308.

1027. **Olschowka, J. A.**, **O'Donohue, T. L.**, **Mueller, G. P.**, and **Jacobowitz, D. M.** (1982b). The distribution of corticotropin releasing factor-like immunoreactive neurons in rat brain. *Peptides* **3**, pp. 995–1015.

1028. **Opp, M. R.** (1995). Corticotropin-releasing hormone involvement in stressor-induced alterations in sleep and in the regulation of waking. *Advances in Neuroimmunology* **5**, pp. 127–143.

1029. **Orchinik, M.**, **Murray, T. F.**, **Franklin, P. H.**, and **Moore, F. L.** (1992). Guanyl nucleotides modulate binding to steroid receptors in neuronal membranes. *Proceedings of the National Academy of Sciences* **89**, pp. 3830–3834.

1030. **Orchinik, M.**, **Murray, T.F.**, and **Moore, F.L.** (1994). Steroid modulation of GABAA receptors in an amphibian brain. *Brain Research* **646**, pp. 258–266.

1031. **Orozco-Cabal, L.**, **Pollandt, S.**, **Liu, J.**, **Shinnick-Gallagher, P.**, and **Gallagher, J. P.** (2006a). Regulation of synaptic transmission by CRF receptors. *Reviews in the Neurosciences* **17**, pp. 279–308.

1032. **Orozco-Cabal, L.**, **Pollandt, S.**, **Liu, J.**, **Vergara, L.**, **Shinnick-Gallagher, P.**, and **Gallagher, J. P.** (2006b). A novel rat medial prefrontal cortical slice preparation to investigate synaptic transmission from amygdala to layer V prelimbic pyramidal neurons. *Journal of Neuroscience Methods* **151**, pp. 148–158.

1033. **Ortlund, E. A.**, **Bridgham, J. T.**, **Redinbo, M. R.**, and **Thornton, J. W.** (2007). Crystal structure of an ancient protein. *Science* **317**, pp. 1544–1547.

1034. **Oswald, M.**, **Drew, R.**, **Murdoch, G.**, and **Robison, B.** (2012). Is behavioral variation along the bold-shy continuum associated with variation in the stress axis in zebrafish? *Physiological and Biochemical Zoology* **85**, pp. 718–728.

1035. **Otsuka, T.**, **Nishii, A.**, **Amemiya, S.**, **Kubota, N.**, **Nishijima, T.**, and **Kita, I.** (2016). Effects of acute treadmill running at different intensities on activities of serotonin and corticotropin-releasing factor neurons, and anxiety-and depressive-like behaviors in rats. *Behavioural Brain Research* **298**, pp. 44–51.

1036. **Ouimette, P. C.**, **Moos, R. H.**, and **Finney, J. W.** (1998). Influence of outpatient treatment and 12-step group involvement on one-year substance abuse treatment outcomes. *Journal of Studies on Alcohol* **59**, pp. 513–522.

1037. **Owen, D.**, and **Matthews, S.** (2006). Prenatal glucocorticoid exposure alters hypothalamic-pituitary-adrenal function in juvenile guinea pigs. *Journal of Neuroendocrinology* **19**, pp. 172–180.

1038. **Palkovits, M.**, **Young III, W. S.**, **Kovacks, K.**, **Toth, Z.**, and **Makara, G. B.** (1998). Alterations in corticotropin releasing hormone gene expression of central amygdala neurons following long-term paraventricular lesions and adrenalectomy. *Neuroscience* **85**, pp. 135–147.

1039. **Palombit, R. A.**, **Seyfarth, R. M.**, and **Cheney, D. L.** (1997). The adaptive value of "friendships" to female baboons: experimental and observational evidence. *Animal Behaviour* **54**, pp. 599–614.

1040. **Pañeda, C.**, **Huitron-Resendiz, S.**, **Frago, L. M.**, **Chowen, J. A.**, **Picetti, R.**, **De Lecea, L.**, and **Roberts, A. J.** (2009). Neuropeptide S reinstates cocaine-seeking behavior and increases locomotor activity through corticotropin releasing factor receptor 1 in mice. *Journal of Neuroscience* **29**, pp. 4155–4161.

1041. **Pang, S.** (1997). Congenital adrenal hyperplasia. *Endocrinology and Metabolism Clinics of North America* **26**, pp. 853–891.

1042. **Pankevich, D. E.**, **Teegarden, S. L.**, **Hedin, A. D.**, **Jensen, C. L.**, and **Bale, T. L.** (2010). Caloric restriction experience reprograms stress and orexegenic pathways and promotes binge eating. *Journal of Neuroscience* **30**, pp. 16399–16407.

1043. **Panossian, A.**, **Wikman, G.**, **Kaur, P.**, and **Asea, A.** (2012). Adaptogens stimulate neuropeptide Y and Hsp72 expression and release in neuroglia cells. *Frontiers in Neuroscience* **6**, p. 6.

1044. **Papaleo, F.**, **Kitchener, P.**, and **Contarino, A.** (2007). Disruption of the CRF/CRF1 receptor stress system exacerbates the somatic signs of opiate withdrawal. *Neuron* **53**, pp. 577–589.

1045. **Park, P. E.**, **Vendruscolo, L. F.**, **Schlosburg, J. E.**, **Edwards, S.**, **Schulteis, G.**, and **Koob, G. F.** (2013). Corticotropin-releasing factor (CRF) and α 2 adrenergic receptors mediate heroin withdrawal-potentiated startle in rats. *International Journal of Neuropsychopharmacology* **16**, pp. 1867–1875.

1046. **Parker, K. J.**, **Schatzberg, A. F.**, and **Lyons, D. M.** (2003). Neuroendocrine aspects of hypercortisolism in major depression. Hormones and Behavior **43**, pp. 60–66.

1047. **Partridge, J. G.**, **Vicini, S.**, **Forcelli, P. A**, **Luo, R.**, and **Cashdan, J.** (2016) Stress increases GABAergic neurotransmission in crf neurons of the central amygdala and bed nucleus stria terminalis. *Neuropharmacology* **107**, pp. 239–250.

1048. **Partridge, J. G.**, **Luo, R.**, **Forcelli, P. A.**, **Cashdan, J. M.**, **Schulkin, J.**, and **Vicni, S.** (2014). Synaptic connections of CRH GABAergic neurons in the central amygdala and bed nucleus stria terminalis. Program No.297.01 2014 Neuroscience Meeting Planner. Washington, DC: Society for Neuroscience. Online.

1049. **Paschos, K., Chouridou, E., Koureta, M., Lambropoulou, M., Kolios, G.,** and **Chatzaki, E.** (2013). The corticotropin releasing factor system in the liver: expression, actions and possible implications in hepatic physiology and pathology. *Hormones* **12**, pp. 236–245.

1050. **Pascual, J.,** and **Heinrichs, S.** (2007). Olfactory neophobia and seizure susceptibility phenotypes in an animal model of epilepsy are normalized by impairment of brain corticotropin-releasing factor. *Epilepsia* **48**, pp. 827–833.

1051. **Pastor, R., Reed, C., Meyer, P., McKinnon, C. Ryabinin, A.,** and **Phillips, T.** (2012). Role of corticotropin-releasing factor and corticosterone in behavioral sensitization to ethanol. *Journal of Pharmacology and Experimental Therapeutics* **341**, pp. 455–463.

1052. **Patchev, V. K.,** and **Almeida, O. F.** (1996). Gonadal steroids exert facilitating and "buffering" effects on glucocorticoid-mediated transcriptional regulation of corticotropin-releasing hormone and corticosteroid receptor genes in rat brain. *Journal of Neuroscience* **16**, pp. 7077–7084.

1053. **Pavcovich, L. A.,** and **Valentino, R. J.** (1997). Regulation of putative neurotransmitter effect of corticotropin-releasing factor: effects of adrenalectomy. *Journal of Neuroscience* **17**, pp. 401–408.

1054. **Pavlov, I. P.** (1928). *Lectures on conditioned reflexes: twenty-five years of objective study of the higher nervous activity*. Trans. W. H. Gantt. New York: Liveright Publishing.

1055. **Peciña, S., Schulkin, J.,** and **Berridge, K. C.** (2006). Nucleus accumbens corticotropin- releasing factor increases cue-triggered motivation for sucrose reward: paradoxical positive incentive effects in stress? *BMC Biology* **4**, pp. **8–16**.

1056. **Pecoraro, N., Dallman, M. F., Warne, J., Ginsberg, A., Laugero, K., La Fleur, S., Houshyar, H., Gomez, F., Bhargava, A.,** and **Akana, S.** (2006). From Malthus to motive: How how the HPA axis engineers the phenotype, yoking needs to wants. *Progress in Neurobiology* **79**, pp. 247–340.

1057. **Pecoraro, N.,** and **Dallman, M. F.** (2005a). c-Fos after incentive shifts: expectancy, incredulity, and recovery. *Behavioral Neuroscience* **119**, pp. 366–387.

1058. **Pecoraro, N., Gomez, F.,** and **Dallman, M.** (2005b) Glucocorticoids dose-dependently remondel energy stores and amplify incentive relativity effects. *Psychoneuroendocrinology* **30**, pp. 815–825.

1059. **Pepels, P. P., Meek, J., Wendelaar Bonga, S. E.,** and **Balm, P. H.** (2002). Distribution and quantification of corticotropin-releasing hormone (CRH) in the brain of the teleost fish Oreochromis mossambicus (tilapia). *Journal of Comparative Neurology* **453**, pp. 247–268.

1060. **Peraglia, F., Florio, F.,** and **Vale, W.** (2005). Placental expression of neurohormones and other neuroactive molecules in human pregnancy. In: Power, M., and Schulkin, J. (eds.). *Birth, distress, and disease: placenta-brain interactions*. Cambridge: Cambrdige University Press (p. 16–73).

1061. **Perkins, A. V.,** and **Linton, E. A.** (1995). Placental corticotrophin-releasing hormone: there by accident or design? *Journal of Endocrinology* **147**, pp. 377–381.

1062. **Perkins, A.,** and **Linton, E.** (1995). Identification and isolation of corticotrophin-releasing hormone-positive cells from the human placenta. *Placenta* **16**, pp. 233–243.

1063. **Perkins, A. V., Eben, F., Wolfe, C.P., Schulte, H. H.,** and **Linton, E. A.** (1993). Plasma measurement of corticotrophin-releasing hormone-binding protein in normal and abnormal human pregnancy. *Journal of Endocrinology* **138**, pp. 149–157.

1064. **Perrin, M. H.,** and **Vale, W. W.** (1999). Corticotropin-releasing factor receptors and their ligand family. *Annals of the New York Academy of Sciences* **885**, pp. 312–328.

1065. **Pesonen, A. K., Räikkönen, K., Kajantie, E., Heinonen, K., Strandberg, T. E.,** and **Järvenpää, A. L.** (2006). Fetal programming ofvtemperamental negative affectivity among children born healthy at term. *Developmental Psychobiology* **48**, pp. 633–643.

1066. **Pessel, C., Moni, S., Zork, N., Brubaker, S., Vink, J., Fuchs, K., Nhan-Chang, C. L., Ananth, C. V.,** and **Gyamfi-Bannerman, C.** (2013). The effect of intramuscular progesterone on the rate of cervical shortening. *American Journal of Obstetrics and Gynecology* **209**, p. 269. e1–7.

1067. **Peters, A., Schweiger, U., Pellerin, L., Hubold, C., Oltmanns, K. M., Conrad, M., Schultes, B., Born, J.,** and **Fehm, H. L.** (2004). The selfish brain: competition for energy resources. *Neuroscience Biobehavioral Review* **28**, pp. 144–178.

1068. **Petraglia, F., Florio, P.,** and **Vale, W. W.** (2005). Placental expression of neurohormones and other neuroactive molecules in human pregnancy. In: Power, M. L., and Schulkin, J., (eds.), *Birth, distress and disease: placenta-brain interactions*. Cambridge: Cambridge University Press. 16–73.

1069. **Petraglia, F., Sutton, S.,** and **Vale, W.** (1989). Neurotransmitters and peptides modulate the release of immunoreactive corticotropin-releasing factor from cultured human placental cells. *American Journal of Obstetrics and Gynecology* **160**, pp. 247–251.

1070. **Petraglia, F., Aguzzoli, L., Florio, P., Baumann, P., Genazzani, A. D., Di Carlo, C.,** and **Romero, R.** (1995). Maternal plasma and placental immunoreactive corticotrophin-releasing factor concentrations in infection-associated term and pre-term delivery. *Placenta* **16**, pp. 157–164.

1071. **Petraglia, F., Giardano, L., Coukos, G.,** et al. (1990). Corticotropin-releasing factor and parturition: plasma and amniotic fluid levels and placental binding sites. *Obstetetrics and Gynecology* **75**, pp. 784–790.

1072. **Petrovich, V.,** and **Swanson, L. W.** (1997). Projections from the lateral part of the central amygdalar nucleus to the postulated fear conditioning circuit. *Brain Research* **763**, pp. 247–254.

1073. **Pfaff, D. W.** (1980). *Estrogens and brain function*. New York: Springer-Verlag.

1074. **Pfaff, D. W.** (1999). *Drive*. Cambridge, MA: MIT Press.

1075. **Pfaffmann, C.,** and **Norgren, R.** (1977). Sensory affect and motivation. *Annals of the New York Academy of Sciences* **290**, pp. 18–34.

1076. **Phelps, E., O'Connor, K., Cunningham, W, Funayama, E., Gatenby, J., Gore, J.,** and **Banaji, M.** (2000). Performance on indirect measures of race evaluation redicts amygdala activation. *Jouranl of Cognitive Neuroscience* **12**, pp. 729–738.

1077. **Phelps, E. A., Delgado, M. R., Nearing, K. I.,** and **LeDoux, J. E.** (2004). Extinction learning in humans: role of the amygdala and vmPFC. *Neuron* **43**, pp. 897–905.

1078. **Phoenix, C. H., Goy, R. W., Gerall, A. A.,** and **Young, W. C.** (1959). Organizing action of prenatally administered testosterone propionate on the tissues mediating mating behavior in the female guinea pig. *Endocrinology* **65**, pp. 369–382.

1079. **Piato, A.**, **Capiotti, K.**, **Tamborski, A.**, **Oses, J.**, **Barcellos, L.**, **Bogo, M.**, **Lara, D.**, **Vianna, M.**, and **Bonan, C.** (2011). Unpredictable chronic stress model in zebrafish (Danio rerio): behavioral and physiological responses. *Progress in Neuro-Psychopharmacology and Biological Psychiatry* **35**, pp. 561–567.

1080. **Piazza, P. V.**, **Deminière, J. M.**, **Le Moal, M.**, and **Simon, H.** (1989). Factors that predict individual vulnerability to amphetamine self-administration. *Science* **245**, pp. 1511–1513.

1081. **Piazza, P. V.**, **Deroche, V.**, **Deminiere, J. M.**, **Maccari, S.**, **Le Moal, M.**, and **Simon, H. E. R. V.** (1993). Corticosterone in the range of stress-induced levels possesses reinforcing properties: implications for sensation-seeking behaviors. *Proceedings of the National Academy of Sciences* **90**, pp. 11738–11742.

1082. **Piazza, P. V.**, **Maccari, S.**, **Deminiere, J. M.**, **Le Moal, M.**, **Mormede, P.**, and **Simon, H.** (1991). Corticosterone levels determine individual vulnerability to amphetamine self-administration. *Proceedings of the National Academy of Sciences* **88**, pp. 2088–2092.

1083. **Piazza, P. V.**, **Marinelli, M.**, **Jodogne, C.**, **Deroche, V.**, **Rougé-Pont, F.**, **Maccari, S.**, **Le Moal, M.** and **Simon, H.** (1994). Inhibition of corticosterone synthesis by metyrapone decreases cocaine-induced locomotion and relapse of cocaine self-administration. *Brain Research* **658**, pp. 259–264.

1084. **Piazza, P.V.**, and **Le Moal, M.** (1997). Glucocorticoids as a biological substrate of reward: physiological and pathophysiological implications. *Brain Research Review* **25**, pp. 359–372.

1085. **Pich, E. M.**, **Heinrichs, S. C.**, **Rivier, C.**, **Miczek, K. A.**, **Fisher, D. A.**, and **Koob, G. F.** (1993). Blockade of pituitary–adrenal axis activation induced by peripheral immunoneutralization of corticotropin-releasing factor does not affect the behavioral response to social defeat stress in rats. *Psychoneuroendocrinology* **18**, pp. 495–507.

1086. **Pinkernelle, J.**, **Abraham, A.**, **Seidel, K.**, and **Braun, K.** (2009). Paternal deprivation induces dendritic and synaptic changes and hemispheric asymmetry of pyramidal neurons in the somatosensory cortex. *Developmental Neurobiology* **69**, pp. 663–673.

1087. **Pinnock, S.**, **Blake, A.**, **Platt, N.**, and **Herbert, J.** (2010). The roles of BDNF, pCREB, and Wnt3a in the latent period preceding activation of progenitor cell mitosis in the adult dentate gyrus by fluoxetine. *PLos ONE* **5**, p. e13652.

1088. **Pinnock, S. B.**, and **Herbert, J.** (2008). Brain-derived neurotropic factor and neurogenesis in adult rat dentate gyrus: interactions with corticosterone. *European Journal of Neuroscience* **27**, pp. 2493–2500.

1089. **Pintor, L.**, **Torres, X.**, **Bailles, E.**, **Navarro, V.**, **de Osaba, M. J. M.**, **Belmonte, A.**, and **Gastó, C.** (2013). CRF test in melancholic depressive patients with partial versus complete relapses: A 2-year follow-up study. *Nordic Journal of Psychiatry* **67**, pp. 177–184.

1090. **Pisarchik, A.**, and **Slominski, A. T.** (2001). Alternative splicing of CRH-R1 receptors in human and mouse skin: identification of new variants and their differential expression. *FASEB Journal* **15**, pp. 2754–2756.

1091. **Pitts, M. W.**, **Todorovic, C.**, **Blank, T.**, and **Takahashi, L. K.** (2009). The central nucleus of the amygdala and corticotropin-releasing factor: insights into contextual fear memory. *Journal of Neuroscience* **29**, pp. 7379–7388.

1092. **Pleil, K. E., Rinker, J. A., Lowery-Gionta, E. G., Mazzone, C. M., McCall, N. M., Kendra, A. M., Olson, D. P., Lowell, B. B., Grant, K. A., Thiele, T. E.,** and **Kash, T. L.** (2015). NPY signaling inhibits extended amygdala CRF neurons to suppress binge alcohol drinking. *Nature Neuroscience* **18**, pp. 545–552.

1093. **Plotsky, P. M., Thrivikraman, K. V., Nemeroff, C. B., Caldji, C., Sharma, S.,** and **Meaney, M. J.** (2005). Long-term consequences of neonatal rearing on central corticotropin-releasing factor systems in adult male rat offspring. *Neuropsychopharmacology* **30**, pp. 2192–2204.

1094. **Plotsky, P. M.,** and **Meaney, M. J.** (1993). Early, postnatal experiences alters hypothalamic corticotropin-releasing factor (CRF) mRNA, median eminence CRF content and stress-induced release in adult rats. *Brain Research Molecular Brain Research* **18**, pp. 195–200.

1095. **Poeggel, G.,** and **Braun, K.** (1996). Early auditory filial learning in degus (Octodon degus): behavioral and autoradiographic studies. *Brain Research* **743**, pp. 162–170.

1096. **Pomrenze, M. B., Millan, E. Z., Hopf, F. W., Keiflin, R., Maiya, R., Blasio, A., Dadgar, J., Kharazia, V., De Guglielmo, G., Crawford, E., Janak, P. H., George, O., Rice, K.C.,** and **Messing, R.O.** (2015). A transgenic rat for investigating the anatomy and function of corticotrophin releasing factor circuits. *Frontiers in Neuroscience* **9**, pp. 1–14.

1097. **Pomrenze, M., Millan, E., Hopf, F., Keiflin, R., Maiya R., Blasio, A., Dadgar, J., Kharazia, V., De Guglielmo, G., Crawford, E., Janak, P., George, O., Rice, K.,** and **Messing, R.** (2015). A transgenic rat for investigating the anatomy and function of corticotrophin releasing factor circuits. *Frontiers in Neuroscience* **9**, pp. 1–14.

1098. **Potter, E., Sutton, S., Donaldson, C., Chen, R., Perrin, M., Lewis, K., Sawchenko, P. E.,** and **Vale, W.** (1994). Distribution of corticotropin-releasing factor receptor mRNA expression in the rat brain and pituitary. *Proceedings of the National Academy of Sciences of the United States of America* **91**, pp. 8777–8781.

1099. **Powell, D. A., Watson, K.,** and **Maxwell, B.** (1994). Involvement of subdivisions of the medial prefrontal cortex in learned cardiac adjustments in rabbits. *Behavioral Neuroscience* **108**, pp. 294.

1100. **Power, M. L.** (2004). Viability as opposed to stability: an evolutionary perspective on physiological regulation. In: Schulkin, J. (ed.), *Allostasis, homeostasis and the costs of adaptation*. Cambridge: Cambridge University Press. pp. 343–364.

1101. **Power, M. L., Bowman, B., Smith, R, Ziegler, T., Layne, D. G., Schulkin, J.,** and **Tardif, S. D.** (2006) The pattern of maternal serum corticotropin-releasing hormone concentration during pregnancy in the common marmoset. *Am J Primatol* **68**, pp. 181–188.

1102. **Power, M. L.,** and **Schulkin, J.** (2005). *Birth distress and disease*. Cambridge: Cambridge University Press.

1103. **Power, M. L.,** and **Schulkin, J.** (2006). Functions of Corticotropin Releasing Hormone in Anthropoid Primates: From Brain to Placenta. *American Journal of Human Biology* **18**, pp. 431–447.

1104. **Power, M. L.,** and **Schulkin, J.** (2008). Anticipatory physiological regulation in feeding biology: cephalic phase responses. *Appetite* **50**, pp. 194–206.

1105. **Power, M. L.,** and **Schulkin, J.** (2012). *The evolution of the placenta and birth*. Baltimore: Johns Hopkins University Press.

1106. **Power, M. L.**, and **Schulkin, J.** (2013). *Evolution of obesity*. Baltimore: John Hopkins University Press.

1107. **Power, M. L.**, and **Schulkin, J.** (2013). Maternal regulation of offspring development in mammals is an ancient adaptation tied to lactation. *Applied and Translational Genomics* 2, pp. 55–63.

1108. **Power, M. L.**, **Williams, L. E.**, **Gibson, S. V.**, **Schulkin, J.**, **Helfers, J.**, and **Zorrilla, E. P.** (2011). Pattern of maternal circulating CRH in laboratory-housed squirrel and owl monkeys. *American Journal of Primatology* 72, pp. 1004–1012.

1109. **Price, J. L.** (1999). Prefrontal cortical networks related to visceral function and mood. *Annals of the New York Academy of Sciences* **877**, pp. 383–396.

1110. **Price, M. L.**, and **Lucki, I.** (2001). Regulation of serotonin release in the lateral septum and striatum by corticotropin-releasing factor. The Journal of Neuroscience **21**, pp. 2833–2841.

1111. **Pringle, R. B.**, **Mouw, N. J.**, **Lukkes, J. L.**, and **Forster, G. L.** (2008). Amphetamine treatment increases corticotropin-releasing factor receptors in the dorsal raphe nucleus. *Neuroscience Research* **62**, pp. 62–65.

1112. **Pucci, M.**, **Micioni Di Bonaventura, M. V.**, **Giusepponi, M. E.**, **Romano, A.**, **Filaferro, M.**, **Maccarrone, M.**, **Ciccocioppo, R.**, **Cifani, C.**, and **D'Addario, C.** (2015). Epigenetic regulation of nociceptin/orphanin FQ and corticotropin-releasing factor system genes in frustration stress-induced binge-like palatable food consumption. *Addiction Biology* **21(6)**, pp. 1168–1185.

1113. **Pugh, C.R.**, **Tremblay, D.**, **Fleshner, M.**, and **Rudy, J. W.** (1997). A selective role for corticosterone in contextual-fear conditioning. *Behavioral Neuroscience* **111**, pp. 503–511.

1114. **Quirarte, G.**, **Teja, I.**, **Casillas, M.**, **Serafin, N.**, **Prado-Alcala, R.**, and **Roozendaal, B.** (2009). Corticosterone infused into the dorsal striatum selectively enhanced memory consolidation of cued water-maze training. *Learning and Memory:* pp. 586–589.

1115. **Quirk, G. J.**, and **Gehlert, D. R.** (2003). Inhibition of the amygdala: key to pathological states? *Annals of the New York Academy of Sciences* **985**, pp. 263–272.

1116. **Quirk, G. J.**, et al. (2004). Corticosterone accelerates extinction of conditioned fear in rats. In: American College of Neuropsychopharmacology 43rd Annual Meeting, Abstract 59.

1117. **Quirk, G. J.**, **Likhtik, E.**, **Pelletier, J. G.**, and **Paré, D.** (2003). Stimulation of medial prefrontal cortex decreases the responsiveness of central amygdala output neurons. *Journal of Neuroscience* **23**, pp. 8800–8807.

1118. **Quirk, G. J.**, **Russo, G. K.**, **Barron, J. L.**, and **Lebron, K.** (2000). The role of ventromedial prefrontal cortex in the recovery of extinguished fear. *Journal of Neuroscience* **20**, pp. 6225–6231.

1119. **Raadsheer, F.**, **Sluiter, A.**, **Ravid, R.**, **Tilders, F.**, and **Swaab, D.** (1993). Localization of corticotropin-releasing hormone (CRH) neurons in the paraventricular nucleus of the human hypothalamus: age-dependent colocalization with vasopressin. *South African Journal of Science* **615**, pp. 50–62.

1120. **Radley, J. F.** (2012). Toward a limbic cortical inhibitory network: implications for hypothalamic-pituitary-adrenal responses following chronic stress. *Frontiers in Behavioral Neuroscience* **6**, pp. 1–10.

1121. **Radley, J.**, and **Sawchenko, P.** (2015). Evidence for involvement of a limbic paraventricular hypothalamic inhibitory network in hypothalamic-pituitary-adrenal axis adaptations to repeated stress. *Journal of Comparative Neurology* **523**, pp. 2769–2787.

1122. **Radley, J. J.**, **Arias, C. M.**, and **Sawchenko, P. E.** (2006b). Regional differentiation of the medial prefrontal cortex in regulating adaptive responses to acute emotional stress. *Journal of Neuroscience* **26**, pp. 12967–12976.

1123. **Radley, J. J.**, **Rocher, A. B.**, **Miller, M.**, **Janssen, W. G.**, **Liston, C.**, **Hof, P. R.**, **McEwen, B. S.**, and **Morrison, J. H.** (2006a). Repeated stress induces dendritic spine loss in the rat medial prefrontal cortex. *Cerebral Cortex* **16**, pp. 313–320.

1124. **Radulovic, J.**, **Ruhmann, A.**, **Lipold, T.**, and **Spiess, J.** (1999). Modulation of learning and anxiety by corticotropin-releasing factor (CRF) and stress: differential roles of CRF receptors 1 and 2. *Journal of Neuroscience* **19**, pp. 5016–5025.

1125. **Rainnie, D. G.**, **Bergeron, R.**, **Sajdyk, T. J.**, **Patil, M.**, **Gehlert, D. R.**, and **Shekhar, A.** (2004). Corticotropin releasing factor-induced synaptic plasticity in the amygdala translates stress into emotional disorders. *Journal of Neuroscience.* **24**, pp. 3471–3479.

1126. **Rajbhandari, A. K.**, **Baldo, B. A.**, and **Bakshi, V. P.** (2015). Predator stress-induced CRF release causes enduring sensitization of basolateral amygdala norepinephrine systems that promote PTSD-like startle abnormalities. *Journal of Neuroscience* **35**, pp. 14270–14285.

1127. **Rao, P. D.**, and **Kanwal, J. S.** (2004). Oxytocin and vasopressin immunoreactivity within the forebrain and limbic-related areas in the mustached bat, Pteronotus parnellii. *Brain, Behavior and Evolution* **63**, pp. 151–168.

1128. **Raper, J.**, **Stephens, S. B.**, **Henry, A.**, **Villareal, T.**, **Bachevalier, J.**, **Wallen, K.**, and **Sanchez, M. M.** (2014). Neonatal amygdala lesions lead to increased activity of brain CRF systems and hypothalamic-pituitary-adrenal axis of juvenile rhesus monkeys. *Journal of Neuroscience* **34**, pp. 11452–11460.

1129. **Rasmussen, D. D.** (1998). Effects of chronic nicotine treatment and withdrawal on hypothalamic proopiomelanocortin gene expression and neuroendocrine regulation. *Psychoneuroendocrinology* **23**, pp. 245–259.

1130. **Reber, S. O.**, **Obermeier, F.**, **Straub, H. R.**, **Falk, W.**, and **Neumann, I. D.** (2006). Chronic intermittent psychosocial stress (social defeat/overcrowding) in mice increases the severity of an acute DSS-induced colitis and impairs regeneration. *Endocrinology* **147**, pp. 4968–4976.

1131. **Refojo, D.**, and **Holsboer, F.** (2009). CRH signaling. *Annals of the New York Academy of Sciences* **1179**, pp. 106–119.

1132. **Refojo, D.**, and **Holsboer, F.** (2009). Molecular specificity for drug targeting in the CNS. *Glucocorticoids and Mood* **1179**, pp. 106–119.

1133. **Refojo, D.**, **Schweizer, M.**, **Kuehne, C.**, **Ehrenberg, S.**, **Thoeringer, C.**, **Vogl, A. M.**, **Dedic, N.**, **Schumacher, M.**, **von Wolff, G.**, **Avrabos, C.**, **Touma, C.**, **Engblom, D.**, **Schutz, G.**, **Nave, K-A.**, **Eder, M.**, **Wotjak, C.T.**, **Sillaber, I.**, **Holsboer, F.**, **Wurst, W.**, and **Deussing, J. M.** (2011). Glutamatergic and dopaminergic neurons mediate anxiogenic and anxiolytic effects of CRHR1. *Science* **333**, pp. 1903–1907.

1134. **Regev, L.**, **Cohen, A.**, **Tsoory, M.**, **Kuperman, Y.**, **Getselter, D.**, **Gil, S.**, and **Chen, A.** (2011). Prolonged and site-specific over-expression of corticotropin-releasing factor revelas differential roles for extended amygdala nucleiu in emotional regulation. *Molecular Psychiatry* **16**, pp. 714–728.

1135. **Regev, L., Tsoory, M., Gil, S.,** and **Chen, A.** (2012). Site-specific genetic manipulation of amygdala corticotropion-releasing factor reveals its imperative role in mediating behavioral response to challenge. *Biological Psychiatry* **71**, pp. 317–326.

1136. **Reiner, A., Medina, L.,** and **Veenman, C.** (1998). Structural and functional evolution of the basal ganglia in vertebrates. *Brain Research Rewards* **28**, pp. 235–285.

1137. **Ren, J., Ding, X., Funk, G.,** and **Greer, J.** (2012). Anxiety-related mechanisms of respiratory dysfunction in a mouse model of rett syndrome. *Journal of Neuroscience* **32**, pp. 17230–17240.

1138. **Rescorla, R. A.** and **Wagner, A. R.** (1972). A theory of Pavlovian conditioning: variations in the effectiveness of reinforcement non-reinforcement. In: Baker W. J., and Prokasy, W. (eds.), *Classical conditioning: current research and theory*. New York, NY: Appleton-Century-Crofts.

1139. **Reynolds, R.** (2015). Trans-generational effects of maternal exposure to social stress on anxiety-like behaviour in rats. *Psychoneuroendocrinology* **61**, pp. 1–78.

1140. **Reynolds, S.,** and **Berridge, K.** (2008). Emotional environments returne the valence of appetitive versus fearful functions in nucleus accumbens. *Nature Neuroscience* **11**, pp. 423–425.

1141. **Reznick, D., Mateos, M.,** and **Springer, M.** (2002). Independent origins and rapid evolution of the placenta in the fish genus Poeciliopsis. *Science* **298**, pp. 1018–1020.

1142. **Rho, J.,** and **Swanson, L.** (1987). Neuroendocrine CRF motoneurons: intrahypothalamic axon terminals shown with a new retrograde-Lucifer-immuno method. *Brain Research* **436**, pp. 143–147.

1143. **Richardson, R., Omachi, K., Kermani, R.,** and **Woods, S. C.** (2002). Intraventricular insulin potentiates the anorexic effect of corticotropin releasing hormone in rats. *American Journal of Physiology-Regulatory, Integrative and Comparative Physiology* **283**, pp. R1321–R1326.

1144. **Richardson, R., Boswell, T., Woods, S.,** and **Wingfield, J.** (2000). Intracerebroventricular corticotropin-releasing factor decreases food intake in white-crowned sparrows. *Physiology and Behavior* **71**, pp. 213–216.

1145. **Richardson, H., Lee, S., O'Dell, L., Koob, G. F.,** and **River, C.** (2008). Alcohol self-administration acutely stimulates the hypothalamic-pituitary-adrenal axis, but alcohol dependence leads to a dampened neuroendocrine state. *European Journal of Neuroscience* **28**, pp. 1641–1653.

1146. **Richter, C. P.** (1942–43). Total self-regulatory functions in animals and human beings. *Harvey Lect.* **38**, pp. 367–371.

1147. **Richter, C.P.** (1957). On the phenomenon of sudden death in animals and man. *Psychosomatic Medicine* **19**, pp. 191–198.

1148. **Richter, R. M.,** and **Weiss, F.** (1999). In vivo crf release in rat amygdala is increased during cocaine withdrawal in self-administering rats. Synapse **32**, pp. 254–261.

1149. **Richter, R. M., Pich, E. M., Koob, G. F.,** and **Weiss, F.** (1995). Sensitization of cocaine-stimulated increase in extracellular levels of corticotropin-releasing factor from the rat amygdala after repeated administration as determined by intracranial microdialysis. *Neuroscience Letters* **187**, pp. 169–172.

1150. **Riley, S. C., Walton, J. C., Herlick, J. M.,** and **Challis, J. R.** (1991). The localization and distribution of corticotropin-releasing hormone in the human placenta and fetal membranes throughout gestation. *Journal of Clinical Endocrinology & Metabolism* **72**, pp. 1001–1007.

1151. **Rilling, J.**, and **Young, L.** (2014). The biology of mammalian parenting and its effect on offspring social development. *Science* **345**, pp. 771–774.

1152. **Rinker, J., Marshall, A., Mazzone, C., Lowery-Gionta, E., Gulati, V., Pleil, K., Kash, T., Navarro, M.**, and **Thiele, T.** (2016, In Press). Extended amygdala to ventral tegmental area corticotropin-releasing factor circuit controls binge ethanol intake. *Biological Psychiatry.*

1153. **Risbrough, V. B., Geyer, M. A., Hauger, R. L., Coste, S., Stenzel-Poore, M., Wurst, W.**, and **Holsboer, F.** (2009). CRF1 and CRF2 receptors are required for potentiated startle to contextual but not discrete cues. *Neuropsychopharmacology* **34**, pp. 1494–1503.

1154. **Rivier, J., Spiess, J.**, and **Vale, W.** (1983). Characterization of rat hypothalamic corticotropin releasing factor. *Proceedings of the National Academy of Sciences of the United States of America* **80**, pp. 4851–4855.

1155. **Robbins, T. W.**, and **Everitt, B. J.** (1999). Drug addiction: bad habits add up. *Nature* **398**, pp. 567–570.

1156. **Robbins, T. W., Cador, M., Taylor, J. R.**, and **Everitt, B. J.** (1989). Limbic-striatal interactions in reward-related processes. *Neuroscience and Biobehavioral Reviews* **13**, pp. 155–162.

1157. **Roberto, M., Cruz, M. T., Gilpin, N. W., Sabino, V., Schweitzer, P., Bajo, M., Cottone, P., Madamba, S. G., Stouffer, D. G., Zorrilla, E. P., Koob, G. F., Siggins, G. R.**, and **Parsons, L. H.** (2010). Corticotropin releasing factor-induced amygdala gamma-aminobutyric acid release plays a key role in alcohol dependence. *Biological Psychiatry* **67**, pp. 831–839.

1158. **Robidoux, J., Simoneau, L., St.Pierre, S., Masse, A.**, and **Lafond, J.** (2000). Characterization of neuropeptide Y-mediated corticotropin releasing factor synthesis and release from human placental trophoblasts. *Endocrinology* **141**, pp. 2795–2804.

1159. **Robinson, B. G., Arbiser, J., Emanuel, R.**, and **Majzoub, J.** (1989). Species-specific placental corticotropin releasing hormone messenger RNA and peptide expression. *Molecular and Cellular Endocrinology* **62**, pp. 337–341.

1160. **Robinson, T. E.**, and **Berridge, K. C.** (2000) The psychology and neurobiology of addiction. *Addiction* **95** Suppl 2, pp. S91–117.

1161. **Robinson, B. G., Tellam, D., Smart, D., Mohammad, Y., Brennand, J., River, J.**, and **Lovejoy, D.** (1999). Cloning and characterization of corticotropin releasing factor and urocortin in Syrian hamster (Mesocricetus auratus). *Peptides* **20**, pp. 1177–1185.

1162. **Robinson, B.G., Arbiser, J.L., Emanuel, R.L., Majzoub, J.A.** (1989). Species-specific placental corticotropin releasing hormone messenger RNA and peptide expression. *Molecular and Cellular Endocrinology* 62(2), 337–341.

1163. **Robinson, B. G., Emanuel, R. L., Frim, D. M.**, and **Majzoub, J. A.** (1988). Glucocorticoid stimulates expression of corticotropin-releasing hormone gene in human placenta. *Proceedings of the National Academy of Sciences of the United States of America* **85**, pp. 5244–5248.

1164. **Rodaros, D., Caruana, D., Amir, S.**, and **Stewart, J.** (2007). Corticotropin-releasing factor projections from limbic forebrain and paraventricular nucleus of the hypothalamus to the region of the ventral tegmental area. *Neuroscience* **150**, pp. 8–13.

1165. **Rodgers, A. B., Morgan, C. P., Bronson, S. L., Revello, S.**, and **Bale, T. L.** (2013). Paternal stress exposure alters sperm microRNA content and reprograms offspring HPA stress axis regulation. *Journal of Neuroscience* **33**, pp. 9003–9012.

1166. **Rogers, J., Raveendran, M., Fawcett, G. L., Fox, A. S., Shelton, S. E., Oler, J. A., Cheverud, J., Muzny, D. M., Gibbs, R. A., Davidson, R. J.,** and **Kalin, N. H.** (2013). CRHR1 genotypes, neural circuits and the diathesis for anxiety and depression. *Molecular Psychiatry* **18**, pp. 700–707.

1167. **Roltsch, E. A., Baynes, B. B., Mayeux, J. P., Whitaker, A. M., Baiamonte, B. A.,** and **Gilpin, N. W.** (2014). Predator odor stress alters corticotropin-releasing factor-1 receptor (CRF1R)-dependent behaviors in rats. *Neuropharmacology* **79**, pp. 83–89.

1168. **Romanowski, C. P., Fenzl, T., Flachskamm, C., Wurst, W., Holsboer, F., Deussing, J. M.,** and **Kimura, M.** (2010). Central deficiency of corticotropin-releasing hormone receptor type 1 (CRH-R1) abolishes effects of CRH on NREM but not on REM sleep in mice. *Sleep* **33**, pp. 427.

1169. **Romanowski, C. P., Fenzl, T., Flachskamm, C., Wurst, W., Holsboer, F., Deussing, J. M.,** and **Kimura, M.** (2010). Central deficiency of corticotropin-releasing hormone receptor type 1 (CRH-R1) abolishes effects of CRH on NREM but not on REM sleep in mice. *Sleep* **33**, pp. 427–436.

1170. **Romeo, R., Karatsoreos, I., Jasnow, A.,** and **McEwen, B.** (2007). Age- and stress-induced changes in corticotropin-releasing hormone mrna expression in the paraventricular nulcues of the hypothalamus. *Neuroendocrinology* **85**, pp. 199–206.

1171. **Romero, L. M., Dickens, M. J.,** and **Cyr, N. E.** (2009). The reactive scope model—a new model integrating homeostasis, allostasis, and stress. *Hormones and Behavior* **55**, pp. 375–389.

1172. **Romero, L. M.,** and **Wikelski, M.** (2001). Corticosterone levels predict survival probabilities of Galapagos marine iguanas during El Nino events. *Proceedings of the National Academy of Sciences of the United States of America* **98**, pp. 7366–7370.

1173. **Romero, R., Yeo, L., Miranda, J., Hassan, S. S., Conde-Agudelo, A.,** and **Chaiworapongsa, T.** (2013). A blueprint for the prevention of preterm birth: vaginal progesterone in women with a short cervix. *Journal of Perinatal Medicine* **41**, pp. 27–44.

1174. **Roozendaal, B., McReynolds, J.,** and **McGaugh, J.** (2004). The basolateral amygdala interacts with the medial prefrontal cortex in regulating glucocorticoid effects on working memory impairment. *Journal of Neuroscience* **24**, pp. 1385–1392.

1175. **Roozendaal, B.** (2000). Glucocorticoids and the regulation of memory consolidation. *Psychoneuroendocrinology* **25**, pp. 213–238.

1176. **Roozendaal, B., Brunson, K. L., Holloway, B. L., McGaugh, J. L.,** and **Baram, T. Z.** (2002). Involvement of stress-released corticotropin-releasing hormone in the basolateral amygdala in regulating memory consolidation. *Proceedings of the National Academy of Sciences of the United States of America* **99**, pp. 13908–13913.

1177. **Roozendaal, B., Schelling, G.,** and **McGaugh, J. L.** (2008). Corticotropin releasing factor in basolateral amygdala enhances memory consolidation via an interaction with the beta-adrenoceptor-cAMP pathway: dependence on glucocorticoid receptor activation. *Journal of Neuroscience* **28**, pp. 6642–6651.

1178. **Roseboom, P. H., Nanda, S. A., Bakshi, V. P., Trentani, A., Newman, S. M.,** and **Kalin, N. H.** (2007). Predator threat induces behavioral inhibition, pituitary-adrenal activation and changes in amygdala CRF-binding protein gene expression. *Psychoneuroendocrinology* **32**, pp. 44–55.

1179. **Rosen, J. B., Pagani, J. H., Rolla, K. L.,** and **Davis, C.** (2008). Analysis of behavioral constraints and the neuroanatomy of fear to the predator odor trimethylthiazoline: a model for animal phobias. *Neuroscience and Biobehavioral Reviews* **32**, pp. 1267–1276.

1180. **Rosen, J.** and **Schulkin, J.** (1998). From normal fear to pathological anxiety. *Psychology Review* **105**, pp. 325–350.

1181. **Rosen, J., Donley, M., Gray, D., West, E., Morgan, M.,** and **Schulkin, J.** (2008). Chronic corticosterone administration does not potentiate unconditioned freezing to the predator odor, trimethylthiazoline. *Behavioral Brain Research* **194**, pp. 32–38.

1182. **Rosen, J., West, E.,** and **Schulkin, J.** (2004). Corticosterone potentiates predator (fox) odor-induced unconditioned fear. Program No. 772.7. In 2004 Abstract Viewer and Itinerary Planner, Society for Neuroscience online.

1183. **Rosen, J. B., Hamerman, E., Sitcoske, M., Glowa, J. R.,** and **Schulkin, J.** (1996). Hyperexcitability: exaggerated fear-potentiated startle produced by partial amygdala kindling. *Behavioral Neuroscience* **110**, pp. 43–50.

1184. **Rosen, J. B., Pishevar, S. K., Weiss, S. R., Smith, M. A., Kling, M. A., Gold, P. W.,** and **Schulkin, J.** (1994). Glucocorticoid potentiation treatment increases the ability of CRH to induce seizures. *Neuroscience Letters* **174**, pp. 113–116.

1185. **Rosen, L. B., Majzoub, J. A.,** and **Adler, G. K.** (1992). Effects of glucocorticoid on corticotropin-releasing hormone gene regulation by second messenger pathways in NPLC and AtT-20 cells. *Endocrinology* **130**, pp. 2237–2244.

1186. **Rosen, T., Schulkin, J., Power, M., Tadesse, S., Norwitz, E., Wen, Z.,** and **Wang, B.** (2015). Comparative immunohistochemistry of placental corticotropin releasing hormone and the transcription factor RelB-NFkB2 between humans and nonhuman primates. *Comparative Medicine* **65**, pp. 1–4.

1187. **Rosenblatt, J. S.** (2003). Outline of the evolution of behavioral and nonbehavioral patterns of parental care among the vertebrates: critical characteristics of mammalian and avian parental behavior. *Scandinavian Journal of Psychology* **44**, pp. 265–271.

1188. **Rosenblum, L. A., Smith, E. L. P., Altemus, M., Scharf, B. A., Owens, M. J., Nemeroff, C. B., Gorman, J.M.** and **Coplan, J. D.** (2002). Differing concentrations of corticotropin-releasing factor and oxytocin in the cerebrospinal fluid of bonnet and pigtail macaques. *Psychoneuroendocrinology* **27**, pp. 651–660.

1189. **Roskoden, T., Hanke, J., Yilmazer-Hanke, D.,** and **Schwegler, H.** (2005). Reduced number of CRF-containing neurons in the central amygdala correlated with enhanced locomotor activity following early postnatal corticosterone treatment in the Wilstar rat. *Behavioural Brain Research* **165**, pp. 221–228.

1190. **Rossant, J.,** and **Cross, J. C.** (2001). Placental development: lessons from mouse mutants. *Nature Reviews Genetics* **2**, pp. 538–548.

1191. **Rougé-Pont, F., Deroche, V., Moal, M. L.,** and **Piazza, P. V.** (1998). Individual differences in stress-induced dopamine release in the nucleus accumbens are influenced by corticosterone. *European Journal of Neuroscience* **10**, pp. 3903–3907.

1192. **Rougé-Pont, F., Piazza, P. V., Kharouby, M., Le Moal, M.,** and **Simon, H.** (1993). Higher and longer stress-induced increase in dopamine concentrations in the nucleus accumbens of animals predisposed to amphetamine self-administration. A microdialysis study. *Brain Research* **602**, pp. 169–174.

1193. **Rouibi, K.,** and **Contarino, A.** (2013). A corticotropin releasing factor receptor 2 mediates the motivational effect of opiate withdrawal. *Neuropharmacology* **73**, pp. 41–47.

1194. **Roy, A., Hodgkinson, C. A., DeLuca, V., Goldman, D.,** and **Enoch, M. A.** (2012). Two HPA axis genes, CRHBP and FKBP5, interact with childhood trauma to increase the risk for suicidal behavior. *Journal of Psychiatric Research* **46**, pp. 72–79.

1195. **Royule, N.**, **Russell, A.**, and **Wilson, A.** (2014). The evolution of flexible parenting. *Science* **345**, pp. 776–778.

1196. **Rozin, P.** (1976). The selection of foods by rats. *Advances in the Study of Behavior* **6**, pp. 21–76.

1197. **Ruiz, R. J.**, **Gennaro, S.**, **O'Connor, C.**, **Dwivedi, A.**, **Gibeau, A.**, **Keshinover, T.**, & and **Welsh, T.** (2015). CRH as a predictor of preterm birth in minority women. *Biological Research for Nursing*: p. 1099800415611248.

1198. **Ruscio, M. G.**, **Sweeny, T.**, **Hazelton, J.**, **Suppatkul, P.**, **Carter, S. C.** (2007). Social environment regulates CRF, corticosterone and vasopressin in juvenile prairie voles. *Hormones and Behavior* **51**, pp. 54–61.

1199. **Russell, J.**, and **Brunton, P.** (2006). Neuroactive steroids attenuate oxytocin stress responses in late pregnancy. *Neuroscience* **138**, pp. 879–889.

1200. **Ryan, K. J.** (1959). Biological aromatization of steroids. *Journal of Biological Chemistry* **234**, pp. 268–272.

1201. **Ryan, P.J.**, **Ma, S.**, **Olucha-Bordonau, F.E.**, **Gundlach, A.L.** (2011), Nucleus incertus – an emerging modulatory role in arousal, stress and memory. *Neuroscience & Biobehavioral Reviews* 35(6), pp. 1326–1341.

1202. **Sabino, V.**, **Kwak, J.**, **Rice, K.C.**, and **Cottone, P.** (2013). Pharmacological characterization of the 20% alcohol intermittent access model in Sardinian alcohol-preferring rats: a model of binge-like drinking. *Alcoholism Clinical and Experimental Research* 37, pp. 635–643.

1203. **Saffran, M.**, and **Schally, A. V.** (1955). The release of corticotrophin by anterior pituitary tissue in vitro. *Canadian Journal of Biochemistry and Physiology* **33**, pp. 408–415.

1204. **Saffran, M.**, **Schally, A. V.**, and **Benfey, B. G.** (1955). Stimulation of the release of corticotropin from the adenohypophysis by a neurohypophysial factor 1. *Endocrinology* **57**, pp. 439–444.

1205. **Saffran, M.**, **Schally, A. V.**, **Segal, M.**, & and **Zimmermann, B.** (1958). Characterization of the corticotrophin releasing factor of the neurohypophysis. In: Zweites Internationales Symposium über Neurosekretion. Berlin: Springer.

1206. **Sagami, Y.**, **Shimada, Y. Tayama, J. Nomura, T.**, **Satake, M.**, **Endo, Y.**, **Shoji, T. Karahashi, K.**, **Hongo, M.**, and **Fukudo, S.** (2004). Effect of a corticotropin releasing hormone receptor antagonist on colonic sensory and motor function inpatients with irritable bowel syndrome. *BMJ* **54**, pp. 958–964.

1207. **Sahgal, A.**, **Wright, C.**, **Edwardson, J. A.**, and **Ketih, A. B.** (1983). Corticotrophin releasing factor is more potent than some corticotrophin-related peptides in affecting passive avoidance behavior in rats. *Neuroscience Letters* **36**, pp. 81–86.

1208. **Sahuque, L.**, **Kullberg, E.**, **McGeehan, A.**, **Kinder, J.**, **Hicks, M.**, **Blanton, M.**, **Janak, P.**, and **Olive, M.** (2006). Anxiogenic and aversive effects of corticotropin releasing factor (CRF) in the bed nucleus of the stria terminalis in the rat: role of CRF receptor subtypes. *Psychopharmacology* **186**, pp. 122–132.

1209. **Sajdyk, T.**, **Shekhar, A.**, and **Gehlert, D.** (2004). Interactions between NPY and CRF in the amygdala to regulate emotionality. *Neuropeptides* **38**, pp. 225–234.

1210. **Sakanaka, M.**, **Shibasaki, T.**, and **Lederis, K.** (1986). Distribution and efferent projections of corticotropin-releasing factor-like immunoreactivity in the rat amygdaloid complex. *Brain Research* **382**, pp. 213–238.

1211. **Saltzman, W., Schultz-Darken, N. J., Scheffler, G., Wegner, F. H.,** and **Abbott, D. H.** (1994). Social and reproductive influences on plasma cortisol in female marmoset monkeys. *Physiology and Behavior* **56**, pp. 801–810.

1212. **Saltzman, W., Schultz-Darken, N. J., Wegner, F. H., Wittwer, D. J.,** and **Abbott, D. H.** (1998). Suppression of cortisol levels in subordinate female marmosets: reproductive and social contributions. *Hormones and Behavior* **33**, pp. 58–74.

1213. **Samaco, R. C., Mandel-Brehm, C., McGraw, C. M., Shaw, C. A., McGill, B. E.,** and **Zoghbi, H. Y.** (2012). Crh and Oprm1 mediate anxiety-related behavior and social approach in a mouse model of MECP2 duplication syndrome. *Nature Genetics* **44**, pp. 206–211.

1214. **Sánchez, M. M., Young, L. J., Plotsky, P. M.,** and **Insel, T. R.** (1999). Autoradiographic and in situ hybridization localization of corticotropin-releasing factor 1 and 2 receptors in nonhuman primate brain. *Journal of Comparative Neurology* **408**, pp. 365–377.

1215. **Sánchez, M. M., Young, L. J., Plotsky, P. M.,** and **Insel, T. R.** (2000). Distribution of corticosteroid receptors in the rhesus brain: relative absence of glucocorticoid receptors in the hippocampal formation. *Journal of Neuroscience* **20**, pp. 4657–4668.

1216. **Sanderson, T., Woods, R., Kemp, C.,** and **Lowry, P.** (2000). Detection of N-terminal Pro-corticotrophin releasing hormone (CRH) and a "novel" CRH in human maternal plasma and placenta. *Placenta* **21**, pp. 218–225.

1217. **Sandman, C. A., Glynn, L. M.,** and **Davis, E. P.** (2013). Is there a viability–vulnerability tradeoff? Sex differences in fetal programming. *Journal of Psychosomatic Research* 75, pp. 327–335.

1218. **Sandman, C. A., Glynn, L., Schetter, C. D., Wadhwa, P., Garite, T., Chicz-DeMet, A.,** and **Hobel, C.** (2006). Elevated maternal cortisol early in pregnancy predicts third trimester levels of placental corticotropin releasing hormone (CRH): priming the placental clock. *Peptides* **27**, pp. 1457–1463.

1219. **Sandman, C. A.,** and **Glynn, L. M.** (2009). Corticotropin-Releasing Hormone (CRH) programs the fetal and maternal brain. *Future Neurology* **4**, pp. 257–261.

1220. **Sandovici, I., Hoelle, K., Angiolini, E., Constancia, M.** (2012). Placental adaptations to the maternal-fetal environment: implications for fetal growth and developmental programming. *Reproductive BioMedicine Online* **25**, pp. 68–69.

1221. **Sanford, C., Soden, M., Baird, M., Miller, S., Schulkin, J., Palmiter, R., Clark, M.,** and **Zweifel, L.** (2016, Under Review). Central amygdala CRF neurons are a unique cell type that facilitate learning about weak threats. *Neuron* **(16)**, pp. 30902–30903.

1222. **Sansoni, V., Forcella, M., Mozzi, A., Fusi, P., Ambrosini, R., Ferini-Strambi, L.,** and **Combi, R.** (2013). Functional characterization of a CRH missense mutation identified in an ADNFLE family. *PloS One* **8**, p. e61306.

1223. **Sapolsky, R. M.** (1990). Adrenocortical function, social rank, and personality among wild baboons. *Biological Psychiatry* **28**, pp. 862–878.

1224. **Sapolsky, R. M., Zola-Morgan, S.,** and **Squire, L. R.** (1991). Inhibition of glucocorticoid secretion by the hippocampal formation in the primate. *Journal of Neuroscience* **11**, pp. 3695–3704.

1225. **Sapolsky, R. M.** (1992). *Stress: the aging brain and the mechanisms of neuron death.* Cambridge, MA: MIT Press.

1226. **Sapolsky, R. M.** (1995). Social subordinance as a marker of hypercortisolism. Some unexpected subtleties. *Annals of the New York Academy of Sciences* **771**, pp. 626–639.

1227. **Sapolsky, R.M.**, **Romero, L.M.**, and **Munck, A.U.** (2000). How do glucocorticoids influence stress responses? Integrating permissive, suppressive, stimulatory, and preparative actions. *Endocrine Reviews* **21**, pp. 55–89.

1228. **Sarnyai, Z. A. N.**, **Dhabhar, F. S.**, **McEwen, B. S.**, and **Kreek, M. J.** (1998). Neuroendocrine-related effects of long-term,'binge' cocaine administration: diminished individual differences in stress-induced corticosterone response. *Neuroendocrinology* **68**, pp. 334–344.

1229. **Sarnyai, Z.**, **Biro, E.**, **Gardi, J.**, **Vecesernyes, M.**, **Julesz, J.**, and **Telegdy, G.** (1995). Brain corticotropin releasing factor mediates "anxiety-like" behavior induced by cocaine withdrawal in rats. *Brain Research* **675**, pp. 89–97.

1230. **Sarnyai, Z.**, **Bíró, É.**, **Gardi, J.**, **Vecsernyés, M.**, **Julesz, J.**, and **Telegdy, G.** (1993). Alterations of corticotropin-releasing factor-like immunoreactivity in different brain regions after acute cocaine administration in rats. *Brain Research* **616**, pp. 315–319.

1231. **Sarnyai, Z.**, **Bíró, É.**, **Penke, B.**, and **Telegdy, G.** (1992a). The cocaine-induced elevation of plasma corticosterone is mediated by endogenous corticotropin-releasing factor (CRF) in rats. *Brain Research* **589**, pp. 154–156.

1232. **Sarnyai, Z.**, **Höhn, J.**, **Szabó, G.**, and **Penke, B.** (1992b). Critical role of endogenous corticotropin-releasing factor (CRF) in the mediation of the behavioral action of cocaine in rats. *Life Sciences* **51**, pp. 2019–2024.

1233. **Sasaki, A.**, **Liotta, A.S.**, **Luckey, M. M.**, **Margioris, A. N.**, **Suda, T.**, and **Krieger, D. T.** (1984). Immunoreactive corticotropoin-releasing factor is present in human maternal serum during the third trimester of pregnancy. *Journal of Clinical Endocrinology & Metabolism* **59**, pp. 812–814.

1234. **Sasaki, A.**, **Shinkawa, O.**, **Margioris, A.N.**, **Liotta, A.S.**, **Sato, S.**, **Murkami, O.**, **Go, M.**, **Shimizu, Y.**, **Hanew, K.**, and **Yoshingaga, K.** (1987). Immunoreactive corticotropin releasing hormone in human plasma during pregnancy, labor, and delivery. *Journal of Clinical Endocrinology & Metabolism* **64**, pp. 224–229.

1235. **Sato, H.**, **Nagashima, Y.**, **Chrousos, G.P.**, **Ichihashi, M.**, and **Funaska, Y.** (2002). The expression of corticotropin-releasing hormone in melanoma. *Pigment Cell Research* **15**, pp. 98–103.

1236. **Sawchenko, P. E.**, **Imaki, T.**, **Potter, E.**, **Kovacs, K.**, **Imaki, J.**, and **Vale, W.** (1993). The functional neuroanatomy of corticotropin-releasing factor. In: Ciba Foundation Symposium 172-Corticotropin-Releasing Factor (p. 5–29).

1237. **Sawchenko, P. E.**, **Li, H. Y.**, and **Ericsson, A.** (2000). Circuits and mechanisms governing hypothalamic responses to stress: a tale of two paradigms. *Progress in Brain Research* **122**, pp. 61–80.

1238. **Sawchenko, P.E.** (1987). Evidence for local site of action for glucocorticoids in inhibiting CRF and vasopressin expression in the paraventricular nucleus. *Brain Research* **403**, pp. 213–223.

1239. **Sawyer, C. H.**, **Everett, J. W.**, and **Markee, J. E.** (1949a). A neural factor in the mechanism by which estrogen induces the release of luteinizing hormone in the rat. *Endocrinology* **44**, pp. 218–233.

1240. **Sawyer, C. H.**, **Markee, J. E.**, and **Townsend, B. F.** (1949b). Cholinergic and adrenergic components in the neurohumoral control of the release of LH in the rabbit. *Endocrinology* **44**, pp. 18–37.

1241. **Schachar-Dadon, A., Leshem, M.,** and **Schulkin, J.** (2009). Adversity before conception will affect adult progeny in rats. *Developmental Psychology* **45**, pp. 9–16.

1242. **Schally, A.** (1977). Aspects of hypothalamic regulation of the pituitary gland with major emphasis on its implications for the control of reproductive processes. Nobel Lecture. Retrieved from http://www.nobelprize.org/nobel_prizes/medicine/laureates/1977/schally-lecture.pdf

1243. **Schally, A. V., Saffran, M.,** and **Zimmermann, B.** (1958). A corticotrophin-releasing factor: partial purification and amino acid composition. *Biochemical Journal* **70**, p. 97.

1244. **Scharrer, B.,** and **Scharrer, E.** (1944). Neurosecretion VI. A comparsion between the intercerberalis cariacum allatum system of the insects and the hypothalamo-hypophyseal system of the vertebrates. *Biological Bulletin* **87**, pp. 242–251.

1245. **Schmidt, L. A., Fox, N. A., Rubin, K. H., Sternberg, E. M., Gold, P. W., Smith, C. C.,** and **Schulkin, J.** (1997). Behavioral and neuroendocrine responses in shy children. *Developmental Psychobiology* **30**, pp. 127–140.

1246. **Schmidt, M., Andrews, R., van der Ark, P., Brown, T., Mannaert, E., Steckler, T., De Hoon, J.,** and **Van Laere, K.** (2010). Dose-dependent effects of the CRF1 receptor antagonist R317573 on regional brain acitivty in healthy male subjects. *Psychopharmacology* **208**, pp. 109–119.

1247. **Schmitz, M. K., Botte, D. A., Sotto, M. N., Borba, E. F., Bonfa, E.,** and **de Mello, S. B. V.** (2015). Increased corticotropin-releasing hormone (CRH) expression in cutaneous lupus lesions. *Lupus* **24**, pp. 854–861.

1248. **Schneider, M. L., Clarke, A. S., Kraemer, G. W., Roughton, E. C., Lubach, G. R., Rimm-Kaufman, S. Schmidt, D.** and **Ebert, M.** (1998). Prenatal stress alters brain biogenic amine levels in primates. *Development and Psychopathology* **10**, pp. 427–440.

1249. **Schneider, M. L., Coe, C. L.,** and **Lubach, G. R.** (1992). Endocrine activation mimics the adverse effects of prenatal stress on the neuromotor development of the infant primate. *Developmental Psychobiology* **25**, pp. 427–439.

1250. **Schulger, J., Bart, G., Green, M., Ho, A.,** and **Kreek, M.** (2003). Corticotropin-releasing factor testing reveals a dose-dependent difference in methadone maintained vs control subjects. *Neuropsychopharmacology* **28**, pp. 985–994.

1251. **Schulkin, J.** (1999). Corticotropin-releasing hormone signals adversity in both the placenta and brain: regulation by glucocorticoids and allostatic overload. *Journal of Endocrinology* **161**, pp. 349–356.

1252. **Schulkin, J.** (1999). *The neuroendocrine regulation of behavior.* Cambridge: Cambridge University Press.

1253. **Schulkin, J.** (2003). *Rethinking homeostasis.* Cambridge, MA: MIT Press.

1254. **Schulkin, J.** (2011). Evolutionary conservation of glucocorticoids and corticotropin releasing hormone: behavioral and physiological adaptations. *Brain Research* **1392**, pp. 27–46.

1255. **Schulkin, J.** (2011). Social allostasis: anticipatory regulation of the internal milieu. *Frontiers in Evolutionary Neuroscience* **2**, pp. 1–15.

1256. **Schulkin, J., Gold, P. W.,** and **McEwen, B. S.** (1998). Induction of corticotropin-releasing hormone gene expression by glucocorticoids: implication for understanding the states of fear and anxiety and allostatic load. *Psychoneuroendocrinology* **23**, pp. 219–243.

1257. **Schulkin, J.**, **McEwen, B.**, and **Gold, P.** (1994). Allostasis, amygdala, and anticipatory angst. *Neuroscience and Biobehavioral Reviews* **18**, pp. 385–396.

1258. **Schulkin, J.**, **Morgan, M. A.**, and **Rosen, J. B.** (2005). A neuroendocrine mechanism for sustaining fear. *Trends in Neurosciences* **28**, pp. 629–635.

1259. **Schultz, W.** (2002). Getting formal with dopamine and reward. *Neuron* **36**, pp. 241–263.

1260. **Schultz, W.** (2007). Multiple dopamine functions at different time courses. *Annual Review of Neuroscience* **30**, pp. 59–88.

1261. **Schultz, W.**, **Preuschoff, K.**, **Camerer, C.**, et al. (2008). Explicit neural signals reflecting reward uncertainty. *Philosophical Transactions of the Royal Society* B **363**, pp. 3801–3811.

1262. **Schulz, C.**, **Paulus, K.**, **Lobmann, R.**, **Dallman, M.**, and **Lehnert, H.** (2010). Endogenous ACTH, not only a-melanocyte-stimulating hormone, reduces food intake mediated by hypothalamic mechanisms. *American Journal of Physiology, Endocrinology and Metabolism* **298**, pp. E237–E244.

1263. **Schumacher, M.**, **Coirini, H.**, **Pfaff, D. W.**, and **McEwen, B. S.** (1990). Behavioral effects of progesterone associated with rapid modulation of oxytocin receptors. *Science* **250**, pp. 691–694.

1264. **Schüssler, P.**, **Kluge, M.**, **Gamringer, W.**, **Wetter, T.C.**, **Yassouridis, A.**, **Uhr, M.**, **Rupprecht, R.**, **Steiger, A.** (2016), Corticotropin-releasing hormone induces depression-like changes of sleep electroencephalogram in healthy women. Psychoneuroendocrinology 74, pp. 302–307.

1265. **Schwartz, C. E.**, **Wright, C. I.**, **Shin, L. M.**, **Kagan, J.**, and **Rauch, S. L.** (2003). Inhibited and uninhibited infants" grown up": adult amygdalar response to novelty. *Science* **300**, pp. 1952–1953.

1266. **Schweitzer, M. H.**, **Suo, Z.**, **Avci, R.**, **Asara, J. M.**, **Allen, M. A.**, **Arce, F. T.**, and **Horner, J. R.** (2007). Analyses of soft tissue from Tyrannosaurus rex suggest the presence of protein. *Science* **316**, pp. 277–280.

1267. **Seasholtz, A.**, **Valverde, R.**, and **Denver, R.** (2002). Corticotropin-releasing hormone-binding protein: biochemistry and function from fishes to mammals. *Journal of Endocrinology* **175**, pp. 89–97.

1268. **Searcy, B.**, **Beckstrom-Sternberg, S.**, **Beckrstrom-Sternberg, J.**, **Stafford, P.**, **Schwendiman, A.**, **Soto-Pena, J.**, **Owen, M.**, **Ramirez, C.**, **Phillips, J.**, **Veldhoen, N.**, **Helbing, C.**, and **Propper, C.** (2012). Thyroid hormone-dependent development in Xenopus laevis: A sensitive screen of thyroid hormone signaling disruption by municipal wastewater treatment plant effluent. *General and Comparative Endocrinology* **176**, pp. 481–492.

1269. **Sears, R. M.**, **Fink, A. E.**, **Wigestrand, M. B.**, **Farb, C. R.**, **de Lecea, L.**, and **LeDoux, J. E.** (2013). Orexin/hypocretin system modulates amygdala-dependent threat learning through the locus coeruleus. *Proceedings of the National Academy of Sciences* **110**, pp. 20260–20265.

1270. **Seckl, J. R.**, and **Walker, B. R.** (2004). 11 beta-hydroxysteroid dehydrogenase type 1 as a modulator of glucocorticoid action: from metabolism to memory. *Trends in Endocrinology and Metabolism* **15**, pp. 418–424.

1271. **Seidel, K.**, **Poeggel, G.**, **Holetschka, R.**, **Helmeke, C.**, and **Braun, K.** (2011). Paternal deprivation affects the development of corticotropin releasing factor-expressing neurons in prefrontal cortex, amygdala, and hippocampus of the biparental octodon degus. *Journal of Neuroendocrinology* **23**, pp. 1166–1176.

1272. **Sellami, A.**, **Agricola, H. J.**, and **Veenstra, J. A.** (2011). Neuroendocrine cells in Drosophila melanogaster producing GPA2/GPB5, a hormone with homology to LH, FSH and TSH. *General and Comparative Endocrinology* **170**, pp. 582–588.

1273. **Seres, J.**, **Bornstein, S. R.**, **Seres, P.**, **Willenberg, H. S.**, **Schulte, K. M.**, **Scherbaum, W. A.**, and **Ehrhart-Bornstein, M.** (2004). Corticotropin-releasing hormone system in human adipose tissue. *Journal of Clinical Endocrinology and Metabolism* **89**, pp. 965–970.

1274. **Shachar-Dadon, A.**, **Schulkin, J.**, and **Leshem, M.** (2009). Adversity before conception will affect adult progeny in rats. *Developmental Psychology* **45**, pp. 9–16.

1275. **Shaham, Y.**, **Erb, S.**, and **Stewart, J.** (2000). Stress-induced relapse to heroin and cocaine seeking in rats: a review. *Brain Research Reviews* **33**, pp. 13–33.

1276. **Shaham, Y.**, **Funk, D.**, **Erb, S.**, **Brown, T. J.**, **Walker, C. D.**, and **Stewart, J.** (1997). Corticotropin-releasing factor, but not corticosterone, is involved in stress-induced relapse to heroin-seeking in rats. *Journal of Neuroscience* **17**, pp. 2605–2614.

1277. **Shalyapina, V.**, **Rakitskaya, V.**, **Petrova, E.**, and **Mironova, V.** (2004). Adaptive behavior in active and passive rats after intranasal administration of corticotropin-releasing hormone. *Neuroscience and Behavioral Physiology* **34**, pp. 193–197.

1278. **Shannon, M.**, **King, T. L.**, and **Kennedy, H. P.** (2007). Allostasis: a theoretical framework for understanding and evaluating perinatal health outcomes. *Journal of Obstetric, Gynecologic, and Neonatal Nursing* **36**, pp. 125–134.

1279. **Shapiro, L. E.**, and **Insel, T. R.** (1990). Infant's response to social separation reflects adult differences in affiliative behavior: A comparative developmetal study in prairie and montane voles. *Developmental Psychobiology* **23**, pp. 375–393.

1280. **Sharpe, A.**, and **Phillips, T. J.** (2009). Central urocortin 3 administration decreases limited-access ethanol intake in non dependent mice. *Behavioural Pharmacology* **20**, pp. 346–351.

1281. **Shekhar, A.**, **Johnson, P. L.**, **Fitz, S. D.**, **Nakazato, A.**, **Chaki, S.**, **Steckler, T.**, and **Schmidt, M.** (2011). A selective, non-peptide CRF receptor 1 antagonist prevents sodium lactate-induced acute panic-like responses. *International Journal of Neuropsychopharmacology* **14**, pp. 355–365.

1282. **Shekhar, A.**, **Truitt, W.**, **Rainne, D.**, and **Sajdyk, T.** (2005). Role of stress, corticotropin releasing factor (CRF) and amygdala plasticity in chronic anxiety. *Stress* **8**, pp. 209–219.

1283. **Sheline, Y. I.**, **Wang, P. W.**, **Gado, M. H.**, **Csernansky, J. G.**, and **Vannier, M. W.** (1996). Hippocampal atrophy in recurrent major depression. *Proceedings of the National Academy of Sciences* **93**, pp. 3908–3913.

1284. **Shemesh, Y.**, **Forkosh, O.**, **Mahn, M.**, **Anpilov, S.**, **Sztainberg, Y.**, **Manashirov, S.**, **Shlapobersky, T.**, **Elliott, E.**, **Tabouy, L.**, **Ezra, G.**, **Adler, E. S.**, **Ben-Efraim, Y. J.**, **Gil. S.**, **Kuperman, Y.**, **Haramati, S.**, **Dine, J.**, **Eder, M.**, **Deussing, J. M.**, **Schneidman, E.**, **Yizhar, O.**, and **Chen, A.** (2016). Ucn3 and CRF-R2 in the medial amygdala regulate complex social dynamics. *Nat Neurosci* **19**(11), pp. 1489–1496.

1285. **Sheng, H.**, **Xu, Y.**, **Chen, Y.**, **Zhang, Y.**, and **Ni, X.** (2012). Corticotropin-releasing hormone stimulates mitotic kinesin-like protein 1 expressionvia a PLC/PKC-depednent signaling pathway in hippocampal neurons. *Molecular and Cellular Endocrinology* **362**, pp. 157–164.

1286. **Sheng, H., Xu, Y., Chen, Y., Zhang, Y., Xu, X., He, C.,** and **Ni, X.** (2012). CRH-R1 and CRH-R2 differentially modulate dendritic outgrowth of hippocampal neurons. *Endocrine* **41**, pp. 458–464.

1287. **Shenoda, B. B., Alexander, G. M.,** and **Ajit, S. K.** (2016). Hsa-miR-34a mediated repression of corticotrophin releasing hormone receptor 1 regulates pro-opiomelanocortin expression in patients with complex regional pain syndrome. *J Transl Med* **3**(14), p. 64.

1288. **Shepard, J. D., Chambers, C. O., Busch, C., Mount, A.,** and **Schulkin, J.** (2009). Chronically elevated corticosterone in the dorsolateral bed nuclei of stria terminalis increases anxiety-like behavior. *Behavioural Brain Research* **203**, pp. 146–149.

1289. **Shepard, J. D., Schulkin, J.,** and **Myers, D. A.** (2006). Chronically elevated corticosterone in the amygdala increases corticotropin releasing factor mRNA in the dorsolateral bed nucleus of stria terminalis following duress. *Behavioural Brain Research* **174**, pp. 193–196.

1290. **Shepard, J. D., Liu, Y., Sassone-Corsi, P.,** and **Aguilera, G.** (2005). Role of glucocorticoids and cAMP-mediated repression in limiting corticotropin-releasing hormone transcription during stress. *Journal of Neuroscience* **25**, pp. 4073–4081.

1291. **Shepard, J. D., Barron, K. W.,** and **Myers, D. A.** (2000). Corticosterone delivery to the amygdala increases corticotropin-releasing factor mRNA in the central amygdaloid nucleus and anxiety-like behavior. *Brain Research* **861**, pp. 288–295.

1292. **Shepard, J. D., Chambers, C. O., Busch, C., Mount, A.,** and **Schulkin, J.** (2010). Behavioral effects of elevated of corticosterone in the medial prefrontal cortex. *Neurosci. Abstr.*

1293. **Sherman, J.,** and **Kalin, N.** (1987). The effects of ICV-CRH on novelty-induced behavior. *Pharmacology, Biochemistry, and Behavior* **26**, pp. 699–703.

1294. **Sherrin, T., Blank, T., Saravana, R., Rayner, M., Spiess, J.,** and **Todorovic, C.** (2009). Region specific gene expression profile in mouse brain after chronic corticotropin releasing factor receptor activation: the novel role for diazepam binding inhibitor in contextual fear conditioning. *Neuroscience* **162**, pp. 14–22.

1295. **Shettleworth, S. J.** (1971). Constraints on learning. In: Lehrman, D. S., Hinde, R. A., and Shaw, E. (eds.), *Advances in the study of behavior*. New York: Academic Press.

1296. **Shibasaki, T., Odagiri, E., Shizume, K.,** and **Ling, N.** (1982). Corticotropin-releasing factor-like activity in human placental extracts. *Journal of Clinical Endocrinology and Metabolism* **55**, pp. 384–386.

1297. **Shimmin, L., Sivamani, N., Heladio, I., Montasser, M., Kim, D., Hanis, C., Boerwinkle, E., Wadhwa, P.,** and **Hixson, J.** (2007). Corticotropin releasing hormone (CRH) gene variation: comprehensive resequencing for variant and molecular haplotype discovery in monosomic hybrid cell lines. *DNA Sequence* **6**, pp. 434–444.

1298. **Siiteri, P. K.,** and **MacDonald, P. C.** (1963). The utilization of circulating dehydroisoandrosterone sulfate for estrogen synthesis during human pregnancy. *Steroids* **2**, pp. 713–730.

1299. **Silberman, Y.,** and **Winder, D. G.** (2013). Corticotropin releasing factor and catecholamines enhance glutamatergic neurotransmission in the lateral subdivision of the central amygdala. *Neuropharmacology* **70**, pp. 316–323.

1300. **Silberman, Y., Fetterly, T. L., Awad, E. K., Milano, E. J., Usdin, T. B.,** and **Winder, D. G.** (2015). Ethanol Produces Corticotropin-Releasing Factor Receptor-Dependent

Enhancement of Spontaneous Glutamatergic Transmission in the Mouse Central Amygdala. *Alcoholism: Clinical and Experimental Research* **39**, pp. 2154–2162.

1301. **Silberman, Y., Matthews, R.**, and **Winder, D.** (2013). A corticotropin releasing factor pathway for ethanol regulation of the ventral tegmental area in the bed nucleus of the stria terminalis. *Journal of Neuroscience* **33**, pp. 950–960.

1302. **Sillaber, I., Rammes, G., Zimmermann, S., Mahal, B., Zieglgänsberger, W., Wurst, W., Holsboer, F.**, and **Spanagel, R.** (2002). Enhanced and delayed stress-induced alcohol drinking in mice lacking functional CRH1 receptors. *Science* **296**, pp. 931–933.

1303. **Silverman, E. S., Breault, D. T., Vallone, J., Subramanian, S., Yilmaz, A. D., Mathew, S., Subramaniam, V., Tantisira, K., Pacek, K., Weiss, S. T.**, and **Majzoub, J. A.** (2004). Corticotropin-releasing hormone deficiency increases allergen-induced airway inflammation in a mouse model of asthma. *Journal of Allergy and Clinical Immunology* **114**, pp. 747–754.

1304. **Singh, L., Boucher, W., Pang, X., Letourneau, R., Seretakis, D., Green, M.**, and **Theoharides, T.** (1999). Potent mast cell degranulation and vascular permeability triggered by urocortin through activation of corticotropin-releasing hormone receptors. *Journal of Pharmacology and Experimental Therapeutics* **288**, pp. 1349–1356.

1305. **Sink, K. S., Walker, D. L., Freeman, S. M., Flandreau, E. I., Ressler, K. J.**, and **Davis, M.** (2013). Effects of continuously enhanced corticotropin releasing factor expression within the bed nucleus of the stria terminalis on conditioned and unconditioned anxiety. *Molecular Psychiatry* **18**, pp. 308–319.

1306. **Sithisarn, T., Bada, H. S., Dai, H., Randall, D. C.**, and **Legan, S. J.** (2011). Effects of perinatal cocaine exposure on open field behavior and the response to corticotropin releasing hormone (CRH) in rat offspring. *Brain Research* **1370**, pp. 136–144.

1307. **Skelton, K. H., Owens, M. J., Gutman, D. A., O'Brien, D., Oren, D., Thrivilaaman, K. V., Easterling, K. W., Holtzmans, G., Knight, D. L.**, and **Nemeroff, C. B.** (2000). The corticotropin-releasing factor (CRF)1 receptor antagonist, R121919, attenuates the behavioral and neuroendocrine response to precipitated morphine or lorazepam withdrawal in the rat. Poster presented at the Second International Congress on Hormones, Brain and Neuropsychopharmacology. Rhodes, Greece.

1308. **Skofitsch, G.**, and **Jacobowitz, D.** (1984). Corticotropin releasing factor-like immunoreactive neurons in the rat retina. *Brain Research Bulletin* **12**, pp. 539–542.

1309. **Skofitsch, G., Jacobowitz, R., Eskay, I.**, and **Zamir, N.** (1985). Distribution of atrial natriuretic factor-like immunoreactive neurons in the rat brain. *Neuroscience* **16**, pp. 917–948.

1310. **Skórzewska, A., Bidziński, A., Lehner, M., Turzyńska, D., Sobolewska, A., Hamed, A., Szyndler, J., Maciejak, P.**, and **Plaznik, A.** (2007). The effects of acute corticosterone administration on anxiety, endogenous corticosterone, and c-Fos expression in the rat brain. *Hormones and Behavior* **52**, pp. 317–325.

1311. **Skórzewska, A., Bidziński, A., Lehner, M., Turzyńska, D., Sobolewska, A., Wisłowska-Stanek, A., Maciejak, P., Szyndler, J.**, and **Płaźnik, A.** (2011). The localization of brain sites of anxiogenic-like effects of urocortin-2. *Neuropeptides* **45**, pp. 83–92.

1312. **Skórzewska, A., Lehner, M., Hamed, A., Wisłowska-Stanek, A., Turzyńska, D., Sobolewska, A.**, and **Płaźnik, A.** (2011). The effect of CRF 2 receptor antagonists on rat conditioned fear responses and c-Fos and CRF expression in the brain limbic structures. *Behavioural Brain Research* **221**, pp. 155–165.

1313. **Skutella, T., Montkowski, A., Stohr, T., Probst, J. C., Landgraf, R., Holsboer, F.,** and **Jirkowski, G. F.** (1994). Corticotropin-releasing hormone (CRH) antisense oligodeoxynucleotide treatment attenuates social defeat-induced anxiety. *Cellular and Molecular Neurobiology* **14**, pp. 579–588.
1314. **Slater, P. G., Noches, V.,** and **Gysling, K.** (2016). Corticotropin-releasing factor type-2 receptor and corticotropin-releasing factor-binding protein coexist in rat ventral tegmental area nerve terminals originated in the lateral hypothalamic area. *European Journal of Neuroscience* **43**, pp. 220–229.
1315. **Slawecki, C., Thorsell, A., Khoury, A., Mathe, A.,** and **Ehlers, C.** (2005). Increased CRF-like and NPY-like immunoreactivity in adult rats exposed to nictone during adolescence: Relation to anxiety-like and depressive-like behavior. *Neuropeptides* **39**, pp. 369–377.
1316. **Slominski, A. T.** (2007). A nervous breakdown in the skin: stress and the epidermal barrier. *Journal of Clinical Investigation* **117**, pp. 3166–3169.
1317. **Slominski, A. T.** (2009). Neuroendocrine activity of the melanocyte. *Experimental Dermatology* **18**, pp. 760–763.
1318. **Slominski, A. T.** (2009). On the role of the corticotropin-releasing hormone signalling system in the aetiology of inflammatory skin disorders. *British Journal of Dermatology* **160**, pp. 229–232.
1319. **Slominski, A. T.** and **Wortsman, J.** (2000). Neuroendocrinology of the Skin. *Endocrine Reviews* **21**, pp. 457–487.
1320. **Slominski, A. T., Zmijewski, M. A., Skobowiat, C., Zbytek, B., Slominski, R. M., &** and **Steketee, J. D.** (2012). Sensing the environment: regulation of local and global homeostasis by the skin's neuroendocrine system. *Advances in Anatomy Embryology and Cell Biology* **212**, pp. 1–115.
1321. **Slominski, A. T., Pisarchik, A. Tobin, D., Mazurkiewicz, J.,** and **Wortsman, J.** (2004). Differential expression of a cutaneous corticotropin releasing hormone system. *Endocrinology* **145**, pp. 941–950.
1322. **Slominski, A. T., Wortsman, J., Tuckey, R.,** and **Paus, R.** (2007). Differential expression of HPA axis homolog in the skin. *Molecular and Cellular Endocrinology* **265**, pp. 143–149.
1323. **Slominski, A. T., Zbytek, B., Pisarchik, A., Slominski, R.M., Zmijewski, M.,** and **Wortsman, J.** (2006). CRH functions as a growth factor/cytokine in the skin. *Journal of Cellular Physiology* **206**, pp. 780–791.
1324. **Slominski, A. T., Zbytek, B., Semak, I., Sweatman, T.,** and **Wortsman, J.** (2005). CRH stimulates POMC activity and corticosterone production in dermal fibroblasts. *Journal of Neuroimmunology* **162**, pp. 97–102.
1325. **Slominski, A. T., Zybtek, B., Nikolakis, G., Manna, P., Skobowiat, C., Smijewski, M., Li, W., Janjetovic, Z., Postlethwaite, A., Zoubloulis, C. C.,** and **Tuckey, R. C.** (2013). Steroidgeneiss in the skin: implications for local immune functions. *Journal of Steroid, Biochemistry, and Molecular Biology* **137**, pp. 107–123.
1326. **Smagin, G. N.,** and **Dunn, A. J.** (2000). The role of CRF receptor subtypes in stress-induced behavioural responses. *European Journal of Pharmacology* **405**, pp. 199–206.

1327. **Smedh, U.**, **Scott, K. A.**, and **Moran, T. H.** (2015). Fourth ventricular CART peptide induces c-fos in the area postrema and nucleus of the solitary tract via a CRF-receptor dependent mechanism. *Neuroscience Letters* **609**, pp. 124–128.

1328. **Smith, G. P.** (1997a). Eating and the American zeitgeist. *Appetite* **29**, pp. 191–200.

1329. **Smith, G. P.** (1997b). *Satiation: from gut to brain*. Oxford: Oxford University Press.

1330. **Smith, G. W.**, **Aubry, J. M.**, **Dellu, F.**, **Contarino, A.**, **Bilezkiianl, M.**,**Gold, L. H.**, **Chen, R.**, et al. (1998). Corticotropin releasing factor receptor 1-deficient mice display decreased anxiety, impaired stress response, and aberrant neuroendocrine development. *Neuron* **20**, pp. 1093–1102.

1331. **Smith, R.** (1998). Alterations in the hypothalamic pituitary adrenal axis during pregnancy and the placental clock that determines the length of parturition. *Journal of Reproductive Immunology* **39**, pp. 215–220.

1332. **Smith, R.** (2007). Parturition. *New England Journal Of Medicine* **356**, pp. 271–283.

1333. **Smith, R.**, **Chan, E. C.**, **Bowman, M. E.**, **Harewood, W. J.**, and **Phippard, A. F.** (1993). Corticotropin releasing hormone in baboon pregnancy. *Journal of Clinical Endocrinology & Metabolism* **76**, pp. 1063–1068.

1334. **Smith, R.**, **Mesiano, S.**, **Chan, E. C.**, **Brown, S.**, and **Jaffe, R. B.** (1998). Corticotropin-releasing hormone directly stimulates dehydroepiandrosterone sulfate excretion by human feltal adrenal cortical cells. *Journal of Clinical Endocrinology & Metabolism* **83**, pp. 2916–2920.

1335. **Smith, R.**, **Paul, J.**, **Maiti, K.**, **Tolosa, J.**, and **Madsen, G.** (2012). Recent advances in understanding endocrinology of human birth. *Trends in Endocrinology & Metabolism* **23**, pp. 516–523.

1336. **Smith, S.**, **De Jonge, L.**, **Pellymounter, M.**, **Nguyen, T.**, **Harris, R.**, **York, D.**, **Redmann, S.**, **Rood, J.**, and **Bray, G.** (2001). Peripheral administration of human corticotropin releasing hormone: a novel method to increase energy expenditure and fat oxidation in man. *Journal of Clinical Endocrinology & Metabolism* **86**, pp. 413–425.

1337. **Smith, R.**, **Smith, J. I.**, **Shen, X.**, **Engel, P. J.**, **Bowman, M. E.**, **McGrath, S. A.**, **Bisits, A. M.**, **McElduff, P.**, **Giles, W. B.**, and **Smith, D. W.** (2009). Patterns of plasma corticotropin-releasing hormone, progesterone, estradiol, and estriol change and the onset of human labor. *J Clin Endocrinol Metab* **94**(6), p. 2066.

1338. **Smith, R.**, **Wickings, E. J.**, **Bowman, M. E.**, **Belleoud, A.**, **Dubreuil, G.**, **Davies, J. J.**, and **Madsen, G.** (1999). Corticotropin-releasing hormone in chimpanzee and gorilla pregnancies. *Journal of Clinical Endocrinology and Metabolism* **84**, pp. 2820–2825.

1339. **Solomon, M.**, **Loftspring, M.**, **de Kloet, A.**, **Ghosal, S.**, **Jankord, R.**, **Flak, J.**, **Wulsin, A.**, **Krause, E.**, **Zhang, R.**, **Rice, T.**, **McKlveen, J.**, **Myers, B.**, **Tasker, J.**, and **Herman, J.** (2015). Neuroendocrine function after hypothalamic depletion of glucocorticoid receptors in male and female mice. *Endocrinology* **156**, pp. 2843–2853.

1340. **Sommer, W. H.**, **Rimondini, R.**, **Hansson, A. C.**, **Hipskind, P. A.**, **Gehlert, D. R.**, **Barr, C. S.**, and **Heilig, M. A.** (2008). Upregulation of voluntary alcohol intake, behavioral sensitivity to stress, and amygdala Crhr1 expression following a history of dependence. *Biological Psychiatry* **63**, pp. 139–145.

1341. **Sommerer, R.**, **Yamanaka, A.**, **Diano, S.**, **Borok, E.**, **Roberts, A.**, **Sakurai, T.**, **Kilduff, T.**, **Horvath, T.**, and **de Lecea, L.** (2004). Interaction between the corticotropin-releasing factor system and hypocretins (orexins): a novel circuit mediating stress response. *Journal of Neuroscience* **24**, pp. 11439–11448.

1342. **Sotres-Bayon, F., Cain, C. K.,** and **LeDoux, J. E.** (2006). Brain mechanisms of fear extinction: historical perspectives on the contribution of prefrontal cortex. *Biological Psychiatry* **15**, pp. 329–336.

1343. **Sower, S., Freamat, M.,** and **Kavanaugh, S.** (2009). The origins of the vertebrate hypothalamic-pituitary-gonadal (HPG) and hypothalamic-pituitary-thyroid (HPT) endocrine systems: New insights from lampreys. *General and Comparative Endocrinology* **161**, pp. 20–29.

1344. **Sparta, D. R., Ferraro, F. M. 3rd, Fee, J. R., Knapp, D. J., Breese, G. R.,** and **Thiele, T. E.** (2009). The alcohol deprivation effect in C57BL/6J mice is observed using operant self-administration procedures and is modulated by CRF-1 receptor signaling. *Alcoholism Clinical and Experimental Research* **33**, pp. 31–42.

1345. **Spencer, S. J., Buller, K. M.,** and **Day, T. A.** (2005). Medial prefrontal cortex control of the paraventricular hypothalamic nucleus response to psychological stress: possible role of the bed nucleus of the stria terminalis. *Journal of Comparative Neurology* **481**, pp. 363–376.

1346. **Spiga, F., Lightman, S., Shekhar, A.,** and **Lowry, C.** (2006). Injections of urocortin 1 into the basolateral amygdala induce anxiety-like behavior and C-Fos expression in brainstem serotonergic neurons. *Neuroscience* **138**, pp. 1265–1276.

1347. **Squire, L. R.** (2004). Memory systems of the brain: a brief history and current perspective. *Neurobiology of Learning and Memory* **82**, pp. 171–177.

1348. **Starling, E. H.** (1923). *The wisdom of the body*. London: H.K. Lewis

1349. **Staub, D., Lunden, A., Cathel, E., Dolben, E.,** and **Kirby, L.** (2012). Morphine history sensitizes postsynaptic GABA receptors in dorsal raphe serotonin neurons in a stress-induced relapse model in rats. *Psychoneuroendocrinology* **37**, pp. 859–870.

1350. **Steckler, T.,** and **Holsboer, F.** (2001). Conditioned activity to amphetamine in transgenic mice expressing and antisense RNA against the glucocorticoid receptor. *Behavioral Neuroscience* **115**, pp. 207.

1351. **Steenbergen, P. J., Richardson, M. K.,** and **Champagne, D. L.** (2011). The use of the zebrafish model in stress research. *Progress in Neuro-Psychopharmacology and Biological Psychiatry* **35**, pp. 1432–1451.

1352. **Steiger, A.** (2002). Sleep and the hypothalamo–pituitary–adrenocortical system. *Sleep Medicine Reviews* **6**, pp. 125–138.

1353. **Stellar, E.** (1954). The physiology of motivation. *Psychological Review* **61**, pp. 5–22.

1354. **Stengel, A., Goebel, M., Million, M., Stenzel-Poore, M., Kobelt, P., Monnikes, H., Tache, Y.,** and **Wang, L.** (2009). Corticotropin-releasing factor-overexpressing mice exhibit reduced neuronal activation in the arcuate nucleus and food intake in response to fasting. *Endocrinology* **150**, pp. 153–160.

1355. **Stenzel-Poore, M., Heinrichs, S., Rivest, S., Koob, G.,** and **Vale, W.** (1994). Overproduction of corticotropin-releasing factor in transgenic mice: genetic model of anxiogenic behavior. *Journal of Neuroscience* **14**, pp. 2579–2584.

1356. **Stenzel-Poore, M. P., Heldwein, K. A., Stenzel, P., Lee, S.,** and **Vale, W. W.** (1992). Characterization of the genomic corticotropin releasing factor (CRF) gene from Xenus laevis: two members of the CRF family exist in amphibians. *Molecular Endocrinology* **6**, pp. 1716–1724.

1357. **Sternberg, E. M., Glowa, J. R., Smith, M. A., Cologero, A. E., Listwak, S. J., Aksentijevich, S., Chrousos, G.P., Wilder, R.L.,** and **Gold, P. W.** (1992).

Corticotropin releasing hormone related behavioral and neuroendocrine responses to stress in Lewis and Fischer rats. *Brain Research* **570**, pp. 54–60.

1358. **Sterrenburg, L., Gaszner, B., Boerrigter, J., Santbergen, L., Bramini, M., Elliott, E., Chen, A., Peeters, B. W. M. M., Roubos, E. W.,** and **Kozicz, T.** (2011). Chronic stress induces sex-specific alterations in methylation and expression of corticotropin-releasing factor gene in the rat. *PLoS One* **6**, p. e28128.

1359. **Stevens, S., Shaw, T., Dykhuizen, E., Lessov, N., Hill, J., Wurst, W.,** and **Stenzel-Poore, M.** (2003). Reduced cerebral injury in the CRH-R1 deficient mice after focal ischemia: a potential link to microglia and astrocytes that express CRH-R1. *Journal of Cerebral Blood Flow and Metabolism* **23**, pp. 1151–1159.

1360. **Stiedl, O., Meyer, M., Jahn, O., Ogren, S.,** and **Spiess, J.** (2005). Corticotropin-releasing factor receptor 1 and central heart rate regulation in mice during expression of conditioned fear. *Journal of Pharmacology and Experimental Therapeutics* **312**, pp. 905–916.

1361. **Stockhorst, U.,** and **Antov, M. I.** (2016). Modulation of fear extinction by stress, stress hormones and estradiol: a review. *Frontiers in Behavioral Neuroscience* **9**, pp. 1–26.

1362. **Stöhr, T., Szuran, T., Welzl, H., Pliska, V., Feldon, J.,** and **Pryce, C. R.** (2000). Lewis/Fischer rat strain differences in endocrine and behavioural responses to environmental challenge. *Pharmacology, Biochemistry, and Behavior* **67**, pp. 809–819.

1363. **Strand, F. L.** (1999). *Neuropeptides: regulators of physiological processes.* Cambridge, MA: MIT Press.

1364. **Stricker-Krongrad, A.,** and **Beck, B.** (2002). Modulation of hypothalamic hypocretin/orexin mRNA expression by glucocorticoids. *Biochemical and Biophysical Research Communication* **296**, pp. 129–133.

1365. **Stuber, G. D., Sparta, D. R., Stamatakis, A. M., van Leeuwen, W. A., Hardjoprajitno, J. E., Cho, S., Tye, K.M., Kempadoo, K.A., Zhang, F., Deisseroth, K.,** and **Bonci, A.** (2011). Excitatory transmission from the amygdala to nucleus accumbens facilitates reward seeking. *Nature* **475**, pp. 377–380.

1366. **Sullivan, G., Apergis, J., Bush, D., Johnson, L., Hou, M.,** and **Ledoux, J.** (2004). Lesions in the bed nucleus of the stria terminalis disrupt corticosterone and freezing responses elicited by a contextual but not by a specific cue-conditioned fear stimulus. *Neuroscience* **128**, pp. 7–14.

1367. **Sullivan, R. M.,** and **Gratton, A.** (1999). Lateralized effects of medial prefrontal cortex lesions on neuroendocrine and autonomic stress responses in rats. Journal of *Neuroscience* **19**, pp. 2834–2840.

1368. **Sullivan, R. M.,** and **Gratton, A.** (2002). Prefrontal cortical regulation of hypothalamic-pituitary-adrenal function in the rat and implications for psychopathology: side matters. *Psychoneuroendocrinology* **27**, pp. 99–114.

1369. **Swanson, L. W.** (1996). *Molecular mechanisms of neuronal communication.* Oxford: Pregamon.

1370. **Swanson, L. W.** (1983). Neuropeptides—new vistas on synaptic transmission. *Trends in Neurosciences* **6**, pp. 294–295.

1371. **Swanson, L.W.,** and **Simmons, D. M.** (1989). Differential steroid hormone and neural influences on peptide mRNA levels in CRH cells of the paraventricular nucleus: a hybridization histochemical study in the rat. *Journal of Comparative Neurology* **285**, pp. 413–435.

1372. **Swanson, L. W.**, and **Petrovich, G. D.** (1998). What is the amygdala? *Trends in Neuroscience* **21**, pp. 323–331.

1373. **Swanson, L. W.**, **Sawchenko, P. E.**, **Lind, R. W.**, and **Rho, J. H.** (1987). The CRH motoneuron: differential peptide regulation in neurons with possible synaptic, paracrine, and endocrine outputs. *Annals of the New York Academy of Sciences* **512**, pp. 12–23.

1374. **Swanson, L.W.**, **Sawchenko, P.**, **River, J.**, and **Vale, W.** (1983). Organization of ovine corticotropin-releasing factor immunoreactive cells and fibers in the rat brain: an immunohistochemical study. *Neuroendocrinology* **36**, pp. 165–186.

1375. **Swanson, L. W.** (1991). Biochemical switching in hypothalamic circuits mediating responses to stress. *Progress in Brain Research* **87**, pp. 181–200.

1376. **Swanson, L. W.**, and **Mogenson, G. J.** (1981). Neural mechanisms for the functional coupling of autonomic, endocrine and somatomotor responses in adaptive behavior. *Brain Research* **228**, pp. 1–34.

1377. **Swanson, L. W.**, and **Simmons, D. M.** (1989). Differential steroid hormone and neural influences on peptide mRNA levels in CRH cells of the paraventricular nucleus: a hybridization histochemical study in the rat. *Journal of Comparative Neurology* **285**, pp. 413–435.

1378. **Sweetser, S.**, **Camilleri, M.**, **Linker, N.**, **Burton, D.**, **Castenada, L.**, **Croop, R.**, **Tong, G.**, **Dockens, R.**, and **Zinsmeister, A.** (2009). Do corticotropin releasing factor-1 receptors influence colonic transit and bowel function in women with irritable bowel syndrome? *American Journal of Physiology, Gastointestinal Liver Physiology* **296**, pp. 1299–306.

1379. **Swerdlow, N. R.**, **Britton, K. T.**, and **Koob, G. F.** (1989). Potentiation of acoustic startle by corticotropin-releasing factor (CRF) and by fear are both reversed by! a-helical CRF. *Neuropsychopharmacology* **2**, pp. 285–292.

1380. **Sztainberg, Y.**, **Kuperman, Y.**, **Justice, N.**, and **Chen, A.** (2011). An anxiolytic role for CRF receptor type 1 in the globus pallidus. *Journal of Neuroscience* **31**, pp. 17416–17424.

1381. **Sztainberg, Y.**, **Kuperman, Y.**, **Tsoory, M.**, **Lebow, M.**, and **Chen, A.** (2010). The anxiolytic effect of environmental enrichment is mediated via amygdalar CRF receptor type 1. *Molecular Psychiatry* **15**, pp. 905–917.

1382. **Tabarin, A.**, **Diz-Chaves, Y.**, **Consoli, D.**, **Monsaingeon, M.**, **Bale, T. L.**, **Culler, M. D.**, **Datta, R.**, **Drago, F.**, **Vale, W. W.**, **Koob, G. F.**, **Zorrilla, E. P.**, and **Contarino, A.** (2007). Role of the corticotropin-releasing factor receptor type 2 in the control of food intake in mice: a meal pattern analysis. *European Journal of Neuroscience* **26**, pp. 2303–2314.

1383. **Tachė, Y.**, **Martinez, V.**, **Million, M.**, and **Wang, L.** (2001). Stress and the gastrointestinal tract III: stress-related alterations of gut motor reactions: brain corticotropin-releasing factor receptors. *American Journal of Physiology, Gastrointestinal and Liver Physiology* **280**, pp. G173–G177.

1384. **Taché, Y.**, **Million, M.**, **Nelson, A. G.**, **Lamy, C.**, and **Wang, L.** (2005). Role of corticotropin-releasing factor pathways in stress-related alterations of colonic motor function and viscerosensibility in female rodents. *Gender Medicine* **2**, pp. 146–154.

1385. **Tachė, Y.**, and **Perdue, M. H.** (2004). Role of peripheral CRF signalling pathways in stress-related alterations of gut motility and mucosal function. *Neurogastroenterology and Motility* **16**, pp. 137–142.

1386. **Takahashi, L. K.** (2014). Olfactory systems and neural circuits that modulate predator odor fear. *Frontiers in Behavioral Neuroscience* **8**, p. 72.

1387. **Takahashi, L. K.**, and **Kim, H.** (1994). Intracranial action of corticosterone facilitates the development of behavioral inhibition in the adrenalectomized preweanling rat. *Neuroscience Letters* pp. 272–276.

1388. **Takahashi, L. K.**, **Kalin, N.**, and **Baker, E.** (1990). Corticotropin-releasing factor antagonist attenuates defensive-withdrawal behavior elicited by odors of stressed conspecifics. *Behavioral Neuroscience* **104**, pp. 386–389.

1389. **Takahashi, L. K.** (2001). Role of CRF(1) and CRF(2) receptors in fear and anxiety. *Neuroscience and Biobehaviorl Reviews* **25**, pp. 627–636.

1390. **Takahashi, L. K.**, **Chan, M. M.**, and **Pilar, M. L.** (2008). Predator odor fear conditioning: current perspectives and new directions. *Neuroscience and Biobehavioral Reviews* **32**, pp. 1218–1227.

1391. **Takahashi, L. K.**, **Kalin, N. H.**, **Vanden-Burgt, J. A.**, and **Sherman, J. E.** (1989). Corticotropin-releasing hormone modulates defensive-withdrawal and exploratory behavior in rats. *Behavioral Neuroscience* **3**, pp. 648–654.

1392. **Takahashi, L. K.**, **Nakashima, B. R.**, **Hong, H.**, and **Watanabe, K.** (2005). The smell of danger: a behavioral and neural analysis of predator odor-induced fear. *Neuroscience and Biobehavioral Review* **29**, pp. 1157–1167.

1393. **Takahashi, L. K.**, and **Rubin, W. W.** (1993). Corticosteroid induction of threat-induced behavioral inhibition in preweanling rats. *Behavioral Neuroscience* **107**, pp. 860–866.

1394. **Takahashi, L. K.**, **Turner, J. G.**, and **Kalin, N. H.** (1998). Prolonged stress-induced elevation in plasma corticosterone during pregnancy in the rat: implications for prenatal stress studies. *Psychoneuroendocrinology* **23**, pp. 571–581.

1395. **Takayanagi, Y.**, **Yoshida, M.**, **Bielsky, I. F.**, **Ross, H. E.**, **Kawamata, M.**, **Onaka, T.**, **Yanagisawa, T.**, **Kimura, T.**, **Matzuk, M. M.**, **Young, L. J.**, and **Nishimori, K.** (2005). Pervasive social deficits, but normal parturition, in oxytocin receptor-deficient mice. *Proceedings of the National Academy of Sciences of the United States of America* **102**, pp. 16096–16101.

1396. **Tan, H.**, **Yi, L.**, **Rote, N.S.**, **Hurd, W. W.**, and **Mesiano, S.** (2012). Progesterone receptor-A and -B have opposite effects on proinflammatory gene expression in human myometrical cells: implications for progesterone actions in human pregnancy and parturition. *Journal of Clinical Endocrinology and Metabolism* **97**, pp. E719–E730.

1397. **Tang, A.**, **Beaton, E.**, **Schulkin, J.**, **Hall, G.**, and **Schmidt, L.** (2014). Revisiting shyness and sociability: a preliminary investigation of hormone-brain-behavior relations. *Frontiers in Psychology*: **5**, pp. 1–11.

1398. **Tanimura, S. M.**, and **Watts, A. G.** (1998). Corticosterone can facilitate as well as inhibit corticotropin-releasing hormone gene expression in the rat hypothalamic paraventricular nucleus. *Endocrinology* **139**, pp. 3830–3836.

1399. **Tariverdian, N.**, **Rucke, M.**, **Szekeres-Bartho, J.**, **Blois, S. M.**, **Karpf, E. F.**, **Sedlmayr, P.**, **Klapp, B. F.**, **Kentenich, H.**, **Siedentopf, F.**, and **Arck, P. C.** (2010). Neuroendocrine circuitry and endometriosis: progesterone derivative dampens corticotropin-releasing hormone-induced inflammation by peritoneal cells in vitro. *Journal of Molecular Medicine* **88**, pp. 267–278.

1400. **Tarjan, E., Denton, D. A., Ferraro, T.,** and **Weisinger, R. S.** (1992). Effect of CRF, ACTH and adrenal steroids on sodium intake and excretion of rabbits. *Kidney International* **37**, pp. S97–101.

1401. **Tasker, J. G., Chen, C., Fisher, M. O., Fu, X., Rainville, J. R.,** and **Weiss, G. L.** (2015). Endocannabinoid regulation of neuroendocrine systems. *International Review of Neurobiology* **125**, pp. 163–201.

1402. **Tasker, J. G., Di, S.,** and **Malcher-Lopes, R.** (2006). Minireview: rapid glucocorticoid signaling via membrane-associated receptors. *Endocrinology* **147**, pp. 5549–5556.

1403. **Taylor, J. S., Van de Peer, Y., Braasch, I.,** and **Meyer, A.** (2001). Comparative genomics provides evidence for an ancient genome duplication event in fish. Philosophical *Transactions of the Royal Society of London B, Biological Sciences* **356**, pp. 1661–1679.

1404. **Taylor, S. E., Klein, L. C., Lewis, B. P., Gruenwald, T. L., Gurung, R. A.,** and **Updegraff, J. A.** (2000). Biobehavioral responses to stress in females: tend and befriend, not fight or flight. *Psychological Reviews* **107**, pp. 411–429.

1405. **Theoharides, T.,** and **Cochrane, D.** (2004). Critical role of mast cells in inflammatory diseases and the effect of acute stress. *Journal of Neuroimmunology* **146**, pp. 1–12.

1406. **Theoharides, T., Donelan, J., Papadopoulou, N., Cao, J., Kempuraj, D.,** and **Conti, P.** (2004). Mast cells as targets of corticotropin-releasing factor and related peptides. *TRENDS in Pharmacological Sciences* **25**, pp. 563–568.

1407. **Theoharides, T. C., Singh, L. K., Boucher, W., Pang, X., Letourneau, R., Webster, E.,** and **Chrousos, G.** (1998).Corticotropin-releasing hormone induces skin mast cell degranulation and incrased vascular permeability, a possible explanation for its proinflammatory effects. *Endocrinology* **139**, pp. 403–413.

1408. **Theoringer, C., Henes, K., Eder, M., Dahloff, M., Wurst, W., Holsbboer, F., Deurssing, J., Moosmang, S.,** and **Wotjak, C.** (2012). Consolidation of remote fear memories involves corticotropin-releasing hormone (CRH) receptor type 1-mediated enhancement of AMPA receptor GluR 1 signaling in the dentate gyrus. *Neuropsychopharmacology* **37**, pp. 787–796.

1409. **Thoeringer, C. K., Henes, K., Eder, M., Dahlhoff, M., Wurst, W., Holsboer, F., Deussing, J. M., Moosmang, S.,** and **Wotjak, C. T.** (2012). Consolidation of remote fear memories involves corticotropin-releasing hormone (CRH) receptor type 1-mediated enhancement of AMPA receptor GluR1 signaling in the dentate gyrus. *Neuropsychopharmacology* **37**, pp. 787–796.

1410. **Thomas, K. L.,** and **Everitt, B. J.** (2001). Limbic-cortical-ventral striatal activation during retrieval of a discrete cocaine-associated stimulus: a cellular imaging study with γ protein kinase C expression. *Journal of Neuroscience* **21**, pp. 2526–2535.

1411. **Thompson, B. L., Schulkin, J.,** and **Rosen, J. B.** (2000). *Chronic corticosterone enhances contextual fear conditioning and CRH mRNA expression in the amygdala.* New Orleans: Society for Neuroscience.

1412. **Thompson, B. L., Erickson, K., Schulkin, J.,** and **Rosen, J. B.** (2004). Corticosterone facilitates retention of contextual fear conditioning and increases CRH mRNA expression in the amygdala. *Behavioural Brain Research* **149**, pp. 209–215.

1413. **Thomson, M.** (2013). The physiological roles of placental corticotropin releasing hormone in pregnancy and childbirth. *Journal of Physiology and Biochemistry* **69**, pp. 559–573.

1414. **Thoreau, H. D.** (1971). *Great short works*. New York: Harper and Row.

1415. **Thornton, J. W., Need, E.,** and **Crews, D.** (2003). Resurrecting the ancestral steroid receptor: ancient origin of estrogen signaling. *Science* **301**, pp. 1714–1717.

1416. **Timpl, P., Spanagel, R., Sillaber, I., Kresse, A., Reul, J., Stalla, G., Blanquet, V., Steckler, T., Holsboer, F.,** and **Wurst, W.** (1998). Impaired stress response and reduced anxiety in mice lacking a functional corticotropin releasing hormone receptor 1. *Nature Genetics* **19**, pp. 162–166.

1417. **Timpl, P., Spanagel, R., Sillaber, I., Kresse, A., Reul, J. M., Stalla, G.K., Blanquet, V., Steckler, T., Holsboer, F.,** and **Wurst, W.** (1998). Impaired stress response and reduced anxiety in mice lacking a functional corticotropin-releasing hormone receptor 1. *Nature Genetics* **19**, pp. 162–166.

1418. **Tan, L., Chand, D., Almeida, R., Xu, M., De Lannoy, L.,** and **Lovejoy, D.** (2012). Modulation of neuroplastic changes and corticotropin-releasing factor-associated behavior by a phylogenetically ancient and conserved peptide family. *General and Comparative Endocrinology* **176**, pp. 309–313.

1419. **To, C. T., Anheuer, Z. E.,** and **Bagdy, G.** (1999). Effects of acute and chronic fluoxetine treatment on CRH-induced anxiety. *Neuroreport* **10**, pp. 553–555.

1420. **Tobin, D.** (2006). Biochemistry of human skin—our brain on the outside. *Chemical Society Reviews* **35**, pp. 52–67.

1421. **Tomiyama, A., Dallman, M.,** and **Epel, E.** (2011). Comfort food is comforting to those most stressed: evidence of the chronic stress response network in high stress women. *Psychoneuroendocrinology* **36**, pp. 1513–1519.

1422. **Torres, G., Horowitz, J. M., Laflamme, N.,** and **Rivest, S.** (1998). Fluoxetine induces the transcription of genes encoding c-fos, corticotropin-releasing factor and its type 1 receptor in rat brain. *Neuroscience* **87**, pp. 463–477.

1423. **Torricelli, M., Novembri, R., Bloise, E., De Bonis, M., Challis, J. R.,** and **Petraglia, F.** (2011a). Changes in placental CRH, urocortins, and CRH-Receptor mRNA expression associated with preterm delivery and chorioamnionitis. *Journal of Clinical Endocrinology & Metabolism* **96**, pp. 534–550.

1424. **Torricelli, M., Novembri, R., Voltolini, C., Conti, N., Biliotti, G., Piccolini, E., Cevenini, G., Smith, R.,** and **Petraglia, F.** (2011b). Biochemical and biophysical predictors of the response to the induction of labor in nulliparous postterm pregnancy. *American Journal of Obstetrics and Gynecology* **204**, pp. 39. e1–e6.

1425. **Toth, M., Flandreau, E. I., Deslauriers, J., Geyer, M. A., Mansuy, I. M., Pich, E. M.,** and **Risbrough, V. B.** (2015). Overexpression of forebrain crh during early life increases trauma susceptibility in adulthood. *Neuropsychopharmacology* **45**, pp. 1681–1690.

1426. **Toth, M., Gresack, J. E., Hauger, R. L., Halberstadt, A. L.,** and **Risbrough, V. B.** (2013). The role of PKC signaling in CRF-induced modulation of startle. *Psychopharmacology* **229**, pp. 579–589.

1427. **Toufexis, D.** (2007). Region- and sex-specific modulation of anxiety behaviours in the rat. *Journal of Neuroendocrinology* **19**, pp. 461–473.

1428. **Toufexis, D. J., Davis, C., Hammond, A.,** and **Davis, M.** (2004). Progesterone attenuates corticotropin-releasing factor-enhanced but not fear-potentiated startle via the activity of its neuroactive metabolite, allopregnanolone. *Journal of Neuroscience* **24**, pp. 10280–10287.

1429. **Toufexis, D. J., Myers, K. M.,** and **Davis, M.** (2006). The effect of gonadal hormones and gender on anxiety and emotional learning. *Hormones and Behavior* **50**, pp. 539–549.

1430. **Tovote, P., Farrokhi, C. B., Gonzales, R. M. K., Schnitzbauer, U., Blanchard, D. C., Blanchard, R. J.,** and **Spiess, J.** (2010). Activation of central CRF receptor 1 by cortagine results in enhanced passive coping with a naturalistic threat in mice. *Psychoneuroendocrinology* **35**, pp. 887–895.

1431. **Tran, L., Schulkin, J.,** and **Greenwood-Van Meerveld, B.** (2014). Importance of CRF receptor-mediated mechanisms of the bed nucleus of the stria terminalis in the processing of anxiety and pain. *Neuropsychopharmacology* **39**, pp. 2633–2645.

1432. **Tran, L., Schulkin, J.,** and **Greenwood-Van Meerveld.** (2014). Epigenetic modulation of chronic anxiety and pain by histone deacytlation. *Molecular Psychiatry* **20**, pp. 1219–1231.

1433. **Treutlein, J., Kissling, C., Frank, J., Wiemann, S., Dong, L., Depner, M., Saam, C., Lascorz, J., Soyka, M., Preuss, U. W., Rujescu, D., Skowronek, M. H., Rietschel, M., Spangel, R., Heinz, A., Mann, K.,** and **Schumann, G.** (2006). Genetic association of the human corticotropin releasing hormone receptor 1 (CRHR1) with binge drinking and alcohol intake patterns in two independent samples. *Molecular Psychiatry* **11**, pp. 594–602.

1434. **Trezza, V.,** and **Compolongo, P.** (2013). The endocannabinoid system as a possible target to treat both the cognitive and emotional features of post-traumatic stress disorder (PTSD). *Frontiers in Behavioral Neuroscience* **7**, pp. 1–6.

1435. **Trivedi, S., Joachim, M., McElrath, T., Kliman, H. J., Allred, E. N., Fichorova, R. N., Onderdonk, A., Hetitor, F., Chaychi, L., Leviton, A.,** and **Majzoub, J.** (2012). Fetal-placental inflammation, but not adrenal activation, is associated with extreme preterm delivery. *American Journal of Obstetrics and Gynecology* **206**, pp. 236.e1–e8.

1436. **Tropper, P. J., Goland, R. S., Wardlaw, S. L., Fox, H. E.,** and **Frantz, A. G.** (1987). Effects of betamethasone on maternal plasma cortiocotropin releasing factor, ACTH and cortisol during pregnancy. *Journal of Perinatal Medicine* **15**, pp. 221–226.

1437. **Trumbo, S. T.** (2002). Hormonal regulation of parental care in insects. *Hormones, Brain and Behavior* **3**, pp. 115–139.

1438. **Tsilioni, I., Russell, I. J., Stewart, J. M., Gleason, R. M.,** and **Theoharides, T. C.** (2016). Neuropeptides CRH, SP, HK-1, and inflammatory cytokines IL-6 and TNF are increased in serum of patients with fibromyalgia syndrome, implicating mast cells. *Journal of Pharmacology and Experimental Therapeutics* **356**, pp. 664–672.

1439. **Uban, K., Comeau, W., Ellis, L., Galea, L.,** and **Weinberg, J.** (2013). Basal regulation of HPA and dopamine systems is altered differentially in males and females by prenatal alcohol exposure and chronic variable stress. *Psychoneuroendocrinology* **38**, pp. 1953–1966.

1440. **Uchoa, E. T., Rorato, R., Ruginsk, S. G., de Carvalho Borges, B., Antunes-Rodrigues, J.,** and **Elias, L. L.** (2015). Corticotrophin-releasing factor receptor 2 mediates the enhanced activation of satiety-related responses through oxytocin neurons in the paraventricular nucleus of the hypothalamus after adrenalectomy. *Neuroscience Letters* **606**, pp. 123–128.

1441. **Ullman, M. T.** (2001). A neurocognitive perspective on language: The declarative/procedural model. *Nature Reviews Neuroscience* **2**, pp. 717–726.

1442. **Ullman, M. T.** (2004). Contributions of memory circuits to language: the declarative/procedural model. *Cognition* **92**, pp. 231–270.

1443. **Ungless, M. A., Whistler, J. L., Malenka, R. C.,** and **Bonci, A.** (2001). Single cocaine exposure in vivo induces long-term potentiation in dopamine neurons. *Nature* **411**, pp. 583–587.

1444. **Uribe-Mariño, A., Gassen, N. C., Wiesbeck, M. F., Balsevich, G., Santarelli, S., Solfrank, B., Dournes, C., Fries, G. R., Masana, M., Labermeier, C., Wang, X. D., Hafner, K., Schmid, B., Rein, T., Chen, A., Deussing, J. M.,** and **Schmidt, M. V.** (2016). Prefrontal cortex corticotropin-releasing factor receptor 1 conveys acute stress-induced executive dysfunction. *Biol Psychiatry* **80**(10), pp. 743–753.

1445. **Vale, W., Grant, G., Rivier, J., Monahan, M., Amoss, M., Blackwell, R., Burgus, R.,** and **Guillemin, R.** (1972). Synthetic polypeptide antagonists of the hypothalamic luteinizing hormone releasing factor. *Science* **176**, pp. 933–934.

1446. **Vale, W., Spiess, J., Rivier, C.,** and **Rivier, J.** (1981). Characterization of a 41-residue ovine hypothalamic peptide that stimulates the secretion of corticotropin releasing hormone and beta-endorphin. *Science* **213**, pp. 1394–1397.

1447. **Valenstein, E.** (1976). The interpretation of behavior evoked by brain stimulation. In: A. Wauquier and E. Rools (eds.), *Brain stimulation reward.* Amsterdam: North-Holland.

1448. **Valentino, R. J., Bangasser, D., Plona, Z., Ding, H.,** and **Seeholzer, S. H.** (2012) The sex biased phosphoproteome: a novel approach towards understanding the molecular basis for sex differences in neuropsychiatric diseases. *Neuropsychopharmacology* **38**, pp. S272–273

1449. **Valentino, R. J., Bangasser, D.,** and **Van Bockstaele, E. J.** (2013). Sex-biased stress signaling: the corticotropin-releasing factor receptor as a model. *Molecular Pharmacology* **83**, pp. 737–745.

1450. **Valentino, R. J., Foote, S. L.,** and **Page, M. E.** (1993). The locus coeruleus as a site for integrating corticotropin-releasing factor and noradrenergic mediation of stress responses. *Annals of the New York Academy of Sciences* **697**, pp. 173–188.

1451. **Valentino, R. J.,** and **Van Bockstaele, E.** (2008). Convergent regulation of locus coeruleus activity as an adaptive response to stress. *European Journal of Pharmacology* **583**, pp. 194–203.

1452. **Valentino, R. J., Pavcovich, L. A.,** and **Hirata, H.** (1995). Evidence for corticotropin-releasing hormone projections from Barrington's nucleus to the periaqueductal gray and dorsal motor nucleus of the vagus in the rat. *Journal of Comparative Neurology* **363**, pp. 402–422.

1453. **Vallée, M., Mayo, W., Dellu, F., Le Moal, M., Simon, H.,** and **Maccari, S.** (1997). Prenatal stress induces high anxiety and postnatal handling induces low anxiety in adult offspring: correlation with stress-induced corticosterone secretion. *Journal of Neuroscience* **17**, pp. 2626–2636.

1454. **Vamvakopoulos, N. C.,** and **Chrousos, G. P.** (1993). Evidence of direct estrogenic regulation of human corticotropin-releasing hormone gene expression. Potential implications for the sexual dimorphism of the stress response and immune/inflammatory reaction. *Journal of Clinical Investigation* **92**, pp. 1896–1902.

1455. **Vamvakopoulos, N. C.,** and **Chrousos, G. P.** (1993). Regulated activity of the distal promoter-like element of the human corticotropin releasing hormone gene and secondary structural features of its corresponding transcripts. *Molecular and Cellular Endocrinology* **94**, pp. 73–78.

1456. **Van Bockstaele, E. J.**, **Chan, J.**, and **Pickel, V. M.** (1996). Input from central nucleus of the amygdala efferents to periocoerulear dendrites, some of which contain tyrosine hydroxylase immunoreactivity. *Journal of Neuroscience Research* **45**, pp. 289–302.

1457. **Van Bockstaele, E. J.**, **Colago, E. E.**, and **Valentino, R. J.** (1998). Amygdaloid corticotropin- releasing factor targets locus coeruleus dendrites: substrate for the coordination of emotional and cognitive limbs of the stress response. *Journal of Neruoendocrinology* **10**, pp. 743–757.

1458. **van der Laan, S.**, **de Kloet, E. R.**, and **Meijer, O. C.** (2009). Timing is critical for effective glucocorticoid receptor mediated repression of the cAMP-induced CRH gene. *PLoS One* **4**, p. e4327.

1459. **Van Enckevort, F. H. J.**, **Pepels, P. P. L. M.**, **Leunissen, J. A. M.**, **Martens, G. J. M.**, **Wendelaar Bonga, S. E.**, and **Balm, P. H. M.** (2000). Oreochromis mossambicus (tilapia) corticotropin-releasing hormone: cDNA sequence and bioactivity. *Journal of Neuroendocrinology* **12**, pp. 177–186.

1460. **Van Essen, D. C.**, **Anderson, C. H.**, and **Felleman, D. J.** (1992). Information processing in the primate visual system: an integrated systems perspective. *Science* **255**, pp. 419–423.

1461. **Van Kempen, T. A.**, **Dodos, M.**, **Woods, C.**, **Marques-Lopes, J.**, **Justice, N. J.**, **Iadecola, C.**, **Pickel, V. M.**, **Glass, M. J.**, and **Milner, T. A.** (2015). Sex differences in NMDA GluN1 plasticity in rostral ventrolateral medulla neurons containing corticotropin-releasing factor type 1 receptor following slow-pressor angiotensin II hypertension. *Neuroscience* **307**, pp. 83–97.

1462. **Van Pett, K.**, **Viau, V.**, **Bittencourt, J. C.**, **Chan, R. K.**, **Li, H. Y.**, **Arias, C.**, **Prins, G. S.**, **Perrin, M.**, **Vale, W.**, and **Sawchenko, P. E.** (2000). Distribution of mRNAs encoding CRF receptors in brain and pituitary of rat and mouse. *Journal of Comparative Neurology* **428**, pp. 191–212.

1463. **Van Wielendaele, P.**, **Dillen, S.**, **Marchal, E.**, **Badisco, L.**, and **Vanden Broeck, J.** (2012). CRF-like diuretic hormone negatively affects both feeding and reproduction in the desert locust, Schistocerca gregaria. *PLoS One* **7**, pp. e31425.

1464. **Van, L. P.**, **Spengler, D. H.**, and **Holsboer, F.** (1990). Glucocorticoid repression of 3', 5'-cyclic-adenosine monophosphate-dependent human corticotropin-releasing-hormone gene promoter activity in a transfected mouse anterior pituitary cell line. *Endocrinology* **127**, pp. 1412–1418.

1465. **Vandenborne, K.**, **De Groef, B.**, **Geelissen, S. M.**, **Kühn, E. R.**, **Darras, V. M.**, and **Van der Geyten, S.** (2005). Corticosterone-induced negative feedback mechanisms within the hypothalamo–pituitary–adrenal axis of the chicken. *Journal of Endocrinology* **185**, pp. 383–391.

1466. **Vazquez, D. M.**, **Bailey, C.**, **Dent, G. W.**, **Okimoto, D. K.**, **Steffek, A.**, **López, J. F.**, and **Levine, S.** (2006). Brain corticotropin-releasing hormone (CRH) circuits in the developing rat: effect of maternal deprivation. *Brain Research* **1121**, pp. 83–94.

1467. **Veenema, A.**, **Reber, S.**, **Selch, S.**, **Obermeier, F.**, and **Neumann, I.** (2008). Early life stress enhances the vulnerability to chronic psychosocial stress and experimental colitis in adult mice. *Endocrinology* **149**, pp. 2727–2736.

1468. **Venihaki, M., Dikes, P., Carrigan, A.,** and **Karalis, K.** (2001). Corticotropin releasing hormone regulates IL-6 expression during inflammation. *Journal of Clinical Investigation* **108**, pp. 1159–1166.

1469. **Verleye, M.,** and **Gillardin, J.** (2004). Effects of etifoxine on stress-induced hypterhtermia, freezing behavior and colonic motor activation in rats. *Physiology and Behavior* **82**, pp. 891–897.

1470. **Viau, V., Soriano, L.,** and **Dallman, M. F.** (2001). Androgens alter corticotropin releasing hormone and arginine vasopressin mRNA within forebrain sites known to regulate activity in the hypothalamic-pituitary-adrenal axis. *Journal of Neuroendocrinolgy* **13**, pp. 442–452.

1471. **Volkow, N. D., Koob, G. F.,** and **McLellan, A. T.** (2016). Neurobiologic advances from the brain disease model of addiction. *New England Journal of Medicine* **374**, pp. 363–371.

1472. **Voltolini, C., Battersby, S., Novembri, R., Toriicelli, M., Severi, F. M., Petraglia, F.,** and **Norman, J. E.** (2014). Urocortin 2 role in placental and myometrial inflammatory mechanisms at parturition. *Endocrinology* **156**, pp. 670–679.

1473. **von Frisch, K.** (1994). The "language" of bees and its utilization in agriculture. *Experientia* **50**, pp. 406–413.

1474. **von Wolff, G., Abrabos, C., Stepan, J., Wurst, W., Deussing, J., Holsboer, F.,** and **Eder, M.** (2011). Voltage-sensitive dye imaging demonstrates an enhancing effect of corticotropin-releasing hormone on neuronal activity propagation through the hippocampal formation. *Journal of Psychiatric Research* **45**, pp. 256–261.

1475. **Vorel, S. R., Liu, X., Hayes, R. J., Spector, J. A.,** and **Gardner, E. L.** (2001). Relapse to cocaine-seeking after hippocampal theta burst stimulation. *Science* **292**, pp. 1175–1178.

1476. **Vrachnis, N., Malamas, F.M., Sifakis, S., Tsikouras, P.,** and **Iliodromiti, Z.** (2012). Immune aspects and myometrial actions of progesterone and CRH in labor. *Clinical and Developmental Immunology* p. 937618.

1477. **Vranjkovic, O., Glasser, P., Gerndt, C. Baker, D.,** and **Mantsch, J.** (2014). Stress-induced cocaine seeking requires a Beta-2 adrenergic receptor-regulated pathway from the ventral bed nucleus of the stria terminalis that regulates CRF actions in the ventral tegmental area. *Journal of Neuroscience* **34**, pp. 12504–12514.

1478. **Vuong, S. M., Oliver, H. A., Scholl, J. L., Oliver, K. M.,** and **Forster, G. L.** (2010). Increased anxiety-like behavior of rats during amphetamine withdrawal is reversed by CRF-2 receptor antagonism. *Behavioural Brain Research* **208**, pp. 278–281.

1479. **Vyas, A., Kim, S. K., Giacomini, N., Boothroyd, J. C.,** and **Sapolsky, R. M.** (2007). Behavioral changes induced by Toxoplasma infection of rodents are highly specific to averson of cat odors. *Proceedings of the National Academy of Sciences of the United States of America* **104**, pp. 6442–6447.

1480. **Vyas, A., Mitra, R., Shankaranarayana Rao, B. S.,** and **Chattarji, S.** (2002). Chronic stress induces contrasting patterns of dendritic remodeling in hippocampal and amygdaloid neurons. *Journal of Neuroscience* **22**, pp. 6810–6818.

1481. **Waddell, J.,** et al. (2004). Central CRF receptor are required for the reinstatement of extinguished conditioned fear. Program No. 328.12. In 2004 Abstract Viewer and Itinerary Planner, Society for Neuroscience online.

1482. **Waddington, C. H.** (1942). Canalization of development and the inheritance of acquired characters. *Nature* **150**, pp. 563–565.

1483. **Wadhwa, P. D., Porto, M., Garite, T. J., Chicz-DeMet, A.,** and **Sandman, C. A.** (1998). Maternal corticotropin-releasing hormone levels in the early third trimester predict length of gestation in human pregnancy. *American Journal of Obstetrics and Gynecology* **179**, pp. 1079–1085.

1484. **Wadhwa, P. D., Culhane, J., Rauh, V., Barve, S., Hogan, V., Sandman, C., Hobel, C., Chicz-Demet, A., Dunkel-Schetter, C., Garite, T,** and **Glynn, L.** (2001). Stress, infection and preterm birth: a biobehavioural perspective. *Paediatric and Perintal Epidemiology* **15**, pp. 17–29.

1485. **Wadhwa, P. D., Dunkel-Schetter, C., Chicz-DeMet, A., Porto, M.,** and **Sandman, C. A.** (1996). Prenatal psychosocial factors and the neuroendocrine axis in human pregnancy. *Psychosomatic Medicine* **58**, pp. 432–446.

1486. **Wadhwa, P. D., Sandman, C. A.,** and **Garite, T. J.** (2001). The neurobiology of stress in human pregnancy: implications for development of the fetal central nervous system. *Progress in Brain Research* **133**, pp. 131–142.

1487. **Wagle, M., Mathur, P.,** and **Guo, S.** (2011). Corticotropin-releasing factor critical for zebrafish camouflage behavior is regulated by light and sensitive to ethanol. *Journal of Neuroscience* **31**, pp. 214–224.

1488. **Wagner, U., Wahle, M., Moritz, F., Wagner, U., Hantzschel, H.,** and **Baerwald, C. G. O.** (2006). Promoter polymorphisms regulating corticotrophin-releasing hormone transcription in vitro. *Hormone and Metabolic Research* **38**, pp. 69–75.

1489. **Walker, D. L.,** and **Davis, M.** (2002). The role of amygdala glutamate receptors in fear learning, fear-potentiated startle, and extinction. *Pharmacology, Biochemistry, and Behavior* **71**, pp. 379–392.

1490. **Walker, D.,** and **Davis, M.** (1997). Double dissociation between the involvement of the bed nucleus of the stria terminalis and the central nucleus of the amygdala in startle increases produced by conditioned versus undocnditioned fear. *Journal of Neuroscience* **17**, pp. 9375–9383.

1491. **Walker, D.,** and **Davis, M.** (2008). Role of the extended amygdala in short-duation versus sustained fear: a tribute to Dr. Lennart Heimer. *Brain Structure and Function* **213**, pp. 29–42.

1492. **Walker, L.C., Lawrence, A.J.** (2016, Epub ahead of print). CRF and the nucleus incertus: a node for integration of stress signals. *Nature Reviews Neuroscience.*

1493. **Walker, D., Toufexis, D. J.,** and **Davis, M.** (2003). Role of the bed nucleus of the stria terminalis versus the amygdala in fear, stress, and anxiety. *European Journal of Pharmacology* **463**, pp. 199–216.

1494. **Walker, D., Yang, Y., Ratti, E., Corsi, M., Trist, D.,** and **Davis, M.** (2008). Differential effects of the CRF-RI antagonist GSK876008 on fear-potentiated, light- and CRF-enhanced startle suggest preferential involvement in sustained vs phasic threat responses. *Neuropsychopharmacology* **34**, pp. 1–10.

1495. **Walker, J. R., Ahmed, S. H., Gracy, K. N.,** and **Koob, G. F.** (2000). Microinjections of an opiate receptor antagonist into the bed nucleus of the stria terminalis suppress heroin self-administration in dependent rats. *Brain Research* **854**, pp. 85–92.

1496. **Wallace, K. J.**, and **Rosen, J. B.** (2000). Predator odor as an unconditioned fear stimulus in rats: elicitation of freezing by trimethylthiazoline, a component of fox feces. *Behavioral Neuroscience* **114**, pp. 912.

1497. **Wallace, K. J.**, and **Rosen, J. B.** (2001). Neurotoxic lesions of the lateral nucleus of the amygdala decrease conditioned fear but not unconditioned fear of a predator odor: comparison with electrolytic lesions. *Journal of Neuroscience* **21**, pp. 3619–3627.

1498. **Walsh, J. J.**, **Friedman, A. K.**, **Sun, H.**, **Heller, E. A.**, **Ku, S. M.**, **Juarez, B.**, **Burnham, V. L.**, **Mazei-Robinson, M. S.**, **Ferguson, D.**, **Golden, S. A.**, **Koo, J. W.**, **Chaudhury, D.**, **Christoffel, D. J.**, **Pomeranz, L.**, **Friedman, J. M.**, **Russo, S. J.**, **Nestler, E. J.**, and **Han, M. H.** (2014). Stress and CRF gate neural activation of BDNF in the mesolimbic reward pathway. *Nature Neuroscience* **17**, pp. 27–29.

1499. **Wanat, M. J.**, **Bonci, A.**, and **Phillips, P. E.** (2013). CRF acts in the midbrain to attenuate accumbens dopamine release to rewards but not their predictors. *Nature Neuroscience* **16**, pp. 383–385.

1500. **Wang, B.**, **Luo, T.**, **Tadesse, S.**, **Tardif, S.**, **Ross, C.**, **Schulkin, J.**, **Power, M.**, **Rosen, T.**, and **Norwitz, E.** (2013). Expression of CRH and CRH regulating proteins in placentas of non-human primates. *Reproductive Sciences* **20**, pp. 136A.

1501. **Wang, B.**, **Parobchak, N.**, and **Rosen, T.** (2012). RelB/NF-κB2 regulates corticotropin-releasing hormone in the human placenta. *Molecular Endocrinology* **26**, pp. 1356–1369.

1502. **Wang, B.**, **Parobchak, N.**, **Rosen, M.**, **Roche, N.**, and **Rosen, T.** (2014). Negative effects of progesterone receptor isoform-A on human placental activity of the noncanonical NF-kB signaling. *Journal of Clinical Endocrinology and Metabolism* **99**, pp. E320–E328.

1503. **Wang, B.**, **Shaham, Y.**, **Zitzman, D.**, **Azari, S.**, **Wise, R.**, and **You, Z.** (2005). Cocaine experience establishes control of midbrain glutamate and dopamine by corticotropin-releasing factor: a role in stress-induced relapse to drug seeking. *Journal of Neuroscience* **25**, pp. 5389–5396.

1504. **Wang, C.**, **Yang, S.**, **Huang, J.**, **Chen, S.**, **Li, Y.**, and **Li, Q.** (2016). Activation of corticotropin releasing factor receptors up regulates collagen production by hepatic stellate cells via promoting p300 expression. *Biological Chemistry* **397(5)**, pp. 437–444.

1505. **Wang, H.**, **Li, S.**, and **Kirouac, G. J.** (2015). Effects of footshocks on anxiety-like behavior and mRNA levels of precursor peptides for corticotropin releasing factor and opioids in the forebrain of the rat. *Neuropeptides* **54**, pp. 1–7.

1506. **Wang, W.**, **Ji, P.**, and **Dow, K.** (2003). Corticotropin-releasing hormone induces proliferation and TNF-a release in cultured rat microglia via MAP kinase signaling pathways. *Journal of Neurochemistry* **84**, pp. 189–195.

1507. **Wang, W.**, **Solc, M.**, **Ji, P.**, and **Dow, K.** (2004). Corticotropin-releasing hormone potentiates neural injury induced by oxygen-glucose deprivation: a possible involvement of microglia. *Neuroscience Letters* **371**, pp. 133–137.

1508. **Wang, X. D.**, **Chen, Y.**, **Wolf, M.**, **Wagner, K. V.**, **Liebl, C.**, **Scharf, S. H.**, **Harbich, D.**, **Mayer, B.**, **Wurst, W.**, **Folsboer, F.**, **Deussing, J. M.**, **Baram, T.Z.**, **Muller, M. B.**, and **Schmidt, M. V.** (2011). Forebrain CRHR1 deficiency attenuates chronic stress-induced cognitive deficits and dendritic remodeling. *Neurobiology of Disease* **42**, pp. 300–310.

1509. **Wang, X., Meng, F. S., Liu, Z. Y., Fan, J. M., Hao, K., Chen, X. Q.,** and **Du, J. Z.** (2013). Gestational hypoxia induces sex-differential methylation of Crhr1 linked to anxiety-like behavior. *Molecular Neurobiology* **48**, pp. 544–555.

1510. **Wang, Y., Tan, W.,** and **Leung, L. K.** (2013). Zeranol upregulates corticotropin releasing hormone expression in the placental cell line JEG-3. *Toxicology Letters* **219**, pp. 218–222.

1511. **Wang, Z.,** and **Young, L. J.** (1997). Ontogeny of oxytocin and vasopressin receptor binding in the lateral septum in prairie and mountain voles. *Developmental Brain Research* **104**, pp. 191–195.

1512. **Warren, W. B., Goland, R. S., Wardlaw, S. L., Stark, R. L., Fox, H. E.,** and **Conwell, I. M.** (1990). Elevated maternal plasma cortictorpin releasing hormone levels in twin gestation. *Journal of Perinatal Medicine* **18**, pp. 39–44.

1513. **Warren, W. B., Patrick, S. L.,** and **Goland, R. S.** (1992). Elevated maternal plasma corticotropin releasing hormone levels in pregnancies complicated by preterm labor. *American Journal of Obstetrics and Gynecology* **166**, pp. 1198–1204.

1514. **Waters, R. P., Rivalan, M., Bangasser, D. A., Deussing, J. M., Ising, M., Wood, S. K., Holsboer, F.** and **Summers, C. H.** (2015). Evidence for the role of corticotropin-releasing factor in major depressive disorder. *Neuroscience and Biobehavioral Reviews* **58**, pp. 63–78.

1515. **Watts, A. G.** (1996). The impact of physiological stimuli on the expression of corticotropin-releasing hormone (CRH) and other neuropeptide genes. *Frontiers in Neuroendocrinology* **17**, pp. 281–326.

1516. **Watts, A. G.** (2005). Glucocorticoid regulation of peptide genes in neuroendocrine CRH neurons: a complexity beyond negative feedback. *Frontiers in Neuroendocrinology* **26**, pp. 109–130.

1517. **Watts, A.G.** (2005). Glucocorticoids and the ups and downs of neuropeptide gene espression. In: Schulkin and Power (eds.), Brith, distress, and disease: placenta-brain interactions (Eds. Schulkin and Power). Cambridge: Cambridge University Press, pp. 202–234.

1518. **Watts, A. G.** (2015). The structure of the neuroendocrine hypothalamus: the neuroanatomical legacy of Geoffrey Harris. *Journal of Endocrinology* **226**, pp. T25–T39.

1519. **Watts, A. G., Tanimura, S.,** and **Sanchez-Watts, G.** (2004). Corticotropin-releasing hormone and arginine vasopressin gene transcription in the hypothalamic paraventricular nucleus of unstressed rats: daily rhythms and their interactions with corticosterone. *Endocrinology* **145**, pp. 529–540.

1520. **Watts, A. G.,** and **Khan, A. M.** (2013). Identifying links in the chain: the dynamic coupling of catecholamines, peptide synthesis, and peptide release in hypothalamic neuroendocrine neurons. *Advances in Pharmacology* **68**, pp. 421–444.

1521. **Watts, A. G.,** and **Sanchez-Watts, G.** (1995). Region-specific regulation of neuropeptide mRNAs in rat limbic forebrain neurons by aldosterone and corticosterone. *Journal of Physiology* **484**, pp. 721–736.

1522. **Weathington, J.,** and **Cooke, B.** (2012). Corticotropin releasing factor receptor binding in the amygdala changes across puberty in a sex-specific manner. *Endocrinology* **153**, pp. 5701–5705.

1523. **Weathington, J., Hamki, A.**, and **Cooke, B.** (2014). Sex and region specific pubertal maturation of the corticotropin releasing factor receptor system in the rat. *Journal of Comparative Neurology* **522**, pp. 1284–1298.

1524. **Weaver, D. R., Liu, C.**, and **Reppert, S. M.** (1996). Nature's knockout: the Mel1b receptor is not necessary for reproductive and circadian responses to melatonin in Siberian hamsters. *Molecular Endocrinology* **10**, pp. 1478–1487.

1525. **Weaver, I. C., Cervoni, N., Champagne, F. A., D'Alessio, A. C., Sharma, S., Seckl, J. R., Dymov, S., Szyf, M.**, and **Meaney, M. J.** (2004). Epigenetic programming by maternal behavior. *Nature Neuroscience* 7, pp. 847–854.

1526. **Weaver, I. C., Champagne, F. A., Brown, S. E., Dymov, S., Sharma, S., Meaney, M. J.**, and **Szyf, M.** (2005). Reversal of maternal programming of stress responses in adult offspring through methyl supplementation: altering epigenetic marking later in life. *Journal of Neuroscience* **25**, pp. 11045–11054.

1527. **Webster, E., Tropy D., Elenkov, I.**, and **Chrousos, G.** (1998). Corticotropin Releasing Hormone and Inflammation. *Annals New York Academy of Sciences* **840**: 21–26.

1528. **Wehr, T. A., Moul, D. E., Barbato, G., Giesen, H. A., Seidel, J. A., Barker, C.**, and **Bender, C.** (1993). Conservation of photoperiod-responsive mechanisms in humans. *American Journal of Physiology, Regulatory, Integrative and Comparative Physiology* **265**, pp. R846–R857.

1529. **Wei, Q., Lu, X. Y., Lui, L., Schafer, G., Shieh, K. R., Burke, S., Robinson, T.E., Watson, S. J., Seasholtz, A. F.**, and **Akil, H.** (2004). Glucocorticoid receptor overexpression in forebrain: a mouse model of increased emotional lability. *Proceedings of the National Academy of Sciences of the United States of America* **101**, pp. 11851–11856.

1530. **Wei, R., Phillips, T.**, and **Sternberg, E.** (2002). Specific up-regulation of CRH or AVP secretion by acetylcholine or lipolysaccharide in inflammatory susceptible Lewis rat fetal hypothalamic cells. *Journal of Neuroimmunology* **131**, pp. 31–40.

1531. **Weinstock, M.** (1997). Does prenatal stress impair coping and regulation of hypothalamic-pituitary-adrenal axis? *Neuroscience and Biobehavioral Reviews* **21**, pp. 1–10.

1532. **Weinstock, M.** (2005). The potential influence of maternal stress hormones on development and mental health of the offspring. *Brain Behavior and Immunity* **19**, pp. 296–308.

1533. **Weisbecker, V.**, and **Goswami, A.** (2010). Brain size, life history, and metabolism at the marsupial/placental dichotomy. *Proceedings of the National Academy of Sciences* **107**, pp. 16216–16221.

1534. **Weiser, M. J., Goel, N., Sandau, U. S., Bale, T. L.**, and **Handa, R. J.** (2008). Androgen regulation of corticotropin-releasing hormone receptor 2 (CRHR2) mRNA expression and receptor binding in the rat brain. *Experimental Neurology* **214**, pp. 62–68.

1535. **Weiss, S. R., Nierenberg, J., Lewis, R.**, and **Post, R. M.** (1992). Corticotropin-releasing hormone: potentiation of cocaine-kindled seizures and lethality. *Epilepsia* **33**, pp. 248–254.

1536. **Weiss, S. R., Post, R. M., Gold, P. W., Chrousos, G., Sullivan, T. L., Walker, D.**, and **Pert, A.** (1986). CRF-induced seizures and behavior: interaction with amygdala kindling. *Brain Research* **372**, pp. 345–351.

1537. **Welberg, L. A.**, and **Seckl, J. R.** (2001). Prenatal stress, glucocorticoids and the programming of the brain. *Journal of Neuroendocrinology* **13**, pp. 113–128.

1538. **Welberg, L. A.**, **Seckl, J. R.**, and **Holmes, M. C.** (2000). Inhibition of 11 beta-hydroxysteroid dehydrogenase, the foeto-placental barrier to maternal glucocorticoids, permanently programs amygdala GR mRNA expression and anxiety-like behavior in the offspring. *European Journal of Neurosciences* **12**, pp. 1047–1054.

1539. **Welberg, L. A.**, **Seckl, J. R.**, and **Holmes, M. C.** (2001). Prenatal glucocorticoid programming of brain corticosteroid receptors and corticotropin releasing hormone: possible implications for behaviour. *Neuroscience* **104**, pp. 71–79.

1540. **Wellman, C. L.** (2001). Dendritic reorganization in pyramidal neurons in medial prefrontal cortex after chronic corticosterone administration. *Journal of Neurobiology* **49**, pp. 245–253.

1541. **Wellman, L. L.**, **Yang, L.**, and **Sanford, L. D.** (2015). Effects of corticotropin releasing factor (CRF) on sleep and temperature following predictable controllable and uncontrollable stress in mice. *Frontiers in Neuroscience* **9**, pp. 1–12.

1542. **Weninger, S.**, **Peters, L.**, and **Majzoub, J.** (2000). Urocortin expression in the Edinger-Westphal nucleus is up-regulated by stress and corticotropin-releasing hormone deficiency. *Endocrinology* **141**, pp. 256–263.

1543. **Weninger, S. C.**, **Dunn, A. J.**, **Muglia, L. J.**, **Dikkes, P.**, **Miczek, K. A.**, **Swiergiel, A. H.**, **Berridge, G. W.**, and **Majzoub, J. A.** (1999). Stress-induced behaviors require the corticotropin-releasing hormone (CRH) receptor, but not CRH. *Proceedings of the National Academy of Sciences of the United States of America* **96**, pp. 8283–8288.

1544. **Westphal, N. J.**, and **Seasholtz, A. F.** (2006). CRH-BP: the regulation and function of a phylogenetically conserved binding protein. *Frontiers in Bioscience* **11**, pp. 1878–1891.

1545. **Whatts, A.** (2011). Structure and function in the conceptual development of mammalian neuroendocrinology between 1920 and 1965. *Brain Research Reviews* **66**, pp. 174–204.

1546. **Whitaker, A.**, and **Gilpin, N.** (2015). Blunted hypothalamic-pituitary adrenal axis response to predator odor predicts high stress reactivity. *Physiology and Behavior* **147**, pp. 16–22.

1547. **Whitman, B. A.**, **Knapp, D. J.**, **Werner, D. F.**, **Crews, F. T.**, and **Breese, G. R.** (2013). The cytokine mRNA increase induced by withdrawal from chronic ethanol in the sterile environment of brain is mediated by CRF and HMGB1 release. *Alcoholism Clinical and Experimental Research* **37**, pp. 2086–2097.

1548. **Wiersielis, K.R.**, **Wicks, B.**, **Simko, H.**, **Cohen, S.R.**, **Khantsis, S.**, **Baksh, N.**, **Waxler, D.E.**, **Bangasser, D.A.** (2016), Sex differences in corticotropin releasing factor-evoked behavior and activated networks. *Psychoneuroendocrinology* **73**, pp. 204–216.

1549. **Wiersma, A.**, **Konsman, J.**, **Knollema, S.**, **Bohus, B.**, and **Koolhaas, J.** (1998). Differential effects of CRH infusion into the central nucleus of the amygdala in the roman high-avoid avoidance and low-avoidance rats. *Psychoneuroendocrinology* **23**, pp. 261–274.

1550. **Wildman, D. E.**, **Chen, C.**, **Erez, O.**, **Grossman, L. I.**, **Goodman, M.**, and **Romero, R.** (2006). Evolution of the mammalian placenta revealed by phylogenetic analysis. *Proceedings of the National Academy of Sciences of the United States of America* **103**, pp. 3203–3208.

1551. **Williams, M. T., Hennessy, M. B.**, and **Davis, H. N.** (1995). CRF administered to pregnant rats alters offspring behavior and morphology. *Pharmacology Biochemistry and Behavior* **52**, pp. 161–167.

1552. **Wills, T. A., Knapp, D. J., Overstreet, D. H.**, and **Breese, G. R.** (2010). Interactions of stress and CRF in ethanol-withdrawal induced anxiety in adolescent and adult rats. *Alcoholism Clinical and Experimental Research* **34**, pp. 1603–1612.

1553. **Wilson, D. S., Clark, A. B., Coleman, K.**, and **Dearstyne, T.** (1994). Shyness and boldness in humans and other animals. *Trends in Ecology and Evolution* **9**, pp. 442–446.

1554. **Wilson, D. N.** (1962). Bifunctional muscles in the thorax of grasshoppers. *Journal of Experimental Biology* **39**, pp. 669–677.

1555. **Wingfield, J. C.**, et al. (1999). Testosterone, aggression and communication: ecological bases of endocrine phenomena. In: Hauser, M. D.m and Konishi, M. (eds.), *The design of animal communication*. Cambridge: Cambridge University Press, pp. 520–529.

1556. **Wingfield, J. C.** (2004). Allostatic load and life cycles: implication for neuroendocrine control mechanisms. In: Sculkin, J. (ed.), *Allostasis, homeostasis and the costs of physiological adaptation*. Cambridge: Cambridge University Press, pp. 303–342.

1557. **Wingfield, J. C.**, and **Romero, L. M.** (2001). Adrenocortical responses to stress and their modulation in free-living vertebrates. In: McEwen, B. (ed.), *Coping with the environment: neural and endocrine mechanisms*. Oxford: Oxford University Press, pp. 211–236.

1558. **Winslow, J. T., Noble, P. L., Lyons, C. K., Sterk, S. M.**, and **Insel, T. R.** (2003). Rearing effects on cerebrospinal fluid oxytocin concentration and social buffering in rhesus monkeys. Neuropsychopharmacology **28**, pp. 910–918.

1559. **Winslow, J. T., Shapiro, L., Carter, C. S.**, and **Insel, T. R.** (1993). Oxytocin and complex social behavior: species comparisons. *Psychopharmacology Bulletin* **29**, pp. 409–414.

1560. **Wise, R. A.** (2005). Forebrain substrates of reward and motivation. *Journal of Comparative Neurology* **493**, pp. 115–121.

1561. **Wise, R.**, and **Bozarth, M.** (1987). A psychomotor stimulant theory of addiction. *Psychological Review* **94**, pp. 469–492.

1562. **Wolf, E. J., Mitchell, K. S., Logue, M. W., Baldwin, C. T., Reardon, A. F., Humphries, D. E.**, and **Miller, M. W.** (2013). Corticotropin releasing hormone receptor 2 (CRHR-2) gene is associated with decreased risk and severity of posttraumatic stress disorder in women. *Depression and Anxiety* **30**, pp. 1161–1169.

1563. **Wolfe, C., Patel, S., Linton, E., Campbell, E., Anderson, J., Dornhorst, A., Lowry, P.**, and **Jones, M.** (1988). Plasma corticotrophinreleasing factor (CRF) in abnormal pregnancy. *British Journal of Obstetrics and Gynaecology* **95**, pp. 1003–1006.

1564. **Wong, E. Y.**, and **Herbert, J.** (2005). Roles of mineralocorticoid and glucocorticoid receptors in the regulation of progenitor proliferation in the adult hippocampus. *European Journal of Neuroscience* **22**, pp. 785–792.

1565. **Wood, S.K., McFadden, K., Griffin, T., Wolfe J.H., Zderic S., Valentino R. J.** (2013). A corticotropin-releasing factor receptor antagonist improves urodynamic dysfunction produced by social stress or partial bladder outlet obstruction in male rats. *American Journal of Physiology* 304(11), pp. 940–950.

1566. **Wood, R. I.**, and **Stanton, S. J.** (2012). Testosterone and sport: current perspectives. *Hormones and Behavior* **61**, pp. 147–155.

1567. **Wood, S. K.**, **McFadden, K. V.**, **Grigoriadis, D.**, **Bhatnagar, S.**, and **Valentino, R. J.** (2012). Depressive and cardiovascular disease comorbidity in a rat model of social stress: a putative role for corticotropin-releasing factor. *Psychopharmacology* **222**, pp. 325–336.

1568. **Wood, S.**, **Walker, H.**, **Valentino, R.**, and **Bhatnager, S.** (2010). Individual differences in reactivity to social stress predict susceptibility and resilience to a depressive phenotype: role of corticotropin releasing factor. *Endocrinology* **151**, pp. 1795–1805.

1569. **Wu, W. X.**, **Unno, S. U. M. I.**, **Giussani, D. A.**, **Mecenas, C. A.**, **McDonald, T. J.**, and **Nathanielsz, P. W.** (1995). Corticotropin-releasing hormone and its receptor distribution in fetal membranes and placenta of the rhesus monkey in late gestation and labor. *Endocrinology* **136**, pp. 4621–4628.

1570. **Wypior, G.**, **Jeschke, U.**, **Kurpisz, M.**, **Szekeres-Bartho, J.** (2011). Expression of CRH, CRH-related peptide and CRH receptor in the ovary and potential CRH signaling pathways. *Journal of Reproductive Immunology* **90**, pp. 67–73.

1571. **Yadawa, A. K.**, and **Chaturvedi, C. M.** (2016). Expression of stress hormones AVP and CRH in the hypothalamus of Mus musculus following water and food deprivation. *General and Comparative Endocrinology* **239**, pp. 13–20.

1572. **Yang, Y.**, **Hahm, E.**, **Kim, Y.**, **Kang, J.**, **Lee, W.**, **Han, I.**, **Myung, P.**, **Kang, H.**, **Park, H.**, and **Cho, D.** (2005). Regulation of IL-18 expression by CRH in mouse micgroglial cells. *Immunology Letters* **98**, pp. 291–296.

1573. **Yang, Y. L.**, **Chao, P. K.**, and **Lu, K. T.** (2006). Systemic and intra-amygdala administration of glucocorticoid agonist and antagonist modulate extinction of conditioned fear. *Neuropsychopharmacology* **31**, pp. 912–924.

1574. **Yao, M.**, **Hu, F.**, and **Denver, R. J.** (2008a). Distribution and corticosteroid regulation of glucocorticoid receptor in the brain of Xenopus laevis. *Journal of Comparative Neurology* **508**, pp. 967–982.

1575. **Yao, M.**, **Schulkin, J.**, and **Denver, R. J.** (2008b). Evolutionary conserved glucocorticoid regulation of CRH. *Endocrinology* **149**, pp. 2352–2360.

1576. **Yao, M.**, **Stenzel-Poore, M.**, and **Denver, R. J.** (2007). Structural and functional conservation of vertebrate corticotropin-releasing factor genes: evidence for a critical role for a conserved cyclic AMP response element. *Endocrinology* **148**, pp. 2518–2531.

1577. **Yao, M.**, **Westphal, N. J.**, and **Denver, R. J.** (2004). Distribution and acute stressor-induced activation of corticotrophin-releasing hormone neurons in the central nervous sysem of Xenopus laevis. *Journal of Neuroendocrinology* **16**, pp. 880–893.

1578. **Yeh, C. M.** (2015). The basal NPO crh fluctuation is sustained under compromised glucocorticoid signaling in diurnal zebrafish. *Frontiers in Neuroscience* **9**, pp. 1–9.

1579. **Yeh, C.**, **Glock, M.**, and **Ryu, S.** (2013). An optimized whole-body cortisol quantification method for assessing stress levels in larval zebrafish. *PLOS One* **8**, p. e79406.

1580. **Yehuda, R.** (2002). Current status of cortisol findings in post-traumatic stress disorder. *Psychiatric Clinics of North America* **25**, pp. 341–368.

1581. **Yen, S. S. C.** (1994). The placenta as the third brain. *Journal of Reproductive Medicine* **39**, pp. 277–80.

1582. **Yilmazer-Hanke, D., Hantsch, M., Hanke, J., Schulz, C., Faber-Zuschratter, H,** and **Schwegler, H.** (2004). Neonatal thyroxine treatment: changes in the number of corticotropin releasing factor (CRF) and neuropeptide Y (NPY) containing neurons and density of tyrosine hydroxylase positive fibers (TH) in the amygdala correlate with anxiety-related behavior of wistar rats. *Neuroscience* **124**, pp. 283–297.

1583. **Yim, I.S., Glynn, L. M., Dunkel-Schetter, C., Hobel, C., Chicz-DeMet, A.,** and **Sandman, C. A.** (2009). Risk of postpartum depressive symptoms with elevated corticotropin-releasing hormone in human pregnancy. *Archives of General Psychiatry* **66**, pp. 162–169.

1584. **Young, L. J.** (2002). The neurobiology of social recognition, approach, and avoidance. *Biological Psychiatry* **51**, pp. 18–26.

1585. **Young, L. J., Lim, M. M., Gingrich, B.,** and **Insel, T. R.** (2001). Cellular mechanisms of social attachment. *Hormones and Behavior* **40**, pp. 133–138.

1586. **Young, L. J., Muns, S., Wang, Z.,** and **Insel, T. R.** (1997). Changes in oxytocin receptor mRNA in rat brain during pregnancy and the effects of estrogen and interleukin-6. *Journal of Neuroendocrinology* **9**, pp. 859–865.

1587. **Young, R. M.** (1970). *Mind, Brain brain and Adaptationadaptation.* Oxford: Oxford University Press.

1588. **Yu, G-Z., Kaba, H., Okutani, S., Takahashi, S.,** and **Higuchi, T.** (1996). The olfactory bulb: a critical site of action for oxytocin on the induction of maternal behavior. *Neuroscience* **72**, pp. 1083–1088.

1589. **Yuan, P. Q., Wu, S. V., Elliott, J., Anton, P. A., Chatzaki, E., Million, M.,** and **Taché, Y.** (2012). Expression of corticotropin releasing factor receptor type 1 (CRF 1) in the human gastrointestinal tract and upregulation in the colonic mucosa in patients with ulcerative colitis. *Peptides* **38**, pp. 62–69.

1590. **Yuan, P., Wu, S., Wang, L.,** and **Tache, Y.** (2010). Corticotropin releasing factor in the rat colon: expression, localization, and upregulation by endotoxin. *Peptides* **31**, pp. 322.

1591. **Zaconeta, A. M., Amato, A. A., Barra, G. B., Casulari da Motta, L. D., de Souza, V. C., Karnikowski, M. G. D. O.,** and **Casulari, L. A.** (2015). Cerebrospinal fluid CRH levels in late pregnancy are not associated with new-onset postpartum depressive symptoms. *Journal of Clinical Endocrinology and Metabolism* **100**, pp. 3159–3164.

1592. **Zalachoras, I., Houtman, R., Atucha, E., Devos, R., Tijssen, A. M., Hu, P., Lockey, P. M., Datson, N. A., Belanoff, J. K., Lucassen, P. J., Joëls, M., de Kloet, E. R., Roozendaal, B., Hunt, H.,** and **Meijer, O. C.** (2013). Differential targeting of brain stress circuits with a selective glucocorticoid receptor modulator. *Proc Natl Acad Sci U S A* **110**(19), pp. 7910–7915.

1593. **Zaidan, H.,** and **Gaisler-Salomon, I.** (2015). Prereproductive stress in adolescent female rats affects behavior and corticosterone levels in second-generation offspring. *Psychoneuroendocrinology* **58**, pp. 120–129.

1594. **Zaidan, H., Leshem, M.,** and **Gaisler-Salomon, I.** (2013). Prereproductive stress to female rats alters corticotropin releasing factor type 1 expression in ova and behavior and brain corticotropin releasing factor type 1 expression in offspring. *Biological Psychiatry* **74**, pp. 680–687.

1595. **Zalachoras, I., Verhoeve, S. L., Toonen, L. J., van Weert, L. T. C. M., van Vlodrop, A. M., Mol, I. M., Meelis, W., de Kloet, E. R.,** and **Meijer, O. C.** (2016). Isoform switching of steroid receptor co-activator-1 attenuates glucocorticoid-induced anxiogenic amygdala CRH expression. *Molecular Psychiatry* **21(12)**, 1733–1739.

1596. **Zbytek, B.,** and **Slominski, A. T.** (2008). CRH mediates inflammation induced by lipopolysaccharide in human adult epidermal keratinocytes. *Journal of Investigative Dermatology* **127**, pp. 730–732.

1597. **Zbytek, B., Wortsman, J.,** and **Slominski, A.** (2006). Characterization of a ultraviolet B-induced corticotropin-releasing hormone-proopiomelanocortin system in human melanocytes. *Molecular Endocrinology* **20**, pp. 2539–2547.

1598. **Zellner, D. A., Loaiza, S., Gonzalez, Z., Pita, J., Morales, J., Pecora, D.,** and **Wolf, A.** (2006). Food selection changes under stress. *Physiology and Behavior* **87**, pp. 789–793.

1599. **Zemkova, H., Tomić, M., Kucka, M., Aguilera, G.,** and **Stojilkovic, S. S.** (2016). Spontaneous and CRH-induced excitability and calcium signaling in mice corticotrophs involves sodium, calcium, and cation-conducting channels. *Endocrinology* **157**, pp. 1576–1589.

1600. **Zhang, R., Asai, M., Mahoney, C. E., Joachim, M., Shen, Y., Gunner, G.,** and **Majzoub, J. A.** (2016, Epub Ahead of Print). Loss of hypothalamic corticotropin-releasing hormone markedly reduces anxiety like behaviors in mice. *Molecular Psychiatry.*

1601. **Zhang, C., Kuo, C. C., Moghadam, S. H., Monte, L., Campbell, S. N., Rice, K. C., Sawchenko, P.E., Masliah, E.,** and **Rissman, R. A.** (2016). Corticotropin-releasing factor receptor-1 antagonism mitigates beta amyloid pathology and cognitive and synaptic deficits in a mouse model of Alzheimer's disease. *Alzheimer's and Dementia* **12**, pp. 527–537.

1602. **Zhang, D. M., Epstein, A. N.,** and **Schulkin, J.** (1993). Medial region of the amygdala: involvement in adrenal-steroid-induced salt appetite. *Brain Research* **600**, pp. 20–26.

1603. **Zhou, F., He, M., Liu, Z., Zhang, L., Gao, B.,** and **Wang, G.** (2013). Expression of corticotropin releasing hormone and its receptor in patients with intrahepstatic cholestasis of pregnancy. *Placenta* **34**, pp. 401–406.

1604. **Zhao, J., Qi, X., Gao, S., Lu, J., van Wamelen, D., Kamphuis, W., Bao, A.,** and **Swaab, D.** (2015). Different stress-related gene expression in depression and suicide. *Journal of Psychiatric Research* **68**, pp. 176–185.

1605. **Zho, Y., Spangler, R., Ho, A.,** and **Kreek, M.** (2003). Increased CRH mRNA levels in the rat amygdala during short-term withdrawal from chronic "binge" cocaine. *Molecular Brain Research* **114**, pp. 73–79.

1606. **Zhou, Y., Spangler, R., LaForge, K. S., Maggos, C. E., Ho, A.,** and **Kreek, M. J.** (1996). Corticotropin-releasing factor and type 1 corticotropin-releasing factor receptor messenger RNAs in rat brain and pituitary during" binge"-pattern cocaine administration and chronic withdrawal. *Journal of Pharmacology and Experimental Therapeutics* **279**, pp. 351–358.

1607. **Zhou, Y., Spangler, R., Schlussman, S. D., Ho, A.,** and **Kreek, M. J.** 2003 Alterations in hypothalamic-pituitary-adrenal axis activity and in levels of proopiomelanocortin and croticotropin-releasing hormone receptor 1 mRNAs in the pituitary and hypothalamus of the rat during chronic 'binge' cocaine and withdrawal. *Brain Res* **964**(2), pp. 187–199.

1608. **Zhao, Z., Jin, X,. Wu, Y., Yang, X,. Xu, Y., Jiang, J. Z., Kim, S. C., Lee, B. H., Yang, C. H.,** and **Zhao, R.** (2013). Amygdaloid corticotropin-releeasing factor is involved in the anxiolytic effect of acupuncture during ethanol wihtdawal in rats. *Journal of Acupuncture and Meridian Studies* **6**, pp. 234–240.

1609. **Zieba, B., Grzegorzewska, M., Branski, P., Domin, H., Wieronska, J., Hess, G.,** and **Smialowska, M.** (2008). The behavioural and electrophysiological effects of CRF in rat frontal cortext. *Neuropeptides* **42**, pp. 513–523.

1610. **Zmijewski, M. A.,** and **Slominski, A. T.** (2010). Emerging role of alternative splicing of CRF1 receptor in CRF signaling. *Acta Biochimica Polonia* **57**, pp. 1–13.

1611. **Zobel, A., Nickel, T., Kunzel, H., Ackl, N., Sonntag, A., Ising, M.,** and **Holsboer, F.** (2000). Effects of high-affinity corticotropin-releasing hormone receptor 1 antagonist R121919 in major depression, pp. the first 20 patients treated. *Journal of Psychiatric Research* **34**, pp. 171–181.

1612. **Zorrilla, E., Heilig, M., Wit, H.,** and **Shaham, Y.** (2013). Behavioral, biological, and chemical perspectives on targeting CRF1 receptor antagonists to treat alcoholism. *Drug Alcohol Dependence* **128**: 175–186.

1613. **Zorrilla, E. P.,** and **Koob, G. F.** (2010). Progress in corticotropin-releasing factor-1 antagonist development. *Drug Discovery Today* **15**, pp. 371–383.

1614. **Zorrilla, E.P., Wee, S., Zhao, Y, Specio, S., Boutrel, B., Koob, G. F.,** and **Weiss, F.** (2012). Extended access cocaine self-administration differentially activates dorsal raphe and amygdala corticotropin releasing factor systems in rats. *Addiction Biology* **17**, pp. 300–308.

1615. **Zoumakis, E., Chatzaki, E., Charlampopoulos, I., Marigioris, A. N., Angelakis, E., Koumantakis, E.,** and **Gravanis, A.** (2001). Cycle and age-related changes in corticotropin- releasing hormone levels in human endometrium and ovaries. *Gynecologic Endocrinology* **15**, pp. 98–102.

1616. **Zoumakis, E., Margioris, A. N., Stournaras, C., Dermitizaki, E., Angelakis, E. Makrigiannakis, A., Koumanatakis, E.,** and **Gravanis, A.** (2000). Corticotrophin-releasing hormone (CRH) interacts with inflammatory prostaglandins and interleukins and affects the decidualization of human endometrial stroma. *Molecular Human Reproduction* **6**, pp. 344–351.

1617. **Zweifel, L. S., Fadok, J. P., Argilli, E., Garelick, M. G., Jones, G. L., Dickerson, T. M., Allen, J. M., Mizumori, S. J., Bonci, A.,** and **Palmiter, R. D.** (2011). Activation of dopamine neurons is critical for aversive conditioning and prevention of generalized anxiety. *Nature Neuroscience* **14**, pp. 620–626.

1618. **Zweifel, L. S., Parker, J. G., Lobb, C. J., Rainwater, A., Wall, V. Z., Fadok, J. P., Darvas, M., Kim, M. J., Mizumori, S. J. Y., Paladini, C. A., Phillips, P. E. M.,** and **Phillips, P. E.** (2009). Disruption of NMDAR-dependent burst firing by dopamine neurons provides selective assessment of phasic dopamine-dependent behavior. *Proceedings of the National Academy of Sciences* **106**, pp. 7281–7288.

1619. **Zybtek, B.,** and **Slominski, A. T.** (2007). CRH mediates inflammation induced by Lipopolysaccharide in human adult epidermal keratinocytes. *Journal of Investigative Dermatology* **127**, pp. 730–732.

# Index

The manufacturer's authorised representative in the EU for product safety is Oxford University Press España S.A. of el Parque Empresarial San Fernando de Henares, Avenida de Castilla, 2 – 28830 Madrid (www.oup.es/en or product.safety@oup.com). OUP España S.A. also acts as importer into Spain of products made by the manufacturer.

www.ingramcontent.com/pod-product-compliance
Lightning Source LLC
LaVergne TN
LVHW011817200926
848484LV00005B/1222

* 9 7 8 0 1 9 8 7 9 3 6 9 4 *